Corrosion Engineering

Volume I

Corrosion Engineering
Volume I

Edited by **Andreas Byers**

NY RESEARCH
PRESS
New York

Published by NY Research Press,
23 West, 55th Street, Suite 816,
New York, NY 10019, USA
www.nyresearchpress.com

Corrosion Engineering: Volume I
Edited by Andreas Byers

International Standard Book Number: 978-1-63238-100-2 (Hardback)

Printed in China.

Contents

Preface

Corrosion is expensive and can be hazardous. It is costly to replace and repair equipment, structures, and other industrial items that have been damaged as a result of corrosion. In order to understand the mechanisms of corrosion, to solve corrosion problems, and improve corrosion resistance of materials, the use of modern evaluation and protection techniques is essential. Therefore no significant progress can be achieved in this area without the use of modern evaluation methods, combined with electrochemical techniques. It is the aim of this text to discuss the recent methods in corrosion evaluation and protection.

Corrosion is a field of chemistry, and a wide range of topics fall under this category. This process causes material degradation such as localized corrosion of metals such as Ni-Fe alloys when it comes in contact with CO. Naturally occurring corrosion is a big problem. For example, stainless steel which is considered to be very resistive of corrosion gets corroded when it comes in contact with natural seawater. Similarly, atmospheric corrosion of copper is a naturally occurring corrosion because tropical humid atmosphere catalyzes the corrosion of copper. The most efficient method to prevent corrosion is electrochemical plating, and it is practiced in many industrial processes. For instance, vehicle parts are plated with a non-corrosive layer of metal. Also electrochemical impedance spectroscopy is a corrosion resistance technique.

I wish to thank all the authors for their contributions for this book and it is my pleasure to acknowledge the assistance of the editorial team and the publishing team who have worked tirelessly for the completion of this book.

Editor

1

Adsorption Properties and Inhibition of C38 Steel Corrosion in Hydrochloric Solution by Some Indole Derivates: Temperature Effect, Activation Energies, and Thermodynamics of Adsorption

M. Lebrini, F. Robert, and C. Roos

Laboratoire Matériaux et Molécules en Milieu Amazonien, UAG-UMR ECOFOG, Campus Trou Biran, 97337 Cayenne, French Guiana

Correspondence should be addressed to C. Roos; christophe.roos@guyane.univ-ag.fr

Academic Editor: Vesna Mišković-Stanković

The corrosion rates in the presence of some indole derivates, namely, 9H-pyrido[3,4-b]indole (norharmane) and 1-methyl-9H-pyrido[3,4-b]indole (harmane), as inhibitors of C38 steel corrosion inhibitor in 1 M HCl solution, were measured by potentiodynamic polarization and electrochemical impedance spectroscopy (EIS) techniques, in the range of temperatures from 25 to 55°C. Results obtained revealed that the organic compounds investigated have inhibiting properties for all temperatures. The inhibition was assumed to occur via adsorption of the indole molecules on the metal surface. Adsorption of indole derivates was found to follow the Langmuir isotherm. The apparent activation energies, enthalpies, and entropies of the dissolution process and the free energies and enthalpies for the adsorption process were determined by potentiodynamic polarization and electrochemical impedance. The fundamental thermodynamic functions were used to collect important information about indole inhibitory behaviour.

1. Introduction

The effect of temperature on the inhibited acid-metal reaction is highly complex, because many changes occur on the metal surface such as rapid etching and desorption of inhibitor and the inhibitor itself may undergo decomposition and/or rearrangement. Temperature effects on acidic corrosion and corrosion inhibition of iron and steel most often in HCl and H_2SO_4 solutions had been the object of a large number of investigations [1–12]. However, it was found that few inhibitors with acid-metal systems have specific reactions which are effective (or more) at high temperature as they are at low temperature [13–15]. Thermodynamic parameters such as adsorption heat, adsorption entropy, and adsorption free energy can be obtained from experimental data of the studies of the inhibition process at different temperatures. The kinetic data such as apparent activation energy and preexponential factor at different inhibitor concentrations are calculated, and the effects of the activation energy and preexponential factor on the corrosion rate of steel were discussed [16–20].

Our earlier results obtained for 9H-pyrido[3,4-b]indole (norharmane) and 1-methyl-9H-pyrido[3,4-b]indole (harmane) as corrosion inhibitors of C38 steel demonstrated that correlation exists between the inhibition efficiency and the chemical structure [21]. It was found that these compounds are good inhibitors in acidic solutions and the inhibition efficiency of these indole-type organic compounds may be explained in terms of electronic properties (the energy of the highest occupied molecular orbital (E_{HOMO}) and the energy of the lowest unoccupied molecular orbital (E_{LUMO})). The aim of this paper is to extend these investigations in order to obtain a better understanding of the mode of inhibitory action of the harmane and norharmane by calculating thermodynamic parameters for both C38 steel dissolution and inhibitor adsorption process in hydrochloric acid solution using potentiodynamic polarization and electrochemical impedance spectroscopy (EIS).

2. Experimental

Electrode and Solution. Corrosion tests have been carried out on electrodes cut from sheets of C38 steel. Steel strips contained 0.36 wt% C, 0.66 wt% Mn, 0.27 wt% Si, 0.02 wt% S, 0.015 wt% P, 0.21 wt% Cr, 0.02 wt% Mo, 0.22 wt% Cu, 0.06 wt% Al, and the remainder iron. The specimens were embedded in epoxy resin leaving a working area of 0.78 cm^2.

Table 1: Corrosion parameters obtained from electrochemical measurements of C38 steel in 1 M HCl containing various concentrations of indole derivates at 25°C.

Inhibitor	Polarisation curves			EIS		
	E_{corr} versus SCE (mV)	I_{corr} (μA·cm^{-2})	IE (%)	R_{ct} (Ω·cm^2)	C_{dl} (μF·cm^{-2})	IE (%)
1 M HCl	−529	284	—	49 ± 0.02	546	
Norharmane						
0.2 mM	−467	67	76	148 ± 0.91	183	67
0.4 mM	−476	61	79	238 ± 1.65	120	79
0.8 mM	−494	52	82	514 ± 1.18	55	90
1.2 mM	−455	28	90	641 ± 1.73	44	92
Harmane						
0.2 mM	−476	65	77	181 ± 0.10	330	73
0.4 mM	−489	55	81	321 ± 0.97	270	85
0.8 mM	−479	34	88	537 ± 0.17	188	91
1.2 mM	−492	24	91	735 ± 0.68	148	93

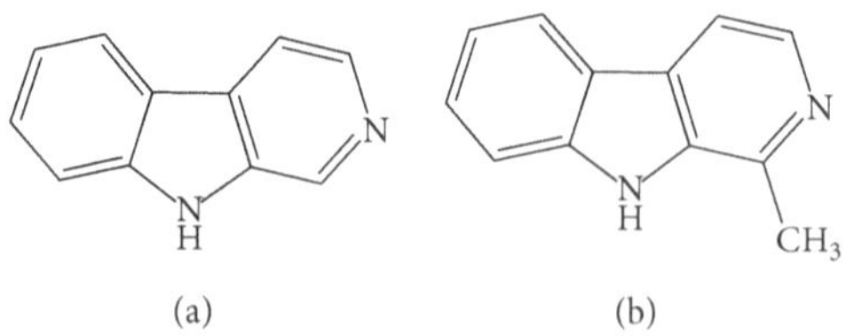

Figure 1: Chemical formulas of the tested inhibitors (a) norharmane and (b) harmane.

The working surface was subsequently ground with 180 and 1200 grit grinding papers, cleaned by distilled water and ethanol. The solutions (1 M HCl) were prepared by dilution of an analytical reagent grade 33% HCl with doubly distilled water. All the tests were performed in the range of temperatures from 25 to 55°C. The tested inhibitors are 9H-pyrido[3,4-b]indole (norharmane) and 1-methyl-9H-pyrido[3,4-b]indole (harmane); their molecular structures are shown in Figure 1. The investigated compounds were commercial products: norharmane (Aldrich, P99%), harmane (Sigma, P98%), the concentration range of inhibitor employed was 0.2 mM M to 1.2 mM. The concentration of inhibitor was determined in accordance with its solubility. Varela et al. reports the solubility of norharmane and harmane in water at 20°C, at pH 7 and 13, using the absorbance at the maxima in the absorption spectra of saturated solutions [22]. The solubility of norharmane was 1.86 mM (pH = 7) and 1.81 mM (pH = 13). For harmane, the solubility was 1.61 mM (pH = 7) and 0.786 mM (pH = 13). At lower pH (Our experimental conditions, pH = 0), where the cationic form is the predominant species, all the compounds present a very high solubility, and it was not possible to obtain data with this technique at lower pH.

Electrochemical Measurements. Electrochemical measurements, including potentiodynamic polarization curves and electrochemical impedance spectroscopy (EIS), were performed in a three-electrode cell. C38 steel specimen was used as the working electrode, a platinum wire as the counter, and a saturated calomel electrode (SCE) as the reference. Before potentiodynamic test and EIS experiments, the electrode was allowed to corrode freely and its open circuit potential (OCP) was recorded as a function of time during 3 h, the time necessary to reach a quasi-stationary value for the open-circuit potential. This steady-state OCP corresponds to the corrosion potential (E_{corr}) of the working electrode. The anodic and cathodic polarisation curves were recorded by a constant sweep rate of 20 mV min^{-1}. Electrochemical impedance spectroscopy (EIS) measurements were carried out, using ac signals of amplitude 5 mV peak to peak in the frequency range of 100 kHz to 10 mHz. Electrochemical measurements were performed through a VSP electrochemical measurement system (Bio-Logic). The above procedures were repeated two times with success for each concentration of the two tested inhibitors. The Tafel fit and EIS data were analysed using graphing and analyzing impedance software, version EC-Lab V9.97.

3. Results and Discussion

3.1. Effect of Concentration. The corrosion rates of metals and alloys in aggressive solutions can be determined using different electrochemical methods. Electrochemical impedance spectroscopy and polarisation curves were thus employed in the present work to study the effect of indole derivates concentration. The percentages of inhibition efficiency and values of associated electrochemical parameters are listed in Table 1. Inspections of the obtained data reveal that the I_{corr} values decrease considerably in the presence of harmane and norharmane with increasing inhibitor concentration. The E_{corr} values shifted to more positive potentials in the presence of indole derivates concentration, although there was not

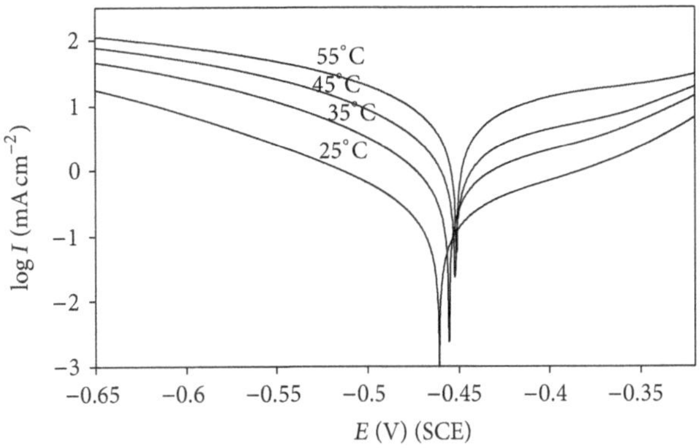

FIGURE 2: Effect of temperature on the cathodic and anodic responses for C38 steel in 1 M HCl.

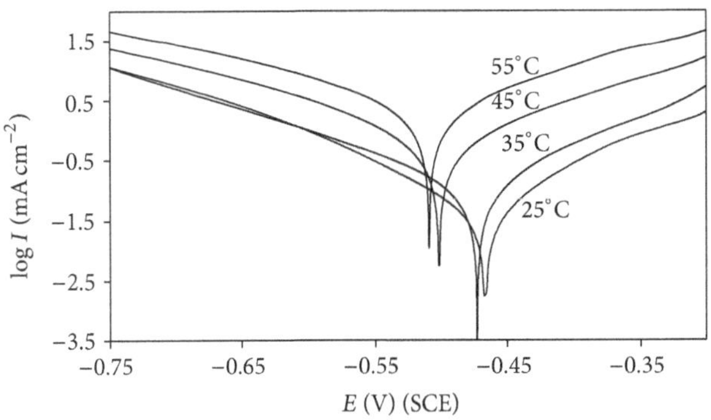

FIGURE 3: Effect of temperature on the cathodic and anodic responses for C38 steel in 1 M HCl + 0.2 mM of norharmane.

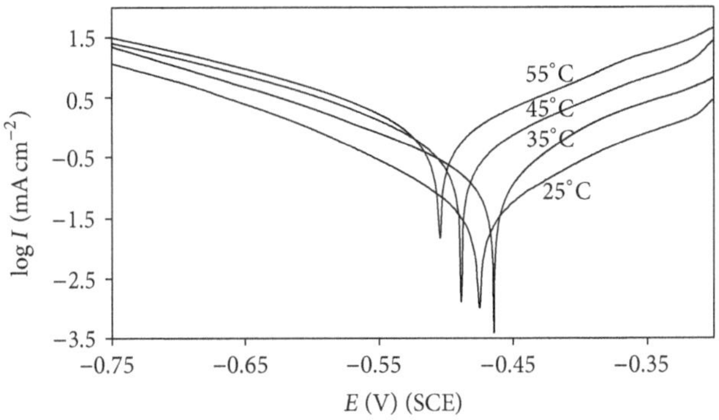

FIGURE 4: Effect of temperature on the cathodic and anodic responses for C38 steel in 1 M HCl + 0.2 mM of harmane.

a specific relation between E_{corr} and inhibitors concentration. This effect may be related to the adsorption of the organic compound at the active sites of the electrode surface, retarding the corrosion reaction. The data also reveal that the addition of harmane or norharmane causes a significant increase in R_{ct} in the additive-free solution and diminishes the value of C_{dl}. The decrease in C_{dl} could be attributed to the adsorption of the inhibitors forming a protective layer on the metal surface [23]. The inhibition efficiencies, calculated from ac impedance study, show the same trend as these obtained from polarization measurements. However, IE (%) increases with inhibitor concentrations for both inhibitors, reaching a maximum value at 1.2 mM. It was found that these compounds are good inhibitors in acidic solutions and the inhibition efficiency of these indole-type organic compounds may be explained in terms of electronic properties. The detailed study of 1 M HCl at 25°C is reported in [21].

3.2. *Effect of Temperature*

3.2.1. Potentiodynamic Polarisation. In order to get more information about the performance of indole derivates, the nature of adsorption isotherm, and thereafter to evaluate the adsorption and activation processes, the influence of temperature is studied by potentiodynamic polarisation. Some of the polarisation curves for C38 steel electrode in 1 M HCl in the absence and in the presence of 0.2 mM of norharmane and harmane in the temperature range 25–55°C are given in Figures 2, 3, and 4, respectively. As it can be seen, raising the temperature increases both anodic and cathodic reactions of C38 steel electrode both in the absence and in the presence of inhibitors. Polarisation curves of the C38 steel electrode in 1 M HCl without and with addition of various concentrations of harmane at 45°C are shown in Figure 5, representative example. As it can be seen, the anodic and cathodic reactions are affected by the addition of inhibitor. Similar observations were found for all concentrations studied in the presence of both inhibitors. Accurate evaluation of Tafel slopes by Tafel extrapolation is often impossible, simply because an experimental polarization curve does not exhibit linear Tafel regions [24–27]. Our experimental polarization curves presented here do not display the expected log/linear Tafel behaviour with anodic branches exhibiting curvature over the complete applied potential range. The curvature of the anodic branch may be attributed to the deposition of the corrosion products or impurities in the steel (e.g., Fe_3C) to form a nonpassive surface film [24]. The various electrochemical parameters were calculated from Tafel plots and summarized in Table 2. The inhibition efficiency and the surface coverage degree (θ) values for steel are also shown in the same table. The surface coverage θ was calculated from (1) [28]. It has been observed from Table 2, that the corrosion current density (I_{corr}) increased with increasing temperature in both uninhibited and inhibited solutions. The addition of norharmane or harmane to the HCl solution shifts the E_{corr} towards more negative values at all of the studied temperatures, although there was not a specific relation between E_{corr} and inhibitors concentration. The IE (%) values decrease with increasing temperature; a decrease in inhibition efficiencies with the increase in temperature might be due to weakening of physical adsorption. Consider

$$\theta = \frac{I_{corr} - I_{corr(inh)}}{I_{corr}}, \qquad (1)$$

where I_{corr} and $I_{corr(inh)}$ are the corrosion current density values in the absence and presence of indole derivates, respectively.

TABLE 2: Electrochemical parameters calculated from Tafel plots and the corresponding inhibition efficiencies at various temperatures studied of C38 steel in 1 M HCl containing different concentrations of indole derivates.

		Norharmane						Harmane					
Temp. (°C)	Con. (mM)	E_{corr} versus SCE (mV)	I_{corr} ($\mu A \cdot cm^{-2}$)	b_a ($mV \cdot dec^{-1}$)	b_c ($mV \cdot dec^{-1}$)	IE (%)	θ	E_{corr} versus SCE (mV)	I_{corr} ($\mu A \cdot cm^{-2}$)	b_a ($mV \cdot dec^{-1}$)	b_c ($mV \cdot dec^{-1}$)	IE (%)	θ
	1 M HCl	−460	232	106	89	—	—	—	—	—	—	—	—
	0.2	−467	67	107	125	76	0.7641	−476	65	110	124	77	0.7711
25	0.4	−476	61	116	122	79	0.7852	−489	55	98	107	81	0.8063
	0.8	−494	52	97	118	82	0.8169	−479	34	113	119	88	0.8803
	1.2	−455	28	93	112	90	0.9014	−492	24	87	107	91	0.9155
	1 M HCl	−454	843	111	109	—	—	—	—	—	—	—	—
	0.2	−473	236	112	144	72	0.7200	−465	212	118	133	75	0.7485
35	0.4	−472	195	109	126	77	0.7687	−470	182	116	136	78	0.7841
	0.8	−470	165	102	131	80	0.8043	−456	154	109	128	82	0.8173
	1.2	−479	98	103	135	88	0.8837	−464	101	103	118	88	0.8802
	1 M HCl	−451	1192	120	109	—	—	—	—	—	—	—	—
	0.2	−501	512	130	146	57	0.5705	−496	413	117	134	65	0.6535
45	0.4	−509	472	112	115	60	0.6040	−488	389	112	125	67	0.6737
	0.8	−505	443	135	135	63	0.6284	−487	335	113	126	72	0.7190
	1.2	−485	352	107	140	70	0.7047	−484	245	110	118	79	0.7945
	1 M HCl	−450	1982	131	116	—	—	—	—	—	—	—	—
	0.2	−509	1023	142	150	48	0.4839	−504	983	136	148	50	0.5040
55	0.4	−497	954	114	156	52	0.5187	−496	852	123	140	57	0.5701
	0.8	−566	867	117	127	56	0.5626	−564	715	110	146	64	0.6393
	1.2	−494	651	109	139	67	0.6715	−515	574	107	132	71	0.7104

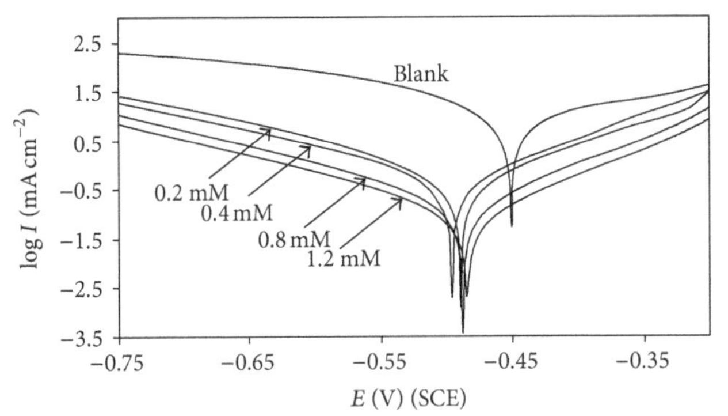

FIGURE 5: Polarisation curves for C38 steel in 1 M HCl containing different concentrations of harmane at 45°C.

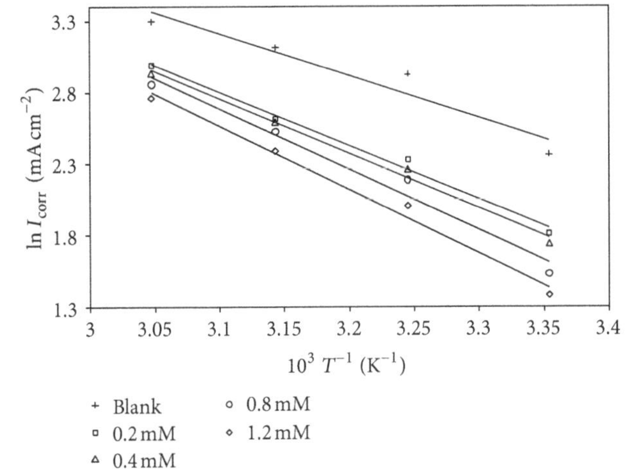

FIGURE 6: Arrhenius plots of corrosion $\ln(I_{\text{corr}})$ versus $1/T$ of 1 M HCl and at different concentrations of harmane.

In order to calculate activation parameters for the corrosion process, Arrhenius (2) and an alternative formulation of Arrhenius (3) were used:

$$I_{\text{corr}} = K \exp\left(\frac{-E_a}{RT}\right), \quad (2)$$

$$I_{\text{corr}} = \frac{RT}{Nh} \exp\left(\frac{\Delta S_a}{R}\right) \exp\left(-\frac{\Delta H_a}{RT}\right), \quad (3)$$

where I_{corr} is the corrosion current density, K the Arrhenius preexponential constant, E_a is the activation energy for the corrosion process, h is the Plank's constant, N is the Avogadro's number, the enthalpy of activation (ΔH_a), and the entropy of activation (ΔS_a).

Arrhenius plots for the corrosion density of C38 steel in the case of harmane are given in Figure 6, representative example. Similar plots are obtained in the case of norharmane. The apparent activation energies (E_a) at different concentrations of indole derivates were determined by linear regression between $\ln I_{\text{corr}}$ and $1/T$ and the result is shown in Table 3. All the linear regression coefficients were close to 1, indicating that the C38 steel corrosion in hydrochloric acid can be elucidated using the kinetic model. Inspection of Table 3 showed that the values of E_a determined in 1 M HCl

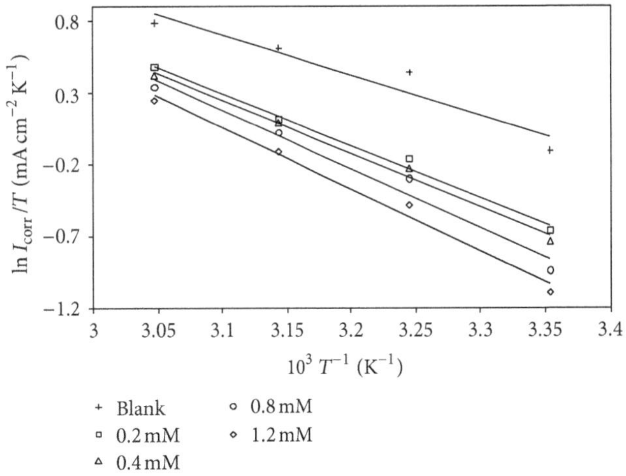

FIGURE 7: Arrhenius plots of corrosion $\ln(I_{\text{corr}}/T)$ versus $1/T$ of 1 M HCl and at different concentrations of harmane.

containing indole derivates are higher than that for uninhibited solution. The increase in the apparent activation energy may be interpreted as physical adsorption that occurs in the first stage [29]. Moreover, the values obtained for norharmane are somewhat higher than those obtained for harmane and confirm the fact that the inhibitor efficiency of norharmane is slightly lower than that of harmane. Szauer and Brand explained that the increase in activation energy can be attributed to an appreciable decrease in the adsorption of the inhibitor on the steel surface with increase in temperature [30]. Figure 7 showed a plot of ln (I_{corr}/T) versus $1/T$ with a slope of ($-\Delta H_a/R$) and an intercept ($\ln R/Nh + \Delta S_a/R$) from which the values of ΔH_a and ΔS_a were calculated, respectively. All estimated kinetic-thermodynamic parameters were tabulated also in Table 3. Inspection of these data revealed that the thermodynamic parameters (ΔH_a and ΔS_a) for dissolution reaction of C38 steel in 1 M HCl in the presence of both inhibitors are higher than that of in the absence of inhibitors. The positive signs of ΔH_a reflect the endothermic nature of the steel dissolution process suggesting that the dissolution of steel is slow in the presence of inhibitors [31]. Furthermore, the values obtained for norharmane are somewhat higher than those obtained for harmane and confirm the fact that the inhibitor efficiency of norharmane is slightly lower than that of harmane. One can notice that E_a and ΔH_a values vary in the same way for both inhibitors (Table 3). The values of ΔS_a were negative both in the absence and presence of inhibitors, implying that the activated complex represented the rate-determining step with respect to the association rather than the dissociation step. It means that a decrease in disorder occurred when proceeding from the reactants to the activated complex [30]. Also, the ΔS_a shift can be explained as a result of the replacement process of water molecules during adsorption of organic inhibitor on the steel surface.

3.2.2. Impedance Measurements. The corrosion behaviour of C38 steel in 1 M HCl solution in the presence of indole derivates was also investigated by EIS in the temperature range 25–55°C. The EIS of C38 steel electrode in 1 M HCl

TABLE 3: Thermodynamic parameters calculated from Tafel plots for the adsorption of indole derivates in 1 M HCl on the C38 steel at different temperatures.

Con. (mM)	Norharmane			Harmane		
	E_a (kJ mol^{-1})	ΔH_a (kJ mol^{-1})	ΔS_a (J mol^{-1} K^{-1})	E_a (kJ mol^{-1})	ΔH_a (kJ mol^{-1})	ΔS_a (J mol^{-1} K^{-1})
HCl	24.42	23.29	−119.48	—	—	—
0.2	31.73	30.60	−99.81	31.20	30.07	−101.88
0.4	32.38	31.25	−98.12	31.80	30.67	−100.43
0.8	33.40	32.27	−95.27	35.18	34.05	−90.53
1.2	37.99	36.85	−81.97	36.89	35.76	−86.63

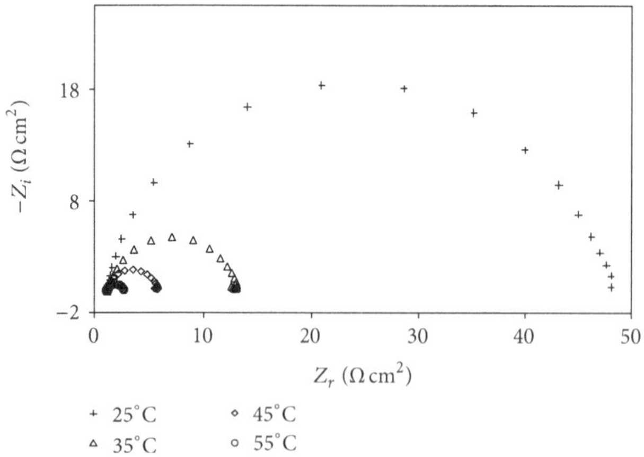

FIGURE 8: Nyquist diagrams for C38 steel in 1 M HCl at different temperatures.

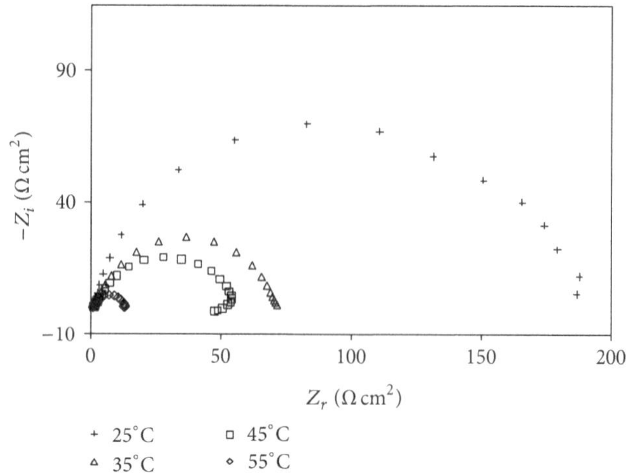

FIGURE 10: Nyquist diagrams for C38 steel in 1 M HCl containing 0.2 mM of harmane at different temperatures.

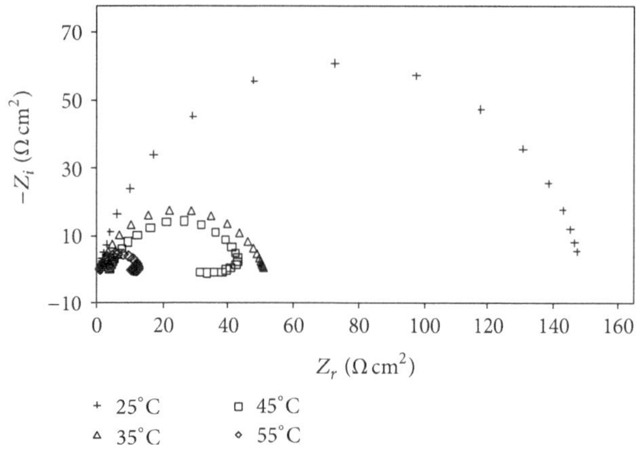

FIGURE 9: Nyquist diagrams for C38 steel in 1 M HCl containing 0.2 mM of norharmane at different temperatures.

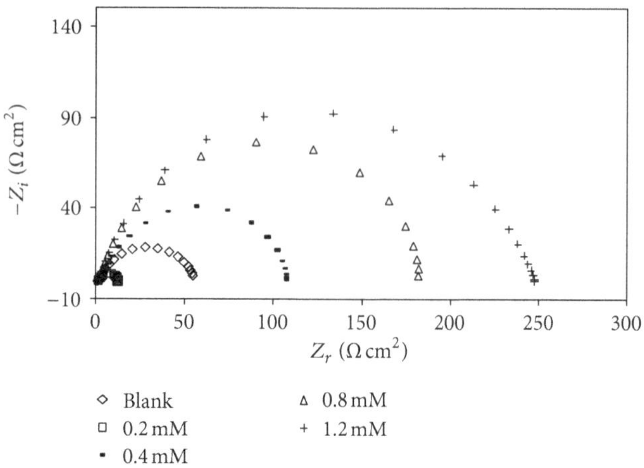

FIGURE 11: Nyquist diagrams for C38 steel in 1 M HCl containing different concentrations of noharmane at 35°C.

solution in the absence and in the presence of of 0.2 mM of noharmane and harmane, at different temperatures, are shown in Figures 8, 9, and 10; representative example. Figure 11 shows the impedance plots of C38 steel electrode in 1 M HCl solution in the absence and presence of different concentrations of norharmane at 35°C. It can be seen that, the size of the capacitive loop increased by increasing the concentration of norharmane. All the impedance spectra were measured at the corresponding open-circuit potentials. After analyzing the shape of the Nyquist plots, it is concluded that the curves obtained up to 45°C approximated by a single capacitive semi-circles, showing that the corrosion process was mainly charge transfer controlled. The general shape of the curves is very similar for all concentrations studied. The diameter of Nyquist plots (R_{ct}) decreases with increasing temperature for both inhibitors. For the most temperatures 45 and 55°C, an inductive loop appeared at low-frequency

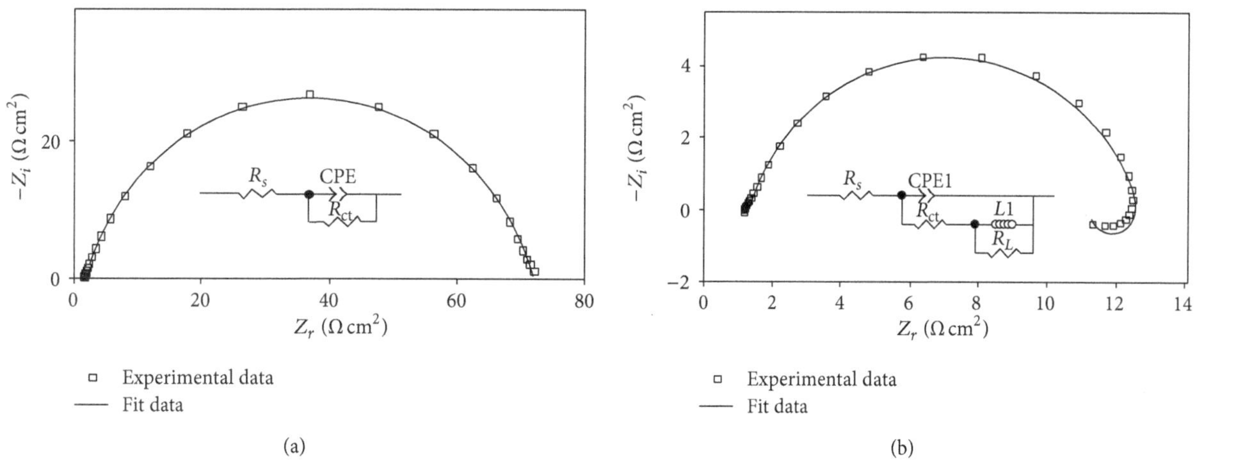

FIGURE 12: Nyquist plots: (□) experimental data and (—) fit data, together with the equivalent circuit used to fit the impedance data, recorded for a C38 electrode (a) in 1 M HCl solution + 0.2 mM harmane at 35°C and (b) in 1 M HCl solution + 0.2 mM norharmane at 55°C.

values. Similar observations were found for all concentrations studied. The presence of the LF inductive loop may be attributed to the relaxation process obtained by adsorption species like Cl^-_{ads} and H^+_{ads} on the electrode surface [32–35]. It may also be attributed to the adsorption of inhibitor on the electrode surface [36] or to the redissolution of the passivated surface at low frequencies [37]. The above impedance diagrams (Nyquist) contain depressed semicircles with the center under the real axis; such behaviors characteristic for solid electrodes and often referred to as frequency dispersion have been attributed to roughness and inhomogeneities of the solid surfaces, impurities [38–41]. Therefore, a constant phase element (CPE) instead of a capacitive element is used to get a more accurate fit of experimental data set. For analysis of the impedance spectra exhibiting one capacitive loop, the equivalent circuit (EC) given in Figure 12(a) was used to fit the experimental spectra while the EC shown in Figure 12(b) was used for the impedance spectra containing one capacitive loop and an inductive loop. Simulation of Nyquist plots with above models shows excellent agreement with experimental data (Figures 12(a) and 12(b); representative example).

The parameters calculated with the application of these models are summarized in Tables 4 and 5. The values discussed show that the organic compounds investigated have inhibiting properties for all temperature values. This conclusion is based on the values of R_{ct} in inhibitors presence. They are greater than those in 1 M HCl. The temperature rise leads to a decrease of R_{ct} values. This is due, on one hand, to the increase of the rate of metal dissolution and, on the other hand, to the shift of the adsorption/desorption equilibrium towards the inhibitor's desorption and hence to the decrease of surface coverage degree. The n value almost decreases with temperature increase and this is interpreted as an evidence for surface inhomogeneity increase. The latter is attributed to the accelerated corrosion and the developed metal surface. At lower temperatures the inhibitors adsorb on the most active surface sites; this results in a decrease of surface inhomogeneity. The desorption, which is favored by high temperatures, will increase with this effect. The C_{dl} value was calculated using (4) [42–44]:

$$C_{dl} = \left(A \cdot R_{ct}^{1-n}\right)^{1/n}. \tag{4}$$

It is clear that the addition of harmane or norharmane causes a significant increase in C_{dl} in the additive-free solution for all temperatures studied. As the double layer capacitance can be connected with the thickness of the layer of adsorbed inhibitor molecules at the interface and this dependence is of reverse proportionality; it follows that the thickness of the adsorption layer will decrease with temperature increase. Similar tendency was found for CPE constant (A) for all temperatures studied. It can be seen from Tables 2, 4, and 5 that the inhibition efficiencies calculated by both methods show the same trend. The IE (%) values decrease with increasing temperature and no difference of numerical values of inhibition efficiencies was observed at 25 and 35°C. However the results obtained at 45 and 55°C show a difference in numerical values obtained by both methods. This variation may be due to the difference between methods, which each method is realised differently with some conditions and assumptions.

To obtain the activation parameters from the impedance technique, values of R_{ct} and those of β_a and β_c obtained at different temperatures were used to calculate values of I_{corr} according to Stern-Geary [45] equation:

$$I_{corr} = \frac{\beta_a \beta_c}{2.303\left(\beta_a + \beta_c\right)}\left(\frac{1}{R_{ct}}\right). \tag{5}$$

Figures 13 and 14 show the Arrhenius plot for the results obtained in 1 M HCl solutions in the presence of harmane, representative example. Similar plots are obtained in the case of norharmane. The results are in good agreement with those obtained from polarization studies and show the same trend. The difference of numerical values of ΔS_a determined from the two techniques may be attributed to the different methods that have been used for the determination of the corrosion rates.

TABLE 4: Values of the elements of equivalent circuit required for fitting the EIS for C38 steel immersed in the presence and absence of norharmane at different temperatures.

	Temp (°C)	R_{ct} (Ω cm^2)	$A10^4$ (Ω^{-1} s^n cm^{-2})	n	C_{dl} (μF cm^{-2})	L (H cm^{-2})	R_L (Ω cm^2)	IE (%)	θ
Blank	25	49 ± 0.02	9.5 ± 0.73	0.853 ± 0.014	546	—	—	—	—
	35	17 ± 0.64	20.5 ± 4.21	0.795 ± 0.421	863	18 ± 1.87	6 ± 1.26	—	—
	45	12 ± 0.94	43.5 ± 3.56	0.745 ± 0.362	1583	15 ± 2.35	3 ± 0.87	—	—
	55	3 ± 0.96	72.8 ± 5.64	0.735 ± 0.093	1834	10 ± 8.54	2 ± 1.25	—	—
0.2 mM	25	148 ± 0.91	2.91 ± 0.02	0.871 ± 0.008	183	—	—	67	0.6689
	35	53 ± 0.43	4.37 ± 0.25	0.836 ± 0.053	209	—	—	60	0.6038
	45	30 ± 0.35	10.88 ± 3.65	0.765 ± 0.059	380	25 ± 4.61	9 ± 1.78	50	0.5000
	55	10 ± 0.65	13.35 ± 1.75	0.815 ± 0.120	501	2 ± 0.40	2 ± 1.37	50	0.5000
0.4 mM	25	238 ± 1.65	1.89 ± 0.19	0.872 ± 0.005	120	—	—	79	0.7941
	35	95 ± 0.89	3.53 ± 0.14	0.829 ± 0.052	175	—	—	78	0.7789
	45	55 ± 3.04	6.19 ± 0.94	0.790 ± 0.032	252	10 ± 1.20	15 ± 1.60	73	0.7273
	55	14 ± 0.93	8.84 ± 2.31	0.831 ± 0.054	362	3 ± 0.21	2 ± 0.32	64	0.6429
0.8 mM	25	514 ± 1.18	0.84 ± 0.03	0.883 ± 0.012	55	—	—	90	0.9047
	35	180 ± 1.04	3.18 ± 0.13	0.825 ± 0.045	173	—	—	82	0.8235
	45	85 ± 1.56	5.01 ± 0.30	0.814 ± 0.023	244	7 ± 0.27	14 ± 0.90	79	0.7857
	55	22 ± 1.55	7.13 ± 0.47	0.809 ± 0.251	267	6 ± 1.60	2 ± 1.10	77	0.7727
1.2 mM	25	641 ± 1.73	0.66 ± 0.02	0.887 ± 0.010	44	—	—	92	0.9236
	35	245 ± 0.55	2.3 ± 0.78	0.803 ± 0.072	114	—	—	91	0.9143
	45	123 ± 0.81	2.9 ± 0.57	0.802 ± 0.140	127	—	—	88	0.8780
	55	25 ± 0.41	4.46 ± 0.17	0.804 ± 0.023	149	6 ± 0.50	3 ± 1.43	80	0.8000

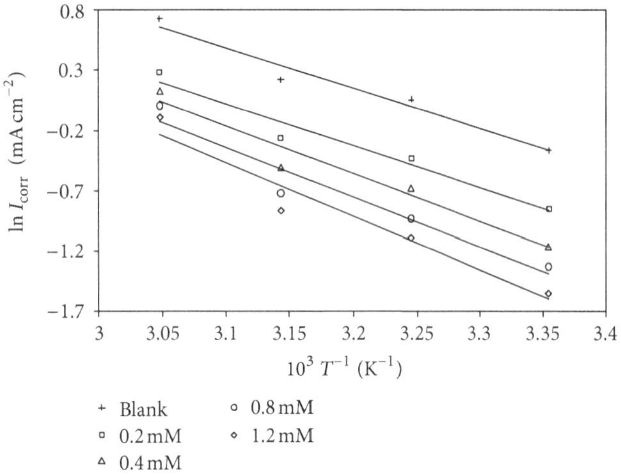

FIGURE 13: Arrhenius plots ln I_{corr} versus $1/T$ of 1 M HCl and at different concentrations of harmane.

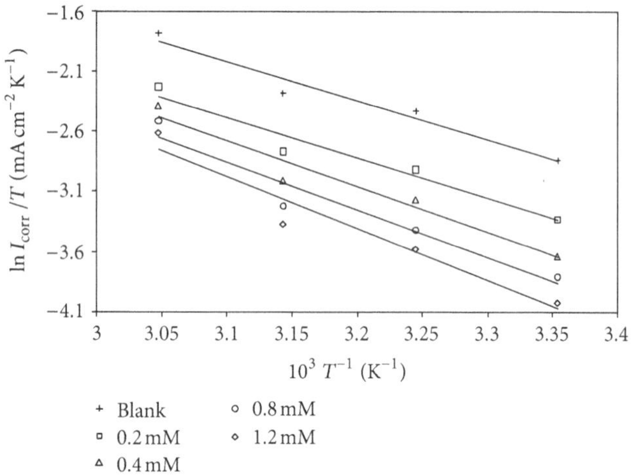

FIGURE 14: Arrhenius plots of corrosion ln(I_{corr}/T) versus $1/T$ of 1 M HCl at different concentrations of harmane.

3.3. Adsorption Isotherm. Inhibition by organic compounds is, mainly, due to their ability to adsorb onto a metal surface to form a protective film. The establishment of isotherms that describe the adsorption behaviour of corrosion inhibitor is important as they provide important information about the nature of metal-inhibitor interaction. It has been reported that the adsorption of the inhibitor molecules depends on a variety of factors such as the presence of functional groups (either electron donating or withdrawing), charge distribution at the donor atom, the π orbital character of donating electrons, the nature of substrate metal and the type of interaction between organic molecules and the metallic surface as well.

The most frequently used isotherms are Langmiur (6), Temkin (7), and Frumkin (8). All these isotherms are tested. In all cases, the degree of surface coverage (θ) is plotted as a function of the inhibitor concentration (C_{inh}). In the present study, values of θ are obtained from I_{corr} and R_{ct}. The significance of the adsorption isotherm was obtained by calculating the correlation coefficients (R^2). As an example,

TABLE 5: Values of the elements of equivalent circuit required for fitting the EIS for C38 steel immersed in the presence and absence of harmane at different temperatures.

	Temp (°C)	R_{ct} (Ω cm^2)	$A10^4$ (Ω^{-1} s^n cm^{-2})	n	C_{dl} (μF cm^{-2})	L (H cm^{-2})	R_L (Ω cm^2)	IE (%)	θ
Blank	25	49 ± 0.02	9.5 ± 0.73	0.853 ± 0.014	546	—	—	—	—
	35	17 ± 0.64	20.5 ± 4.21	0.795 ± 0.421	863	18 ± 1.87	6 ± 1.26	—	—
	45	12 ± 0.94	43.5 ± 3.56	0.745 ± 0.362	1583	15 ± 2.35	3 ± 0.87	—	—
	55	3 ± 0.96	72.8 ± 5.64	0.735 ± 0.093	1834	10 ± 8.54	2 ± 1.25	—	—
0.2 mM	25	181 ± 0.10	4.82 ± 0.19	0.866 ± 0.042	330	—	—	73	0.7293
	35	73 ± 0.61	6.92 ± 0.44	0.816 ± 0.501	353	—	—	71	0.7123
	45	50 ± 1.23	9.73 ± 0.27	0.805 ± 0.062	468	25 ± 8.41	8 ± 3.02	70	0.7000
	55	16 ± 0.81	19.5 ± 4.03	0.719 ± 0.501	503	11 ± 0.34	3 ± 0.8	69	0.6875
0.4 mM	25	321 ± 0.97	3.71 ± 0.37	0.870 ± 0.036	270	—	—	85	0.8474
	35	129 ± 0.65	5.29 ± 0.09	0.804 ± 0.074	275	—	—	84	0.8372
	45	82 ± 1.54	7.95 ± 1.02	0.784 ± 0.128	375	25 ± 1.74	9 ± 1.5	82	0.8171
	55	21 ± 2.31	10.31 ± 2.13	0.824 ± 0.054	455	21 ± 0.58	4 ± 1.3	76	0.7619
0.8 mM	25	537 ± 0.17	2.52 ± 0.27	0.873 ± 0.051	188	—	—	91	0.9088
	35	218 ± 0.56	3.96 ± 0.72	0.795 ± 0.104	211	—	—	90	0.9037
	45	136 ± 1.53	4.37 ± 0.73	0.811 ± 0.091	226	—	—	89	0.8897
	55	27 ± 1.20	8.64 ± 2.03	0.780 ± 0.035	299	5 ± 0.12	2 ± 0.76	81	0.8148
1.2 mM	25	735 ± 0.68	1.92 ± 0.12	0.883 ± 0.036	148	—	—	93	0.9333
	35	293 ± 0.54	3.28 ± 0.27	0.814 ± 0.037	192	—	—	93	0.9283
	45	182 ± 0.53	4.33 ± 0.07	0.826 ± 0.051	254	—	—	92	0.9176
	55	32 ± 0.66	7.36 ± 1.91	0.783 ± 0.102	260	11 ± 1.54	5 ± 0.8	84	0.8438

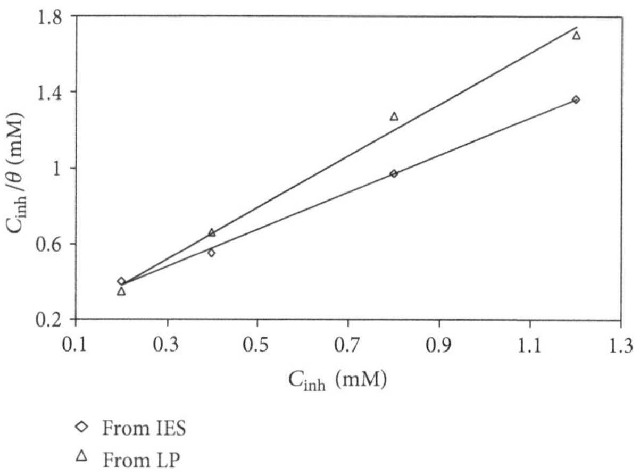

FIGURE 15: Langmuir adsorption plots for C38 steel in 1 M HCl containing different concentrations of norharmane at 55°C.

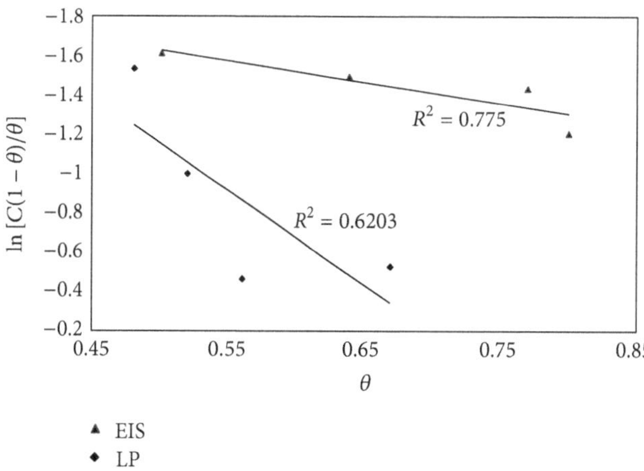

FIGURE 16: Frumkin adsorption plots for C38 steel in 1 M HCl containing different concentrations of norharmane at 55°C.

Figures 15, 16, and 17 represent fitting of impedance and polarization data obtained for C38 electrode in 1 M HCl containing various concentrations of norharmane at 55°C to Langmuir, Temkin, and Frunkin isotherms. The best correlation between the experimental results, obtained from the three tested isotherm functions, was obtained using Langmuir isotherm (the strong correlation $R^2 = 0.999$ for both methods). Similar observations were found for all temperatures studied. The simplest, being the Langmuir isotherm, is based on assumption that all adsorption sites are equivalent and that particle binding occurs independently from nearby sites being occupied or not [46]. Consider

$$\frac{C_{inh}}{\theta} = \frac{1}{K} + C_{inh} \quad \text{(Langmuir isotherm)}, \tag{6}$$

$$\left(\frac{\theta}{1-\theta}\right) \exp(-2a\theta) = KC_{inh} \quad \text{(Frumkin isotherm)}, \tag{7}$$

$$\exp(-2a\theta) = KC_{inh} \quad \text{(Temkin isotherm)}, \tag{8}$$

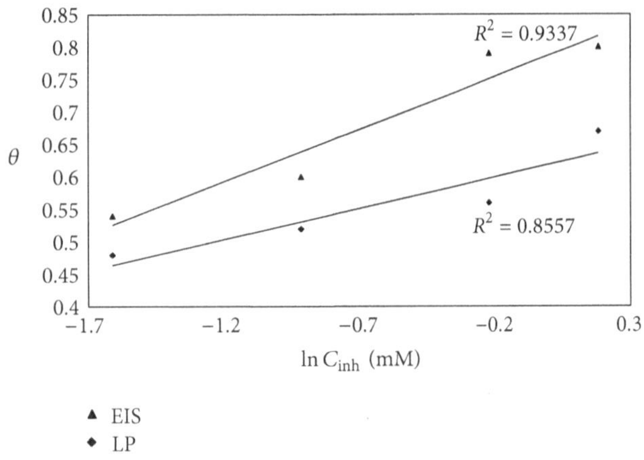

FIGURE 17: Temkin adsorption plots for C38 steel in 1 M HCl containing different concentrations of norharmane at 55°C.

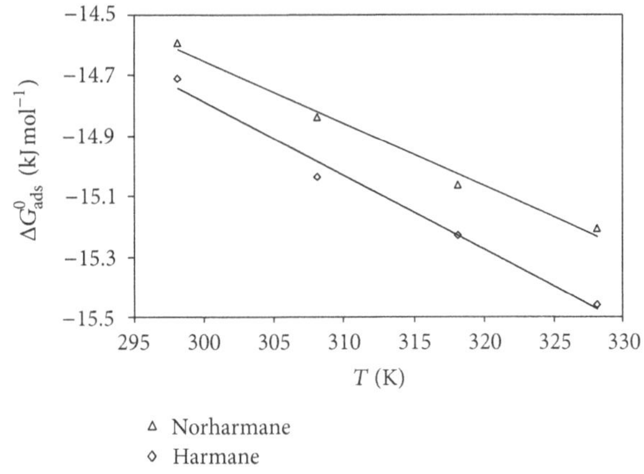

FIGURE 19: Variation of ΔG^0_{ads}, evaluated from LP, versus T on C38 steel in 1 M HCl containing indole derivates.

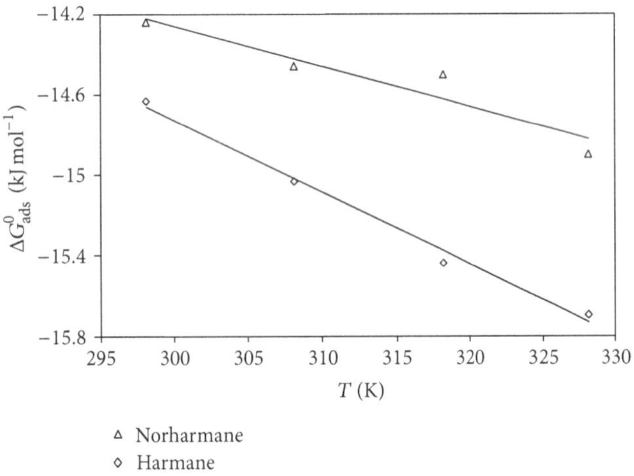

FIGURE 18: Variation of ΔG^0_{ads}, evaluated from IES, versus T on C38 steel in 1 M HCl containing indole derivates.

where "K" is the binding constant of the adsorption reaction and "a" is the lateral interaction term describing the molecular interactions in the adsorption layer and the heterogeneity of the surface.

The well known thermodynamic adsorption parameters are the standard free energy of adsorption (ΔG^0_{ads}), the heat of adsorption (ΔH^0_{ads}), and the standard entropy of adsorption (ΔS^0_{ads}). These quantities can be calculated depending on the estimated values of K from adsorption isotherms at different temperatures.

The constant of adsorption, K, is related to the standard free energy of adsorption, ΔG^0_{ads}, with the following equation:

$$K = \frac{1}{55.5} \exp\left(\frac{-\Delta G^0_{ads}}{RT}\right). \quad (9)$$

The value 55.5 in the above equation is the concentration of water in solution in mol L^{-1} [47]. The ΔG^0_{ads} values at all studied temperatures can be calculated. Then, the obtained ΔG^0_{ads} values plotted versus T (Figures 18 and 19) gave the heat of adsorption ΔH^0_{ads} and the standard adsorption entropy ΔS^0_{ads} in accordance with the basic equation:

$$\Delta G^0_{ads} = \Delta H^0_{ads} - T\Delta S^0_{ads}. \quad (10)$$

All thermodynamic adsorption parameters estimated from both methods for indole derivates on C38 steel from 1 M of HCl solution are listed in Tables 6, 7, and 8. The entire values of thermodynamic adsorption obtained by both methods are in good agreement and follow the same trend. The negative values of ΔG^0_{ads} ensure the spontaneity of the adsorption process and stability of the adsorbed layer on the steel surface. The calculated ΔG^0_{ads} values indicate the physical nature of adsorption of norharmane and harmane. The dependence of ΔG^0_{ads} on experimental temperature can be summarized in two cases as follows.

(a) Value of ΔG^0_{ads} becomes less negative with increasing temperature indicating the occurrence of exothermic process at which adsorption was unfavourable with increasing reaction temperature as the result of the inhibitor desorption from the steel surface [48].

(b) Value of ΔG^0_{ads} becomes more negative with increasing temperature indicating the occurrence of endothermic process at which increasing temperature facilitates inhibitor adsorption.

The first case (a) is observed for the adsorption of the studied inhibitor species on C38 steel surface from HCl solution depending on the applied temperature range (Tables 7 and 8), indicating the occurrence of exothermic adsorption processes. Also, it is noteworthy feature that the calculated value of ΔG^0_{ads} at 25°C in the case of harmane is greater than that in the case of norharmane, which means that the electrostatic interaction between the harmane and the steel surface is stronger. This is in good agreement with the range of the inhibition efficiency values obtained at 25°C from both impedance and polarisation measurements. However in the temperature range 35–55°C, there is an observable contradiction between the results of inhibition efficiency values and the finding of the calculated value of ΔG^0_{ads}. The negative sign of (ΔH^0_{ads}) reveals that the adsorption of inhibitor molecules is

TABLE 6: Thermodynamic parameters calculated from IES for the adsorption of indole derivates in 1 M HCl on the C38 steel at different temperatures.

Con. (mM)	Norharmane			Harmane		
	E_a (kJ mol^{-1})	ΔH_a (kJ mol^{-1})	ΔS_a (J mol^{-1} K^{-1})	E_a (kJ mol^{-1})	ΔH_a (kJ mol^{-1})	ΔS_a (J mol^{-1} K^{-1})
HCl	28.00	26.87	−131.04	—	—	—
0.2	32.78	31.65	−120.35	29.03	27.91	−131.68
0.4	33.34	32.22	−116.44	32.82	31.69	−121.55
0.8	38.00	36.87	−105.77	34.05	32.92	−119.20
1.2	38.77	37.64	−104.36	37.05	35.92	−110.91

TABLE 7: Thermodynamic parameters calculated from IES for the adsorption of indole derivates on C38 steel in 1 M HCl at different temperatures.

Temp. (°C)	Norharmane			Harmane		
	ΔG^0_{ads} (kJ mol^{-1})	ΔH^0_{ads} (kJ mol^{-1})	ΔS^0_{ads} (J mol^{-1} K^{-1})	ΔG^0_{ads} (kJ mol^{-1})	ΔH^0_{ads} (kJ mol^{-1})	ΔS^0_{ads} (J mol^{-1} K^{-1})
25	−18.00			−14.63		
35	−14.46	−8.18	20.2	−15.03	−3.94	36
45	−14.50			−15.44		
55	−14.90			−15.69		

TABLE 8: Thermodynamic parameters calculated from LP for the adsorption of indole derivates on C38 steel in 1 M HCl at different temperatures.

Temp. (°C)	Norharmane			Harmane		
	ΔG^0_{ads} (kJ mol^{-1})	ΔH^0_{ads} (kJ mol^{-1})	ΔS^0_{ads} (J mol^{-1} K^{-1})	ΔG^0_{ads} (kJ mol^{-1})	ΔH^0_{ads} (kJ mol^{-1})	ΔS^0_{ads} (J mol^{-1} K^{-1})
25	−14.59			−14.71		
35	−14.84	−8.45	20.7	−15.04	−7.43	24.5
45	−15.06			−15.23		
55	−15.21			−15.46		

an exothermic process. Generally, an exothermic adsorption process suggests either physisorption or chemisorption while endothermic process is attributed to chemisorption [17]. Moreover, the values obtained for harmane are somewhat higher than those obtained for norharmane. This result indicated that harmane is more strongly adsorbed on the steel surface. This is in good agreement with the results obtained from impedance and polarisation measurements. The results show a great difference between the value of ΔH^0_{ads} for harmane obtained from EIS (−3.88 kJ mol^{-1}) and from LP (−7.40 kJ mol^{-1}). This difference may be due to the accuracy of linear regression between ΔG^0_{ads} and T, where the linear regression coefficient (R^2) obtained from EIS is 0.9905 and from LP is 0.9254. The ΔS^0_{ads} values in the presence of indole derivates are large and positive, meaning that an increasing in disordering takes place in going from reactants to the metal-adsorbed species reaction complex [49].

The integrated version of the Van't Hoff equation expressed by (11) can be also deduced the ΔH^0_{ads} value [50]:

$$\ln K = -\frac{\Delta H^0_{ads}}{RT} + \text{constant}. \quad (11)$$

Figures 20 and 21 show the plot of $\ln K$ versus $1/T$ from both methods which gives straight lines with slopes of $(-\Delta H^0_{ads}/R)$ and intercepts of $(-\Delta S^0_{ads}/R + \ln 1/55.5)$. The calculated ΔH^0_{ads} using the Van't Hoff equation are −8.25 (from IES) and −8.41 (from LP) kJ mol^{-1} for norharmane. For harmane, the calculated ΔH^0_{ads} are −3.88 (from IES) and −7.40 (from LP) kJ mol^{-1}. These results confirm the exothermic behaviour of the adsorption of the inhibitors on the steel surface, therefore, the physisorption process. Values of

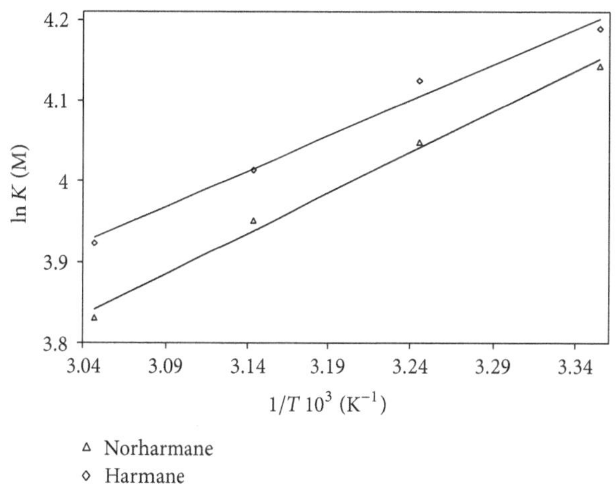

FIGURE 20: Van't Hoff plots for the C38 steel/indoles/1 M HCl system evaluated from LP.

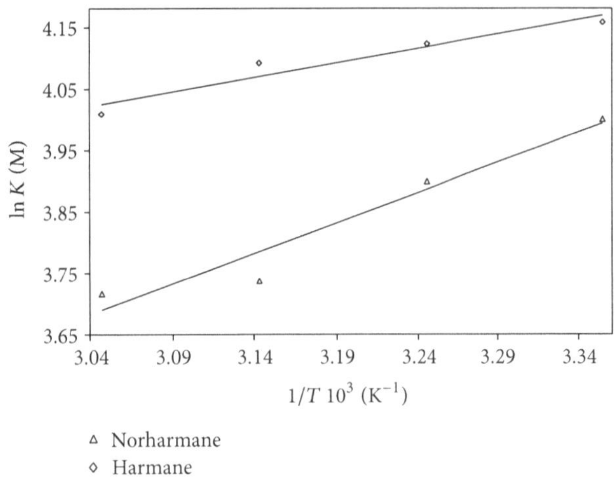

FIGURE 21: Van't Hoff plots for the C38 steel/indoles/1 M HCl system evaluated from IES.

ΔH^0_{ads} obtained by both (9) and (10) are in good agreement. Moreover, the deduced ΔS^0_{ads} values (-20.01 J mol^{-1} K^{-1} from IES and -20.82 J mol^{-1} K^{-1} from LP) and (-36.14 J mol^{-1} K^{-1} from IES and -24.63 J mol^{-1} K^{-1} from LP) for norharmane and harmane, respectively, are very close to those obtained in Tables 7 and 8.

4. Conclusion

Norharmane and harmane were found to be inhibitors for C38 steel corrosion in 1 M HCl solution and their inhibition efficiency increased with increasing concentration. Indole derivates are found to affect both the anodic and cathodic processes and their inhibition efficiencies decrease with increasing temperature. Impedance measurements show that the interface studied (in presence of the compounds investigated) is described by a relatively simple structural model, which shows the presence of one time constant and dispersed capacitance. For the most temperatures 45 and 55°C, an inductive loop appeared at low-frequency values. The structural parameters calculated show decrease of the resistance R_{ct} and increase of the capacitance. The inhibition is attributed to physisorption of the heterocyclic compound on the steel surface and blocking its active sites. The activation parameters of corrosion process such as activation energy, E_a, activation enthalpy, ΔH_a, and activation entropy, ΔS_a, were calculated from the dependence of corrosion rates on the temperature from polarization and impedance measurements. Also, the thermodynamic adsorption parameters were calculated and commented on from polarization and impedance measurements. The adsorption of indole derivates on the mild steel surface in 1 M HCl obeyed the Langmuir adsorption isotherm model at all studied temperatures.

References

[1] A. Khamis, M. M. Saleh, and M. I. Awad, "Synergistic inhibitor effect of cetylpyridinium chloride and other halides on the corrosion of mild steel in 0.5 M H_2SO_4," *Corrosion Science*, vol. 66, pp. 343–349, 2013.

[2] R. Sharmila, N. Selvakumar, and K. Jeyasubramanian, "Evaluation of corrosion inhibition in mild steel using cerium oxide nanoparticles," *Materials Letters*, vol. 91, pp. 78–80, 2013.

[3] L. Fragoza-Mar, O. Olivares-Xometl, M. A. Domínguez-Aguilar, E. A. Flores, P. Arellanes-Lozada, and F. Jiménez-Cruz, "Corrosion inhibitor activity of 1,3-diketone malonates for mild steel in aqueous hydrochloric acid solution," *Corrosion Science*, vol. 61, pp. 171–184, 2012.

[4] G. Quartarone, L. Ronchin, A. Vavasori, C. Tortato, and L. Bonaldo, "Inhibitive action of gramine towards corrosion of mild steel in deaerated 1.0 M hydrochloric acid solutions," *Corrosion Science*, vol. 64, pp. 82–89, 2012.

[5] M. K. Pavithra, T. V. Venkatesha, M. K. Punith Kumar, and H. C. Tondan, "Inhibition of mild steel corrosion by Rabeprazole sulfide," *Corrosion Science*, vol. 60, pp. 104–111, 2012.

[6] S. E. Nataraja, T. V. Venkatesha, K. Manjunatha, B. Poojary, M. K. Pavithra, and H. C. Tandon, "Inhibition of the corrosion of steel in hydrochloric acid solution by some organic molecules containing the methylthiophenyl moiety," *Corrosion Science*, vol. 53, no. 8, pp. 2651–2659, 2011.

[7] M. A. Hegazy, H. M. Ahmed, and A. S. El-Tabei, "Investigation of the inhibitive effect of p-substituted 4-(N,N,N-dimethyldodecylammonium bromide)benzylidene-benzene-2-yl-amine on corrosion of carbon steel pipelines in acidic medium," *Corrosion Science*, vol. 53, no. 2, pp. 671–678, 2011.

[8] M. M. Solomon, S. A. Umoren, I. I. Udosoro, and A. P. Udoh, "Inhibitive and adsorption behaviour of carboxymethyl cellulose on mild steel corrosion in sulphuric acid solution," *Corrosion Science*, vol. 52, no. 4, pp. 1317–1325, 2010.

[9] I. Ahamad and M. A. Quraishi, "Mebendazole: new and efficient corrosion inhibitor for mild steel in acid medium," *Corrosion Science*, vol. 52, no. 2, pp. 651–656, 2010.

[10] A. Popova, E. Sokolova, S. Raicheva, and M. Christov, "AC and DC study of the temperature effect on mild steel corrosion in acid media in the presence of benzimidazole derivatives," *Corrosion Science*, vol. 45, no. 1, pp. 33–58, 2003.

[11] H. L. Wang, H. B. Fan, and J. S. Zheng, "Corrosion inhibition of mild steel in hydrochloric acid solution by a mercapto-triazole compound," *Materials Chemistry and Physics*, vol. 77, no. 3, pp. 655–661, 2003.

[12] A. Popova, M. Christov, and A. Vasilev, "Inhibitive properties of quaternary ammonium bromides of N-containing heterocycles on acid mild steel corrosion—part II: EIS results," *Corrosion Science*, vol. 49, no. 8, pp. 3290–3302, 2007.

[13] M. M. Singh and A. Gupta, "Inhibition of mild steel corrosion in formic acid solution," *Bulletin of Electrochemistry*, vol. 12, no. 9, pp. 511–516, 1996.

[14] M. H. Wahdan, A. A. Hermas, and M. S. Morad, "Corrosion inhibition of carbon-steels by propargyltriphenylphosphonium bromide in H_2SO_4 solution," *Materials Chemistry and Physics*, vol. 76, no. 2, pp. 111–118, 2002.

[15] F. Bentiss, M. Lebrini, and M. Lagrenée, "Thermodynamic characterization of metal dissolution and inhibitor adsorption processes in mild steel/2,5-bis(n-thienyl)-1,3,4-thiadiazoles/hydrochloric acid system," *Corrosion Science*, vol. 47, no. 12, pp. 2915–2931, 2005.

[16] M. M. Osman, "Hexadecyl trimethyl ammonium bromide as an effective inhibitor for the corrosion of steel in sulphuric acid solution," *Anti-Corrosion Methods and Materials*, vol. 45, no. 3, pp. 176–180, 1998.

[17] W. Durnie, R. D. Marco, A. Jefferson, and B. Kinsella, "Development of a structure activity relationship for oil field corrosion

inhibitors," *Journal of the Electrochemical Society*, vol. 146, no. 5, pp. 1751–1756, 1999.

[18] L. Tang, G. Mu, and G. Liu, "The effect of neutral red on the corrosion inhibition of cold rolled steel in 1.0 M hydrochloric acid," *Corrosion Science*, vol. 45, no. 10, pp. 2251–2262, 2003.

[19] T. Zhao and G. Mu, "The adsorption and corrosion inhibition of anion surfactants on aluminium surface in hydrochloric acid," *Corrosion Science*, vol. 41, no. 10, pp. 1937–1944, 1999.

[20] M. H. Wahdan, "The synergistic inhibition effect and thermodynamic properties of 2-mercaptobenzimidazol and some selected cations as a mixed inhibitor for pickling of mild steel in acid solution," *Materials Chemistry and Physics*, vol. 49, no. 2, pp. 135–140, 1997.

[21] M. Lebrini, F. Robert, H. Vezin, and C. Roos, "Electrochemical and quantum chemical studies of some indole derivatives as corrosion inhibitors for C38 steel in molar hydrochloric acid," *Corrosion Science*, vol. 52, no. 10, pp. 3367–3376, 2010.

[22] A. P. Varela, M. D. Miguel, A. L. Maqanita, H. D. Burrows, and R. S. Becker, "Beta-carboline photosensitizers. 3. Studies on ground and excited state partitioning in AOT/water/cyclohexane microemulsions," *The Journal of Physical Chemistry*, vol. 99, no. 43, pp. 16093–16100, 1995.

[23] M. Lebrini, M. Lagrenée, H. Vezin, M. Traisnel, and F. Bentiss, "Experimental and theoretical study for corrosion inhibition of mild steel in normal hydrochloric acid solution by some new macrocyclic polyether compounds," *Corrosion Science*, vol. 49, no. 5, pp. 2254–2269, 2007.

[24] D. A. Jones, *Principles and Prevention of Corrosion*, MaCmillan, New York, NY, USA, 1992.

[25] H. J. Flitt and D. P. Schweinsberg, "A guide to polarisation curve interpretation: deconstruction of experimental curves typical of the $Fe/H_2O/H^+/O_2$ corrosion system," *Corrosion Science*, vol. 47, no. 9, pp. 2125–2156, 2005.

[26] H. J. Flitt and D. Paul Schweinsberg, "Evaluation of corrosion rate from polarisation curves not exhibiting a Tafel region," *Corrosion Science*, vol. 47, no. 12, pp. 3034–3052, 2005.

[27] F. Mansfeld, "Tafel slopes and corrosion rates obtained in the pre-Tafel region of polarization curves," *Corrosion Science*, vol. 47, no. 12, pp. 3178–3186, 2005.

[28] E. Khamis, "Effect of temperature on the acidic dissolution of steel in the presence of inhibitors," *Corrosion*, vol. 46, no. 6, pp. 476–484, 1990.

[29] S. Martinez and I. Stern, "Thermodynamic characterization of metal dissolution and inhibitor adsorption processes in the low carbon steel/mimosa tannin/sulfuric acid system," *Applied Surface Science*, vol. 199, no. 1–4, pp. 83–89, 2002.

[30] T. Szauer and A. Brand, "Mechanism of inhibition of electrode reactions at high surface coverages—II," *Electrochimica Acta*, vol. 26, no. 9, pp. 1219–1224, 1981.

[31] N. M. Guan, L. Xueming, and L. Fei, "Synergistic inhibition between *o*-phenanthroline and chloride ion on cold rolled steel corrosion in phosphoric acid," *Materials Chemistry and Physics*, vol. 86, no. 1, pp. 59–68, 2004.

[32] M. A. Amin, S. S. Abd El-Rehim, E. E. F. El-Sherbini, and R. S. Bayyomi, "The inhibition of low carbon steel corrosion in hydrochloric acid solutions by succinic acid—part I: weight loss, polarization, EIS, PZC, EDX and SEM studies," *Electrochimica Acta*, vol. 52, no. 11, pp. 3588–3600, 2007.

[33] H. J. W. Lenderrink, M. V. D. Linden, and J. H. W. De Wit, "Corrosion of aluminium in acidic and neutral solutions," *Electrochimica Acta*, vol. 38, no. 14, pp. 1989–1992, 1993.

[34] M. Kedam, O. R. Mattos, and H. Takenouti, "Reaction model for iron dissolution studied by electrode impedance I . Experimental results and reaction model," *Journal of the Electrochemical Society*, vol. 128, no. 2, pp. 257–266, 1981.

[35] M. A. Veloz and I. Gonzalez, "Electrochemical study of carbon steel corrosion in buffered acetic acid solutions with chlorides and H_2S," *Electrochimica Acta*, vol. 48, no. 2, pp. 135–144, 2002.

[36] M. S. Morad, "An electrochemical study on the inhibiting action of some organic phosphonium compounds on the corrosion of mild steel in aerated acid solutions," *Corrosion Science*, vol. 42, no. 8, pp. 1307–1326, 2000.

[37] E. M. Sherif and S. M. Park, "Effects of 1,4-naphthoquinone on aluminum corrosion in 0.50 M sodium chloride solutions," *Electrochimica Acta*, vol. 51, no. 7, pp. 1313–1321, 2006.

[38] Z. B. Stoynov, B. M. Grafov, B. Savova-Stoynova, and V. V. Elkin, *Electrochemical Impedance*, Nauka, Moscow, Russia, 1991.

[39] F. B. Growcock and R. J. Jasinski, "Time resolved impedance spectroscopy of mild steel in concentrated hydrochloric acid," *The Journal of the Electrochemical Society*, vol. 136, no. 8, pp. 2310–2314, 1989.

[40] P. Li, J. Y. Lin, K. L. Tan, and J. Y. Lee, "Electrochemical impedance and X-ray photoelectron spectroscopic studies of the inhibition of mild steel corrosion in acids by cyclohexylamine," *Electrochimica Acta*, vol. 42, no. 4, pp. 605–615, 1997.

[41] D. A. Lopez, S. N. Simison, and S. R. de Sanchez, "The influence of steel microstructure on CO_2 corrosion. EIS studies on the inhibition efficiency of benzimidazole," *Electrochimica Acta*, vol. 48, no. 7, pp. 845–854, 2003.

[42] S. Martinez and M. Metikos-Hukovic, "A nonlinear kinetic model introduced for the corrosion inhibitive properties of some organic inhibitors," *Journal of Applied Electrochemistry*, vol. 33, no. 12, pp. 1137–1142, 2003.

[43] X. Wu, H. Ma, S. Chen, Z. Xu, and A. Sui, "General equivalent circuits for Faradaic electrode processes under electrochemical reaction control," *The Journal of the Electrochemical Society*, vol. 146, no. 5, pp. 1847–1853, 1999.

[44] H. Ma, X. Cheng, G. Li et al., "The influence of hydrogen sulfide on corrosion of iron under different conditions," *Corrosion Science*, vol. 42, no. 10, pp. 1669–1683, 2000.

[45] M. Stern and A. L. Geary, "Electrochemical polarization I . A theoretical analysis of the shape of polarization curves," *The Journal of the Electrochemical Society*, vol. 104, no. 1, pp. 56–63, 1957.

[46] H. A. Sorkhabi, B. Shaabani, and D. Seifzadeh, "Effect of some pyrimidinic Shciff bases on the corrosion of mild steel in hydrochloric acid solution," *Electrochimica Acta*, vol. 50, no. 16-17, pp. 3446–3452, 2005.

[47] J. Flis and T. Zakroczymski, "Impedance study of reinforcing steel in simulated pore solution with tannin," *The Journal of the Electrochemical Society*, vol. 143, no. 8, pp. 2458–2464, 1996.

[48] L. Tang, G. Mu, and G. Liu, "The effect of neutral red on the corrosion inhibition of cold rolled steel in 1.0 M hydrochloric acid," *Corrosion Science*, vol. 45, no. 10, pp. 2251–2262, 2003.

[49] G. Banerjee and S. N. Malhotra, "Contribution to adsorption of aromatic amines on mild steel surface from HCl solutions by impedance, UV, and Raman spectroscopy," *Corrosion*, vol. 48, no. 1, pp. 10–15, 1992.

[50] D. Do, *Adsorption Analysis: Equilibria and Kinetics*, Imperial College Press, New York, NY, USA, 1998.

2

Corrosion Behavior of Fe40Al Alloy with Additions of Ti, Ag, and Cr in Molten KCl + $ZnCl_2$

R. Ademar,[1] J. G. Gonzalez-Rodriguez,[1] J. Uruchurtu,[1] J. Porcayo-Calderon,[2] V. M. Salinas-Bravo,[2] G. Dominguez-Patiño,[3] and A. Bedolla-Jacuinde[4]

[1] *U.A.E.M.-C.I.I.C.A.P. Av. Universidad 1001, Col. Chamilpa, 62500-Cuernavaca, Mor., Mexico*
[2] *I.I.E., Gerencia de Procesos Termico, Av. Reforma 120, Temixco, Mor., Mexico*
[3] *U.A.E.M.-Facultad de Biología, Av. Universidad 1001, Col. Chamilpa, 62500-Cuernavaca, Mor., Mexico*
[4] *Instituto de Investigaciones Metalúrgicas, Universidad Michoacana de San Nicolás de Hidalgo. Edificio "U", Ciudad Universitaria, 58066—Morelia, Michoacán, Mexico*

Correspondence should be addressed to J. G. Gonzalez-Rodriguez, ggonzalez@uaem.mx

Academic Editor: W. Ke

The effect of 2.5 at.% Cr, Ti, and Ag on the corrosion behavior of Fe40Al intermetallic alloy in KCl-$ZnCl_2$ (1 : 1 M) at 670°C has been evaluated by using electrochemical techniques. Techniques included potentiodynamic polarization curves, linear polarization resistance (LPR), and electrochemical impedance spectroscopy (EIS) measurements. Results have shown that additions of both Cr and Ti were beneficial to the alloy, since they decreased its corrosion rate, whereas additions of Ag was detrimental, since its additions increased the corrosion rate, although the alloy was passivated by adding Ag or Cr. The best corrosion performance was obtained with the addition of Cr, whereas the highest corrosion rate was obtained by adding Ag. This is explained in terms of the stability of the corrosion products formed film.

1. Introduction

Sodium and potassium impurities present in the form of chloride or sulfates are very corrosive constituents under certain combustion conditions such as waste incinerators and biomass-fired boilers [1, 2]. Early failure of the thermal components frequently occurs due to the complex reactions between the metallic materials and the hostile combustion environment. Incineration has become a viable technology for disposing of various types of wastes, including municipal, hospital, chemical, and hazardous. Problems with process equipment resulting from fireside corrosion have been frequently encountered in incinerators. The major problem is the complex nature of the feed (waste) as well as corrosive impurities which form low-melting point compounds with heavy and alkali metal chloride which prevents the formation of protective oxide scales and then causes an accelerated degradation of metallic elements [1]. In particular, under reducing conditions such as those typical of the operation of waste gasification plants or even under localized reducing conditions, which frequently arise in the case of incorrect operation of waste incineration systems, it is difficult to form protective oxide scales such as Cr_2O_3, SiO_2, and Al_2O_3 on the surface of structural materials. Thus, the corrosion attack can be further enhanced under reducing atmospheres in the presence of salt deposits [2].

The effect of individual KCl, NaCl, and their mixtures with heavy metal chlorides or sulfates on the corrosion behavior of a series of alloy systems has been studied in detail so far [3–9]. It is generally realized that Cr is not as effective element for corrosion resistance of Fe-base and Ni-based alloys due to chloride attack. In contrast, alumina- (Al_2O_3) forming alloys exhibit promising candidate materials considering the better high temperature resistance of alumina over chromia and their properties such as lower cost, low density, high strength, and good wear resistance [10–13].

In order to obtain more information on the corrosion mechanisms taking place, electrochemical techniques are more appropriated than gravimetric methods. Since a molten salt acts as an electrolyte, electrochemical techniques

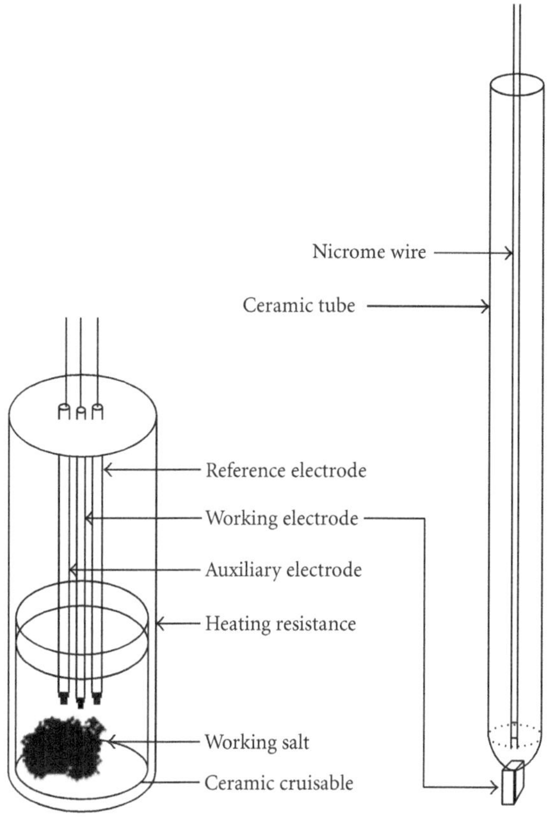

FIGURE 1: Schematic diagram showing the experimental setup.

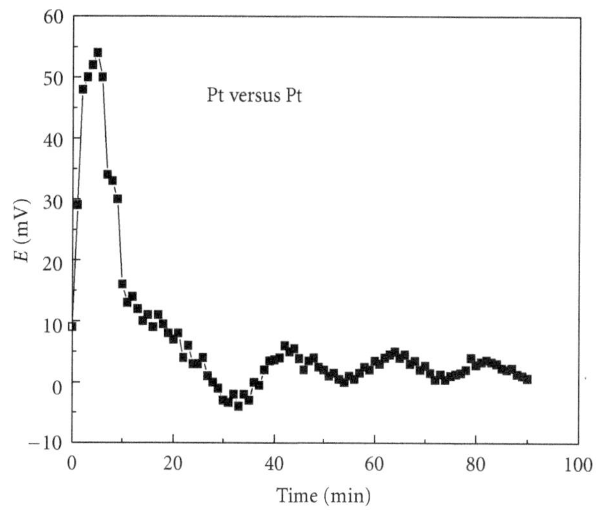

FIGURE 2: Potential fluctuations of a platinum reference electrode versus another platinum reference electrode.

such as potentiodynamic polarization curves and linear polarization resistance (LPR) measurements can be used just like Ducati et al. [14] showed in one of the earliest works on this topic. Zeng et al. used electrochemical impedance spectroscopy (EIS) studies for the corrosion of two-phase Cu-15Al alloy in the eutectic (Li, K)$_2$CO$_3$ at 650°C [15] and for Ni_3Al alloy in (Li, Na, K)$_2$SO$_4$ at 650°C [16]. They found that the complex plane of the electrochemical impedance suggested that the fast corrosion rate of these alloys was due to the formation of a nonprotective oxide, and that the corrosion process was controlled by the diffusion of oxidants. Shirvani et al. [17] used polarization curves to study the corrosion performance of the slurry Si-modified aluminide coating on the nickel base superalloy In-738LC exposed to Na_2SO_4—20 wt. % NaCl melt at 750°C. Zhu et al. [18] used Tafel extrapolation, linear polarization resistance, chronopotentiometry, and EIS to determine corrosion rates of iron and iron-based alloys in molten carbonate melts for both cathode and anode in molten carbonate fuel cell (MCFC) environments. Cuevas-Arteaga and coworkers [19–21] have successfully used electrochemical techniques to study the corrosion of materials used in sulfate+vanadate environments and obtained information about both the type and mechanisms of corrosion on these environments. This work deals with an electrochemical study of the effect of adding 2.5 at. % Cr, Ti, and Ag to a Fe40Al intermetallic alloy corroded in KCl-$ZnCl_2$ (1 : 1 M) at 670°C.

2. Experimental Procedure

Cast ingots of binary Fe-40 at. % Al and ternary Fe40Al-X (X = 2.5 at. % Cr, Ti or Ag) alloys were fabricated using a high-frequency vacuum induction furnace. Iron, Aluminum, Chromium of high purity (99.9%) were placed in an alumina crucible placed inside a graphite crucible in order to be induction melted under vacuum. The molten Fe-40Al, Fe40Al-X alloys were poured into a rectangular copper mould. The produced ingots were cut by a diamond wheel cutter in small rectangular parallelepiped pieces of surface area ranging from 1.5 to 2.5 cm^2. Electrochemical techniques included potentiodynamic polarization curves, linear polarization resistance (LPR), and electrochemical impedance spectroscopy (EIS) measurements. Details of the experimental setup for the electrochemical cell used in this work are given elsewhere [22]. The body of the cell was a 15 mL ceramic crucible. The most important elements were a reference and auxiliary electrodes, made of a 0.5 mm diameter platinum (Pt) wire inside a mullite tube and filled with commercial refractory, ceramic cement as shown in Figure 1. The amount of salt in each run was 0.5 g for an exposed area of 1.0 cm^2 under static conditions, without stirring the molten salt. The electrical contact was made by welding an 80 wt. % Cr—20 Ni, Nicrome wire to the specimen. The stability of the platinum reference electrode was measured by measuring its free corrosion potential versus another identical platinum reference electrode and the results are shown in Figure 2. We can see that during the first ten minutes the potential value shifted towards more active values for up to 55 mV more positive, but after this, the potential value dropped and reached a more or less stable value after approximately one hour or so. Polarization curves were obtained by polarizing the specimens from −500 to +1000 mV_{Pt} with respect to the free corrosion potential value, E_{corr}, at a scanning rate of 1.0 mV/s. Corrosion current density values, I_{corr}, were calculated by using the Tafel

Table 1: Electrochemical parameters obtained from the polarization curves.

Alloy	E_{corr} (mV)	I_{corr} (mA/cm^2)	E_{pas} (mV)	I_{pas} (mA/cm^2)	E_{pit} (mV)
Fe-40Al	−330	0.3	—	—	—
Fe-40Al-2.5Cr	−975	0.8	−562	50	−337
Fe-40Al-2.5Ag	−98	9	87	7	360
Fe-40Al-2.5Ti	156	0.4	—	—	—

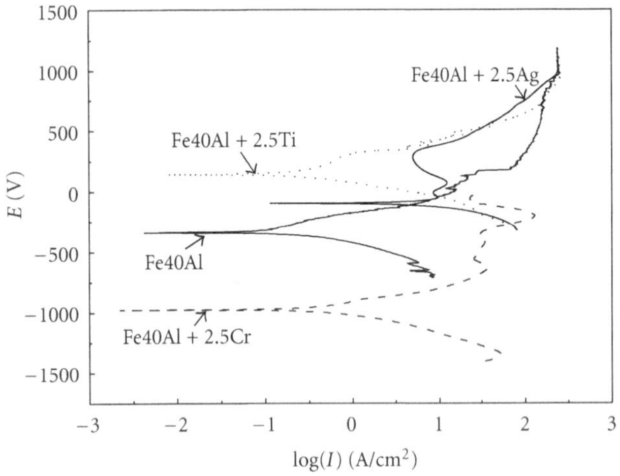

Figure 3: Polarization curves for Fe40Al alloyed with 2.5% Cr, Ti, and Ag in the KCl + $ZnCl_2$ mixture at 670°C.

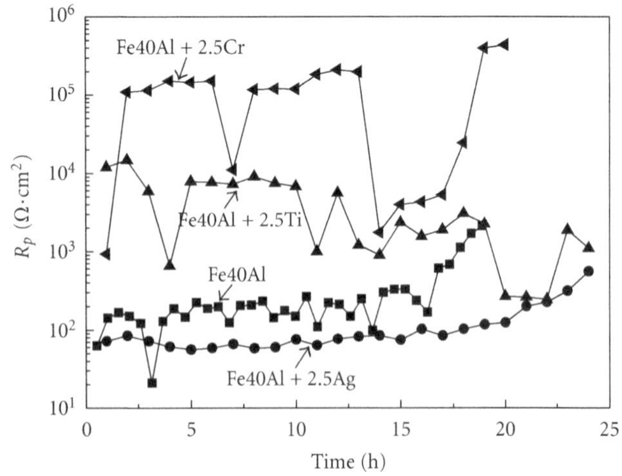

Figure 4: Change in the R_p value with time for Fe40Al alloyed with 2.5% Cr, Ti, and Ag in the KCl + $ZnCl_2$ mixture at 670°C.

extrapolation method, which is valid for the sweep rates used here according to [21] and taking an extrapolation zone of ± 250 mV around the E_{corr} value once it was stable. LPR measurements were carried out by polarizing the specimen from −10 to +10 mV with respect to E_{corr}, at a scanning rate of 1 mV/s every 20 minutes during 24 hours. EIS measurements were done in the frequency interval of 0.05 to 30000 Hz at the E_{corr} value by using a PC4-300 Gamry potentiostat. The amplitude of the input sine-wave voltage was 10 mV as recommended by previous works done on this topic [15, 16]. Working salt consisted of 500 mg/cm^2 of an eutectic mixture of KCl-$ZnCl_2$, 1 : 1 M, analytical grade for each test. The testing temperature was 670°C in static air condition. Prior the tests, the surfaces of specimens were prepared by the standard technique of grinding with SiC from 240 to 600 grit emery paper, washed with water, and degreased with acetone. Some specimens taken after the LPR measurements were mounted in bakelite and polished to analyze the corroded surface in a scanning electron microscope (SEM) with an accelerating voltage of 10 keV. Microchemical analysis was carried out with energy dispersive of X-ray analyzer (EDX) attached to the SEM.

3. Results and Discussion

Polarization curves for Fe40Al alloyed with 2.5% Cr, Ti, or Ag are shown in Figure 3. This figure shows that unalloyed Fe40Al intermetallic alloy shows an active behavior only, with an E_{corr} value close to −330 mV and a corrosion rate, expressed in terms of the I_{corr} value, close to 0.3 mA/cm^2. The anodic current density increases as the potential is made more anodic, but around 100 mV a kind of anodic limiting current density is reached, maybe because of the formation of a layer of corrosion products. With the addition of both Ti and Ag, the E_{corr} is made more active, reaching values of 156 and −98 mV, respectively. Similarly, the corrosion current density value increases in all cases, especially with the addition of 2.5Ag. Thus, the addition of Ag increases the corrosion current density value and shifted the E_{corr} value towards the active region, which shows that this element is detrimental for the base alloy. On the other hand, a passive region was found with Cr and Ag at a passivating potential value, E_{pas}, of −562 and 87 mV, respectively, reaching a much higher passive current density, I_{pas}, with Cr (around 35 mA/cm^2) than that reached with Ti (close to 5 mA/cm^2). Table 1 summarizes all these parameters, which shows the beneficial effect of both Cr and Ti.

The change in the linear polarization resistance value with time, R_p, for the different alloys in the KCl + $ZnCl_2$ mixture are shown in Figure 4. It can be seen that by alloying with either Ti or Cr the R_p value increases for up to 3 orders of magnitude in the case of Cr, and thus, a decrease in the corrosion rate. This can be due to the formation of an Al_2O_3 layer, probably reinforced with Cr_2O_3. In both cases, the R_p value shows a decrease, but after that it increases once again, indicating a disruption of the external protective

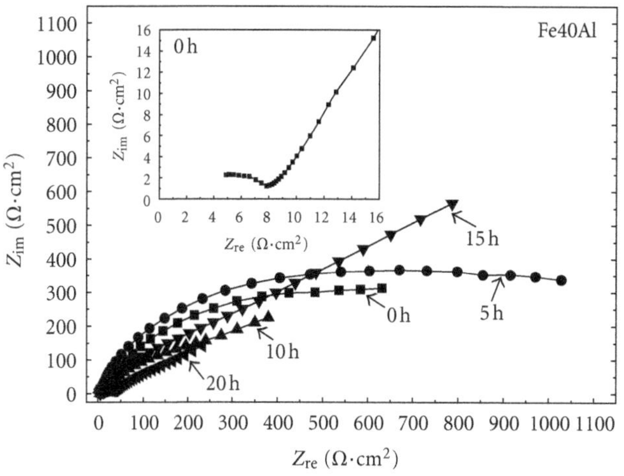

FIGURE 5: Nyquist diagrams for Fe40Al in the KCl + $ZnCl_2$ mixture at 670°C.

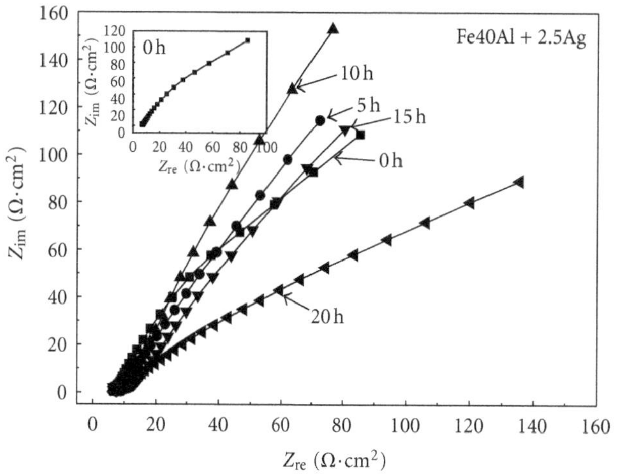

FIGURE 7: Nyquist diagrams for Fe40Al + 2.5Ag in the KCl + $ZnCl_2$ mixture at 670°C.

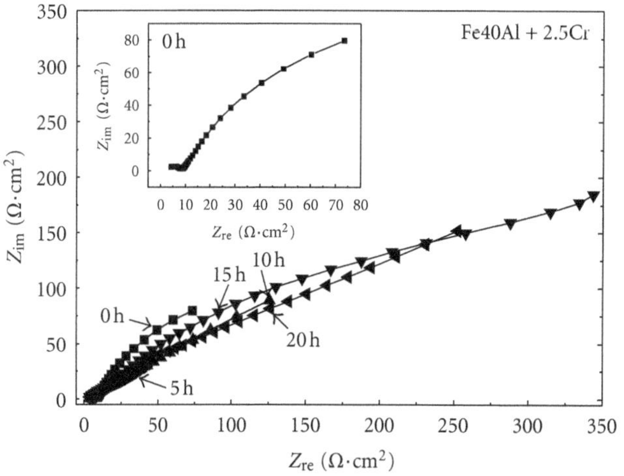

FIGURE 6: Nyquist diagrams for the Fe40Al + 2.5Cr alloy corroded in KCl + $ZnCl_2$ at 670°C.

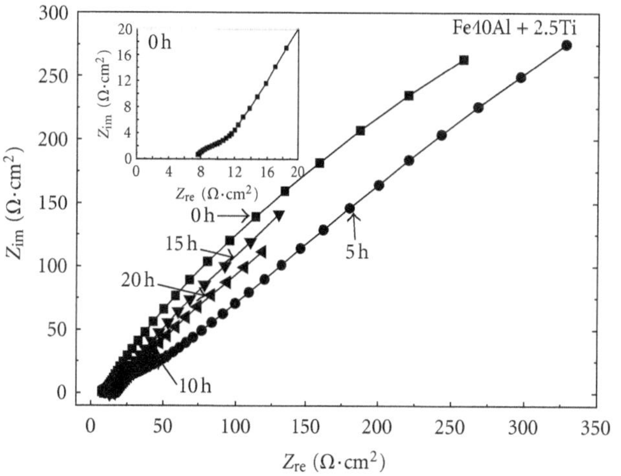

FIGURE 8: Nyquist diagrams for Fe40Al + 2.5Ti in the KCl + $ZnCl_2$ mixture at 670°C.

corrosion products layer, but a rapid re-establishment of it. Unlike this, the addition of 2.5Ag brings a decrease in the R_p value, and thus, an increase in the corrosion rate. The R_p value for unalloyed Fe40Al and Fe40Al + 2.5Ag intermetallic alloys shows a trend to increase as time increases, indicating a decrease in the corrosion rate and an increase in the protectiveness of the external layer, although this effect is marginal, not sufficient to reach the R_p values shown by the alloys containing either Cr or Ti. This means that the oxides formed in presence of Ag are more unstable as compared to those formed in presence of either Cr or Ti, leading to an increase in the corrosion rate for the alloy containing Ag.

Nyquist diagrams for unalloyed Fe40Al exposed to the KCl + $ZnCl_2$ mixture is shown in Figure 5, where it can be seen that for times shorter than 5 hours approximately, data describe a single capacitive-like semicircle, with its center in the real axis and its diameter increasing with time indicating that the corrosion process is under control by charge transfer from the metal to the electrolyte through the double electrochemical layer. For longer times than 5 hours or so, however, in addition to the high-frequency capacitive semicircle there is what seems an uncompleted low-frequency semicircle. The diameter of the low-frequency semicircle is much bigger than that for the high-frequency semicircle and it has a trend to decrease as time elapses. The emergence of a large capacitive loop at the low-frequency part may indicate the formation of a protective scale on the alloy surface.

For the alloy with addition of 2.5Cr, Figure 6, the data describes, at all exposure times, a semicircle at high and intermediate frequency values followed by a second semicircle at low-frequency values, with its diameter increasing with time. This time, the high-frequency semicircle diameter is lower than that for unalloyed Fe40Al base alloy, Figure 5. A similar behavior is observed for alloys with the addition of either Ag or Ti (Figures 7 and 8). However, the low-frequency

Table 2: Parameters used to simulate the EIS data for Fe-40Al-2.5Ag alloy.

Time (h)	R_s (Ohm cm^2)	R_{ct} (Ohm cm^2)	Y_{dl} (Ohm cm^{-2}s^{-n})	n_{dl}	Y_f (Ohm cm^{-2}s^{-n})	n_f	R_f (Ohm cm^2)
0	7.2	6.2	8.9×10^{-3}	0.80	9.2×10^{-3}	0.78	77.6
5	8.1	8.5	8.1×10^{-3}	0.77	9.9×10^{-3}	0.75	89.4
10	9.7	9.6	3.2×10^{-3}	0.71	3.8×10^{-3}	0.72	91.6
15	6.7	9.5	2.2×10^{-3}	0.51	7.5×10^{-3}	0.70	82.5
20	7.5	11.8	6.2×10^{-3}	0.32	7.4×10^{-3}	0.40	75.4

Table 3: Parameters used to simulate the EIS data for Fe- 40Al-2.5Ti alloy.

Time (h)	R_s (Ohm cm^2)	R_{ct} (Ohm cm^2)	Y_{dl} (Ohm cm^{-2}s^{-n})	n_{dl}	Y_f (Ohm cm^{-2}s^{-n})	n_f	R_f (Ohm cm^2)
0	7.2	10.7	1.3×10^{-3}	0.67	3.1×10^{-3}	0.59	9057
5	9.2	46.8	5.6×10^{-3}	0.43	2.5×10^{-3}	0.51	4736
10	9.4	11.8	3.9×10^{-2}	0.24	2.2×10^{-2}	0.59	8016
15	14	34.7	5.5×10^{-3}	0.66	7.4×10^{-3}	0.56	7784
20	12	13.5	5.7×10^{-3}	0.60	7.7×10^{-3}	0.54	2369

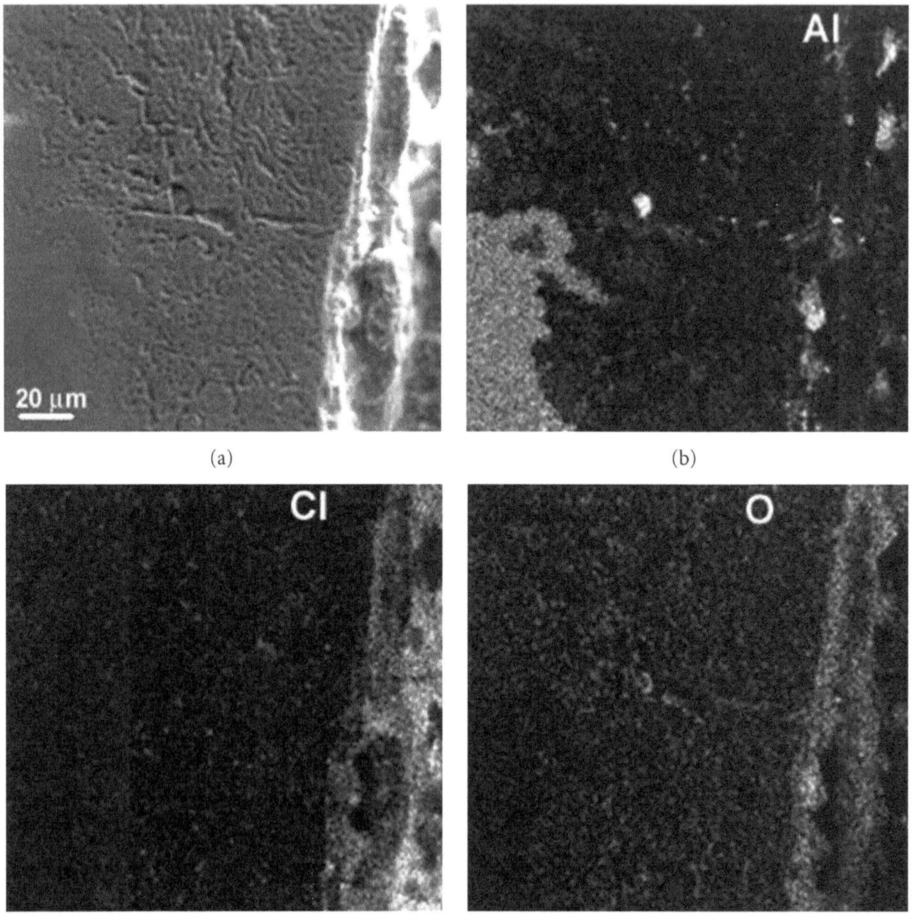

(a) (b) (c) (d)

Figure 9: Microphotograph of Fe40Al corroded in the KCl + $ZnCl_2$ mixture at 670°C together with X-ray mappings of Al, Cl, and O.

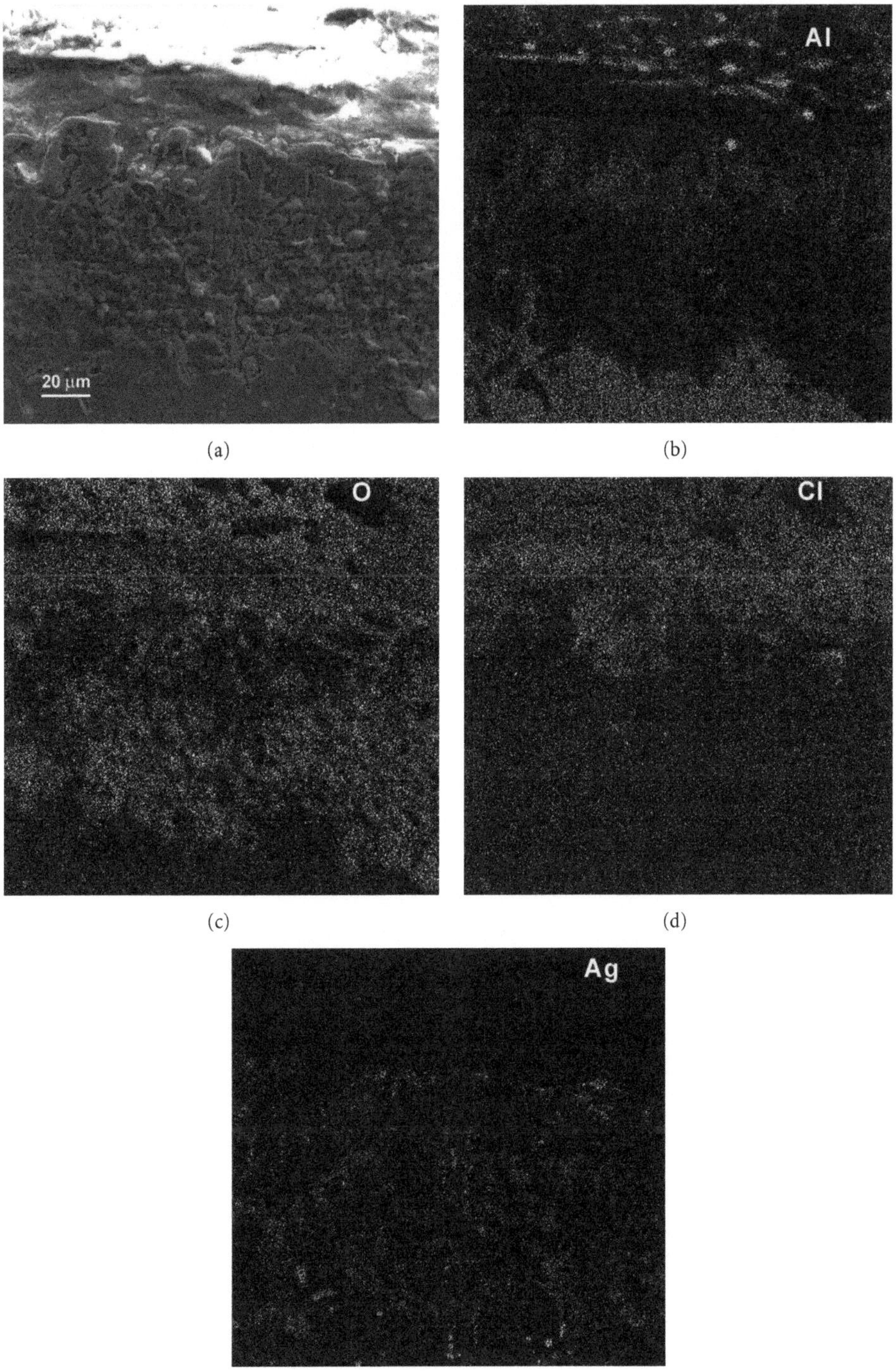

(a) (b) (c) (d) (e)

FIGURE 10: Microphotograph of Fe40Al + 2.5Ag corroded in the KCl + $ZnCl_2$ mixture at 670°C together with X-ray mappings of Al, Cl, O, and Ag.

semicircle diameter this time shows an erratic behavior, since sometimes it increases with time, and some other times it decreases. This explains why the addition of 2.5Ag increases the corrosion rate of the base alloy, since Nyquist diagrams show the formation of unstable oxides due to the disruption of the external layer formed by the corrosion products or to the dissolution of any protective oxide layer formed on top of the alloy, which is dissolved by the molten salt, Figures 3 and 4.

In order to have a better picture of this behavior, some micrographs of corroded specimens were examined in the scanning electronic microscope. Figure 9 shows a micrograph of corroded Fe40Al base alloy together with the corresponding X-ray mappings of Al, Cl, and O. The

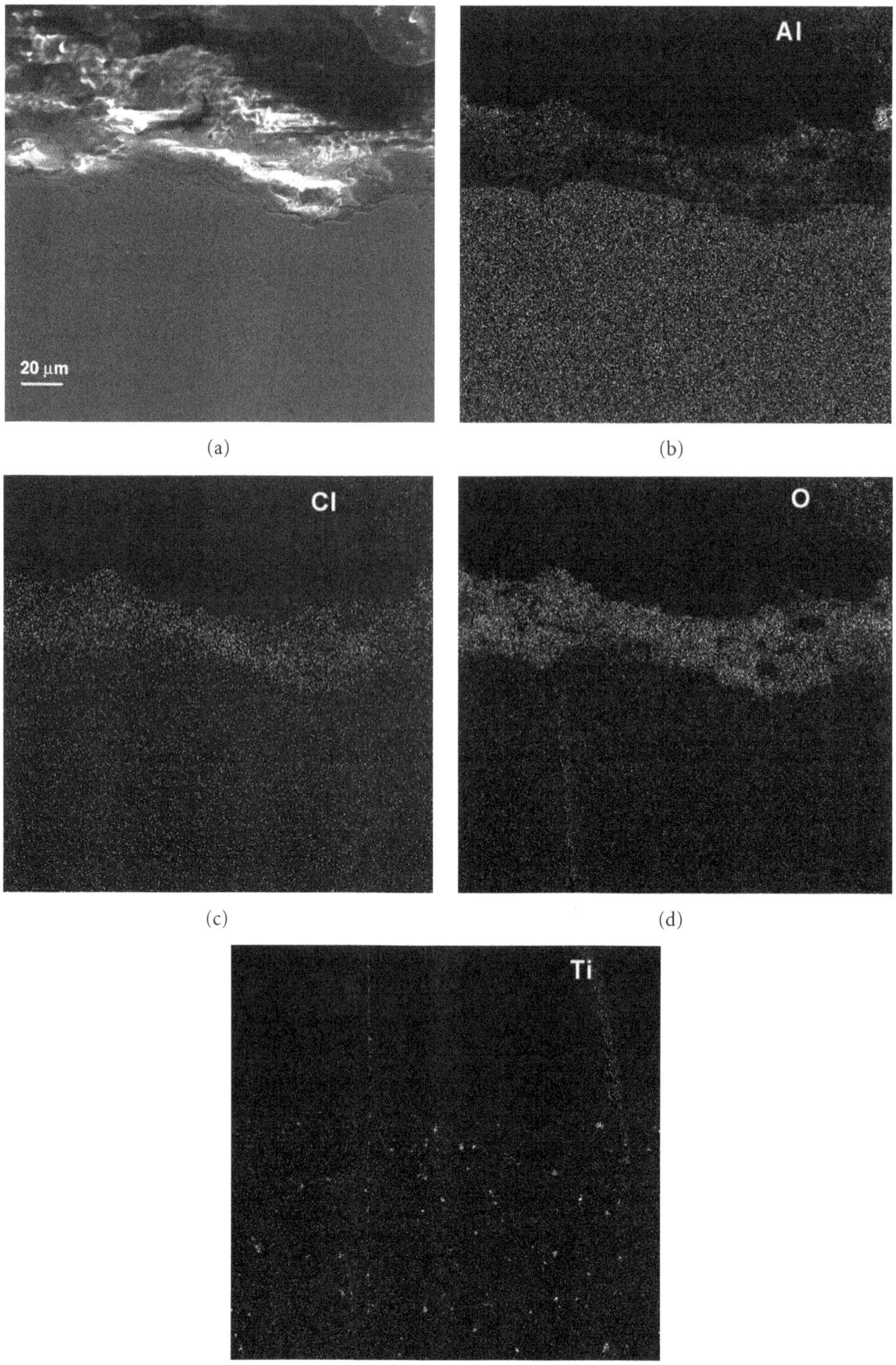

(a) (b) (c) (d) (e)

Figure 11: Microphotograph of Fe40Al + 2.5Ti corroded in the KCl + $ZnCl_2$ mixture at 670°C together with X-ray mappings of Al, Cl, O and Ti.

presence of O and Al on the external layer, although the Al-layer was not continuous, leads us to the assumption that there was some Al_2O_3 layer formed on top of the alloy. In addition to this, there is Cl present, which indicates that the molten salt has penetrated and dissolved the external, protective Al_2O_3 layer, leading to a continuous degradation of the alloy. Similar features were present the alloy with addition of Ag and Ti, Figures 10 and 11, respectively, where it can be seen the presence of an Al-containing oxide on the external surface, dissolved by chloride ions, although the oxide present in the Ag-containing alloy seems to be more porous than that present in the Ti-containing alloy. It can be seen that in the Ti-containing alloy the formed oxide is external and the presence of any Ti-containing oxide,

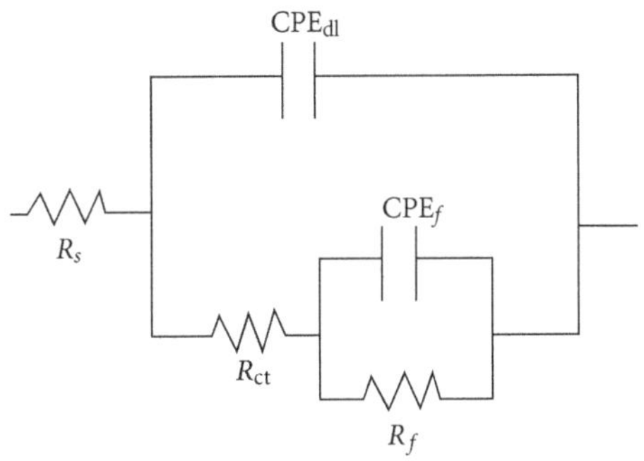

FIGURE 12: Electric circuit used to simulate the EIS data.

perhaps TiO_2 is in very negligible. On the other hand, on the Ag-containing alloy, it can be seen that Ag reacted more with O than Ti, the main corrosion product seems to be silver oxides with, such as AgO and Ag_2O and, maybe, some silver chloride, $AgCl_2$; maybe this is the reason why this alloy is less corrosion resistant than the one containing Ti.

EIS results can be represented by equivalent circuit shown on Figure 12, where R_s is the solution or electrolyte resistance, CPE_{dl} the double-layer capacitance, R_{ct} the charge transfer resistance, CPE_f the capacitance of any formed film on the surface, R_f its resistance. The impedance of a CPE is described by the expression [23]:

$$Z_{CPE} = Y^{-1}(j\omega)^{-n}, \quad (1)$$

where Y is the admittance, j is $\sqrt{-1}$ and ω is $2\pi f$ and f the used frequency. As an example of this, Tables 2 and 3 give the parameters used to simulate the EIS data for Fe40Al with 2.5Ag and 2.5Ti, respectively. It can be seen that, generally speaking, the R_{ct} values for both alloys are smaller than the corrosion products resistance, R_f, indicating that the corrosion resistance of both alloys is given by this film. However, the R_f value for the Ti-containing alloy shows an erratic behavior as time elapses, that is, the film thickness decreases and increases, indicating that the film is constantly detached from the alloy surface. In addition to this, the values of n_{dl} and n_f deviate from 1, indicating a strong dispersion effect which is common in molten salt corrosion [24]. Moreover, their values at the later stage were smaller than those at the initial stage, which may be related to the partial loss of the molten salt due to evaporation or to the roughness of the surface oxide layer.

4. Conclusions

A study of the effect of 2.5 at. % Cr, Ag, and Ti on the corrosion behavior of Fe40Al in molten KCl-$ZnCl_2$ (1 : 1 M) at 670°C has been carried out by using electrochemical techniques. Polarization curves have shown that by adding either Cr or Ag the alloy was passivated, although the passive current density was higher with the later. Addition of either Cr or Ti were beneficial, since the lowest corrosion rate was obtained by their addition, whereas the highest corrosion rate, even lower than that obtained with the unalloyed Fe40Al base alloy, was obtained by adding 2.5Ag, being the addition of this element detrimental for the alloy. EIS data showed that the corrosion rate was controlled by the formation of an external corrosion products external film, which was dissolved by the Cl-containing molten salt. The stability of this external film provides the corrosion resistance of the alloy.

References

[1] H. J. Grabke, E. Reese, and M. Spiegel, "The effects of chlorides, hydrogen chloride, and sulfur dioxide in the oxidation of steels below deposits," *Corrosion Science*, vol. 37, no. 7, pp. 1023–1043, 1995.

[2] W. M. Lu, T. J. Pan, K. Zhang, and Y. Niu, "Accelerated corrosion of five commercial steels under a $ZnCl_2$–KCl deposit in a reducing environment typical of waste gasification at 673–773 K," *Corrosion Science*, vol. 50, no. 7, pp. 1900–1906, 2008.

[3] Y. S. Li, M. Spiegel, and S. Shimada, "Corrosion behaviour of various model alloys with NaCl-KCl coating," *Materials Chemistry and Physics*, vol. 93, no. 1, pp. 217–223, 2005.

[4] Y. Shibata, "Cyclic oxidation behavior of Ni3Al–0.1B base alloys containing a Ti, Zr, or Hf addition," *Oxidation of Metals*, vol. 26, no. 3, pp. 201–207, 1986.

[5] N. Hiramatsu, Y. Uematsu, T. Tanaka, and M. Kinugasa, "Effects of alloying elements on NaCl-induced hot corrosion of stainless steels," *Materials Science and Engineering A*, vol. 120, no. 11, pp. 319–328, 1989.

[6] Y. S. Li, M. Sanchez-Pasten, and M. Spiegel, "Fundamental studies on alkali chloride induced corrosion during combustion of biomass," *Materials Science Forum*, vol. 461, no. 3, pp. 1047–1052, 2004.

[7] F. Wang, Y. Shu, and W. P. Pan, "High-temperature corrosion of A210–C carbon steel in simulated coal-combustion atmospheres," *Oxidation of Metals*, vol. 59, no. 2, pp. 201–208, 2003.

[8] C. J. Wang and Y. C. Chang, "NaCl-induced hot corrosion of Fe–Mn–Al–C alloys," *Materials Chemistry and Physics*, vol. 76, no. 2, pp. 151–157, 2002.

[9] Y. S. Li, Y. Niu, and W. T. Wu, "Accelerated corrosion of pure Fe, Ni, Cr and several Fe-based alloys induced by $ZnCl_2$–KCl at 450°C in oxidizing environment," *Materials Science and Engineering A*, vol. 345, no. 3, pp. 964–970, 2003.

[10] F. H. Stott and G. C. Wood, "Internal oxidation," *Materials Science and Technology*, vol. 4, no. 12, pp. 1072–1078, 1988.

[11] G. Han and W. D. Cho, "High-temperature corrosion of Fe_3Al in 1% Cl_2Ar," *Oxidation of Metals*, vol. 58, no. 3-4, pp. 391–413, 2002.

[12] F. Lang and Z. Yu, "Corrosion behavior of Fe–40Al sheet in N_2–11.2O_2–7.5CO_2atmospheres with various SO_2 contents at 1273 K," *Intermetallics*, vol. 11, no. 2, pp. 135–141, 2003.

[13] Y. Kawahara, "High temperature corrosion mechanisms and effect of alloying elements for materials used in waste incineration environment," *Corrosion Science*, vol. 44, no. 2, pp. 223–232, 2002.

[14] U. Ducati, G. L. Coccia, and G. Caironi, "Electrochemical evaluation of hot corrosion resistance of metallic materials," *Materials Chemistry and Physics*, vol. 8, no. 2, pp. 135–145, 1983.

[15] C. L. Zeng, P. Y. Guo, and W. T. Wu, "Electrochemical impedance spectra for the corrosion of two-phase Cu–15Al alloy in eutectic (Li, K)$_2CO_3$ at 650°C in air," *Electrochimica Acta*, vol. 49, no. 9, pp. 1445–1450, 2004.

[16] C. L. Zeng, W. Wang, and W. T. Wu, "Electrochemical impedance models for molten salt corrosion," *Corrosion Science*, vol. 43, no. 4, pp. 787–801, 2001.

[17] K. Shirvani, M. Saremi, A. Nishikata, and T. Tsuru, "Electrochemical study on hot corrosion of Si-modified aluminide coated In–738LC in Na_2SO_4–20 wt.% NaCl melt at 750°C," *Corrosion Science*, vol. 45, no. 5, pp. 1011–1021, 2003.
[18] B. Zhu, G. Lindbergh, and D. Simonsson, "Electrochemical impedance studies of the initial-stage corrosion of 310 S stainless steel beneath thin film of molten $(0.62Li, 0.38K)_2CO_3$ at 650°C," *Corrosion Science*, vol. 41, no. 9, pp. 1497–1513, 1999.
[19] C. Cuevas-Arteaga, J. Uruchurtu-Chavarín, J. Porcayo-Calderon, G. Izquierdo-Montalvo, and J. Gonzalez, "Study of molten salt corrosion of HK–40 m alloy applying linear polarization resistance. And conventional weight loss techniques," *Corrosion Science*, vol. 46, no. 11, pp. 2663–2679, 2004.
[20] C. Cuevas-Arteaga, J. Uruchurtu-Chavarín, J. G. González, G. Izquierdo-Montalvo, J. Porcayo-Calderón, and U. Cano, "Corrosion evaluation of Alloy 800 in sulfate/vanadate molten salts," *Corrosion*, vol. 60, no. 4, pp. 520–527, 2004.
[21] C. Cuevas-Arteaga, "Corrosion study of HK–40 m alloy exposed to molten sulfate/vanadate mixtures using the electrochemical noise technique," *Corrosion Science*, vol. 50, no. 3, pp. 650–663, 2008.
[22] J. G. Gonzalez-Rodriguez, O. L. Arenas, J. Porcayo-Calderon, V. M. Salinas-Bravo, M. Casales, and A. Martinez-Villafañe, "An electrochemical study of the effect of B on the corrosion of atomized Fe40Al intermetallics in molten Na_2SO_4," *High Temperature Materials and Processes*, vol. 25, no. 2, pp. 293–301, 2006.
[23] J. R. MacDonald, "Equivalent circuits for the binary electrolyte in the Warburg region," *Journal of Electroanalytical Chemistry*, vol. 223, no. 9, pp. 182–189, 1987.
[24] C. S. Ni, L. Y. Lu, C. L. Zeng, and Y. Niu, "Electrochemical impedance studies of the initial-stage corrosion of 310 S stainless steel beneath thin film of molten $(0.62Li, 0.38K)_2CO_3$ at 650°C," *Corrosion Science*, vol. 53, pp. 1018–1024, 2011.

3

Preliminary Study of Corrosion Status on Bronzes Excavated from Qin Dynasty Tombs at Xinfeng Town in China

Qian-li Fu,[1] Pu-jun Jin,[2] Xue Ling,[3] Shang-xin Zhang,[1] Wei-gang Sun,[4] and Yin Xia[1]

[1] *Emperor Qin's Terra-Cotta Warriors & Horses Museum, Lin Tong, Shaanxi Province, Xi'an 710600, China*
[2] *School of Materials Science and Engineering, Shaanxi Normal University, Chang An South Road 199#, Shaanxi Province, Xi'an 710062, China*
[3] *School of Cultural Heritage, Northwest University, Tai Bai North Road 229#, Shaanxi Province, Xi'an 710069, China*
[4] *Institute of Archaeology of Shaanxi, Yan Ta Region, Shaanxi Province, Xi'an 710043, China*

Correspondence should be addressed to Pu-jun Jin, jinpj@snnu.edu.cn

Academic Editor: Rokuro Nishimura

From 2007 to 2008, many bronze wares of Qin Dynasty were excavated from tombs at Xinfeng town. Being an important finding, these bronze wares attracted people's attention, especially for their conservation. Therefore, the corrosive products were explored by using Scanning Electron Microscope with Energy Dispersive X-ray Detector (SEM/EDS), X-Ray diffraction (XRD), and Raman spectroscopy (RM), which provided much valuable information on the conservation of these bronze wares. According to tested results, the corrosive products of bronzes were found to be comprised of cuprite (Cu_2O), covellite (CuS), lead carbonate ($PbCO_3$), and malachite ($CuCO_3 \cdot Cu(OH)_2$). Meantime, the multilayer corrosive structure was found in some samples due to the cracks in Cu_2O layer which had formed many microchannels to promote the material migration.

1. Introduction

From 2007 to 2008, about six hundred Qin Dynasty tombs were unearthed by archaeologists at Xinfeng town in Xi'an city, Shanxi province, China. It was reported as a significant archaeological excavation to dig out the largest number and scale of the Qin tombs in the central Shaanxi plain [1].

The site located south shore secondary tableland of Wei He River which was a domicile named as "Xiyi" where ceramics craftman had once lived to build Qin Shi Huang Mausoleum [2].

A lot of bronzes were unearthed from these tombs, such as bronze ding tripod, bronze dou, bronze hu, and arrow, many of which were broken and in severe corrosion [2].The bronzes are ternary alloy with the element of Cu, Sn, and Pb that would be in different corrosive situation after a long buried time [3]. In general, people had always research the corrosive mechanism of ancient bronze by analyzing on corrosive products or doing the simulated experiments [4, 5]. The analysis of corrosive products will supply important information to protect them and also provide important reference to research their corrosive mechanism. This paper studied the corrosive products from the bronze fragments by Scanning electron microscope with energy dispersive X-ray detector (SEM/EDAX), X-Ray diffraction (XRD), and Raman spectroscopy (RM) and revealed the corrosive situation and mechanism on them.

2. Samples and Methods

2.1. Test Samples. In this paper, eight samples form different bronzes were studied, as shown in Table 1. All of them were collected from bronze fragile pieces, whose section was dealt with coarse grinding and polishing before the experiment.

2.2. Instrumentation. The cross-section morphologies were examined by a digital optical microscope (Keyence, VHX-600 K, Japan), which has large depth of field and was equipped with a 3CCD camera about 54 million pixels. X-ray diffraction (XRD) was obtained in a Rigaku D/Max-3C X-ray diffractometer, equipped with a Cu Kα radiation

Table 1: The record of test samples collected from bronze fragile piece at Xinfeng town.

Number	Tombs/Implements	Sample site	Time	Location
Sample 1	M274/bronze *hu*	Abdomen	Qin	Xinfeng town
Sample 2	M83/bronze *ding*	Bottom	Qin	Xinfeng town
Sample 3	M691/bronze *ding*	Bottom	Qin	Xinfeng town
Sample 4	M301/bronze *ding*	Bottom	Qin	Xinfeng town
Sample 5	M55/bronze *ding*	Bottom	Qin	Xinfeng town
Sample 6	M268/bronze *hu*	Abdomen	Qin	Xinfeng town
Sample 7	M04/bronze *hu*	Abdomen	Qin	Xinfeng town
Sample 8	M256/bronze *pen*	Edge	Qin	Xinfeng town

Section of sample 1 Section of sample 2 Section of sample 3 Section of sample 4

Section of sample 5 Section of sample 6 Section of sample 7 Section of sample 8

Figure 1: Section photomicrograph of bronze samples.

source ($\lambda = 0.154056$ nm) in the 2θ range from 10° to 70°, tube voltage 40 KV, current 40 mA. A FEI-Quanta 200 environment scanning electron microscopy (SEM) equipped with an X-ray energy dispersive spectroscopy (EDS) was used to research samples, which were examined in the high-vacuum mode and low high-vacuum mode at an accelerating voltage about 20 kV. Roman spectra were attained by a microscopic confocal Raman Spectrometer(Renishaw in via Plus, England) employing a 514 nm laser beam excited by argon ion laser with information collection time about 10 s and accumulation time about 3~5 s.

3. Results

3.1. Optical Microscope Analysis. The cross section of samples indicated that these bronze fragile pieces were a serious corrosion, as shown in Figure 1. Many samples had similarly corrosive situation, such as sample 1, 3, 4, 5, 6, 7, and 8, which had outer green rust layer and inner red rust layer [6]. Specially, the section of sample 2 demonstrated a multilayer dust, which had the existence of alternate red and green corrosive product layer.

3.2. Backscattered Electron Phase and Component Analysis. After samples were inlaid into Acrylic resin, they were dealt with surface grinding and polishing. Then sample 1, 2, 3, and 6 were analyzed by SEM/EDS, whose electron images were attained by backscattered electron and elementary content were detected by X-ray energy dispersive spectroscopy (EDS), data as shown in Table 2.

There are four scanning area of sample1, such as Area-1 to Area-4, as shown in Figure 2. The result indicates that Area-1 and Area-2 were outer layer of corrosion with complex component, especially with high content of C and O element, which revealed that the surface layer of sample had been corroded. Meantime, the backscattered electron images of Area-3 take on a white look which due to $PbCO_3$ according to the high content of some element as 20.53% C 12.4% O and 67.07% Pb. The analyzed Area-4 revealed that a significant amount of 83.63% Cu and 16.37% Sn, which should be remnant α-phase in bronze alloy.

For sample 2, four areas were analyzed about Area-5 to Area-8, as shown in Figure 2. The analyzed Area-5 is middle area of red corrosive product layer whose component is complex with 36.46% C, 43.55% O, 14.13% Cu, 5.29% Pb and 0.57% Si, which mean that the red corrosive product layer (Cu2O) contains impurities. The complex composition of 27.48% C, 26.67% O, 3.74% Cu, and 42.11% Pb indicates that lead carbonate is main corrosive product in the other analyzed Area-6.

Five areas of sample3 were analyzed, Area-7 is α-phase in bronze alloy with 86.61% Cu and 13.39% Sn; Area-8 is

TABLE 2: The content of composite element of samples (wt%).

	Element																
Area	C	O	Cu	Sn	Pb	Ca	Si	Al	S	As	K	Fe	Mg	Ni	Cl	Na	P
Area-1	18	52.79	26.08		2.67	0.46											
Area-2	19.17	31.19	18.46	22.35	7.15		1.25	0.42									
Area-3	20.53	12.4			67.07												
Area-4			83.63	16.37													
Area-5	36.46	43.55	14.13		5.29		0.57										
Area-6	27.48	26.67	3.74		42.11												
Area-7			86.61	13.39													
Area-8	8.69	5.86	54.28	31.17													
Area-9		20.07	79.93														
Area-10	16.61	6.08	57.82						19.5								
Area-11	11.34	22.08	27.26	16.55	16.66					2.42					3.69		
Area-12	26.07	4.77	69.16														
Area-13	15.82	45.55			2.04		21.79	7.94			5.89					0.97	
Area-14	23.24	34.74	5.99		24.49	1.47	4.43	2.11				1.57	0.63	1.34			

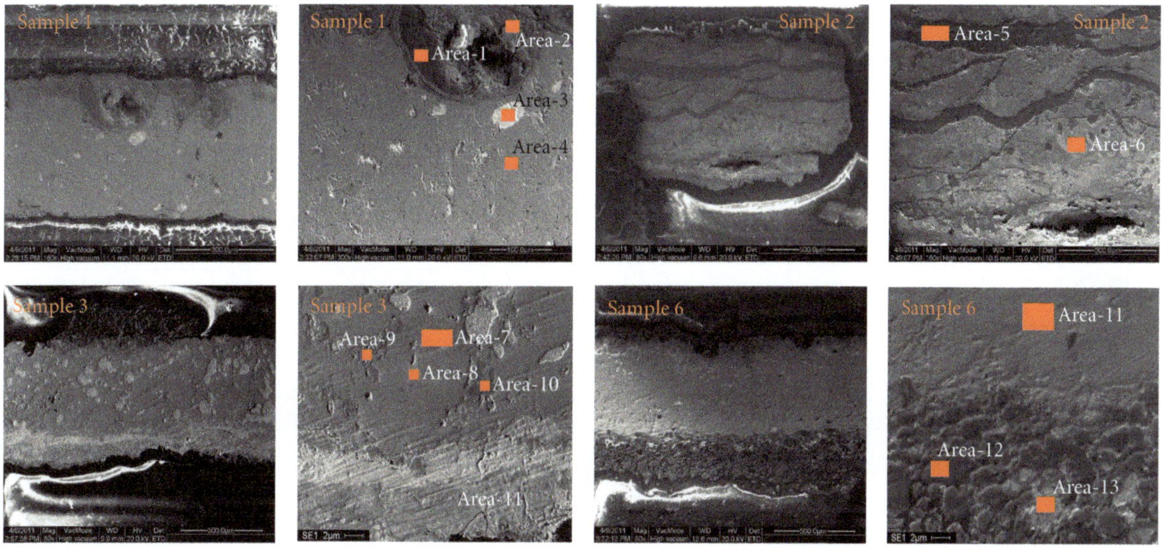

FIGURE 2: SEM photograph of sample cross section and drawing of site of energy spectrum analysis area.

the remnant α-δphase with 8.69% C, 5.86% O, 54.28% Cu, and 31.17% Sn; Area-9 is dull color area where is composed of 20.07% O and 79.93% Cu, which suggest the existence of copper oxide [7]; Area-10 is dark area with 19.5% S in which there are some copper sulfide; Area-11 shows white with complex composition and 16.66% Pb.

Sample 6 was analyzed in three areas from Area-12 to Area-14, as seen in Figure 2. In the Area-14, the high content of 24.49% Pb element was detected by EDS which is more than that in Area-13. In fact, Area-13 has complex component of 15.82% C, 45.55% O, 2.04% Pb, 21.79% Si, 7.94% Al, 5.89% K, and 0.97% Na which means this area is soil. On the other hand, Pb element wasn't found in the analyzed Area-12. It could draw a conclusion that element Pb has the trend of migration from the bronze alloy to the circumambient soil.

3.3. Secondary Electron Image of Sample 2. The secondary electron image of sample 2 showed obviously multilayer structure about corrosive layer, as seen in Figure 3. The zonal cuprite could be seen in the SEM image in which many cracks also were observed clearly.

3.4. Analysis of XRD. The XRD pattern of sample (Figure 4) showed that the peaks at 29.6°, 36.5°, 42.4°, and 61.5° can be indexed to Cu_2O (JCPDS Card No. 77-0199). Other diffraction peaks at 23.4°, 25.5°, 29.7°, 34.5°, 36.7°, 43.3°, and 49.7° can be indexed to $PbCO_3$ (JCPDS Card No. 03-0358).

3.5. Raman Spectrum. Different color corrosive product of cross section of sample 2 and sample 6 was analyzed by Micro-Raman spectroscopy. Figures 5(a) and 6(a) show

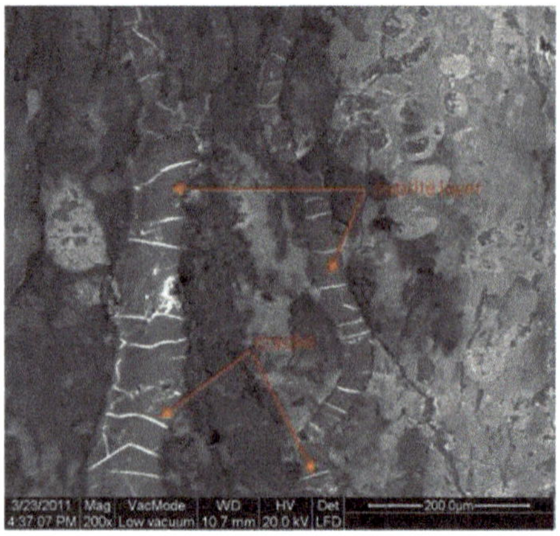

FIGURE 3: Morphology photograph of cross section of sample no. 2.

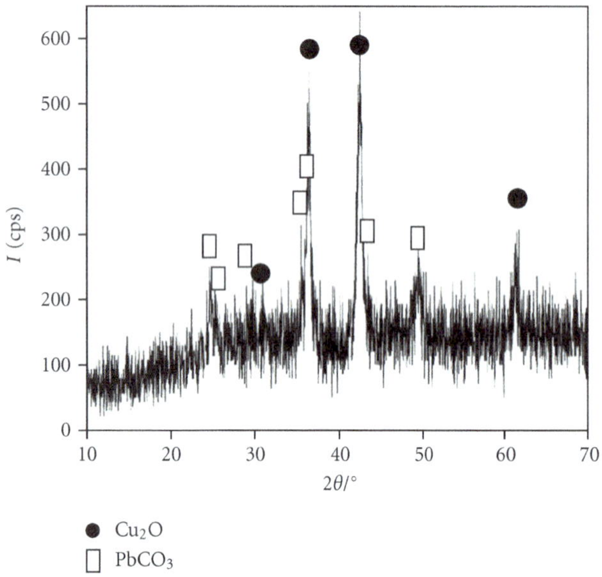

FIGURE 4: X-ray diffraction pattern of sample 1.

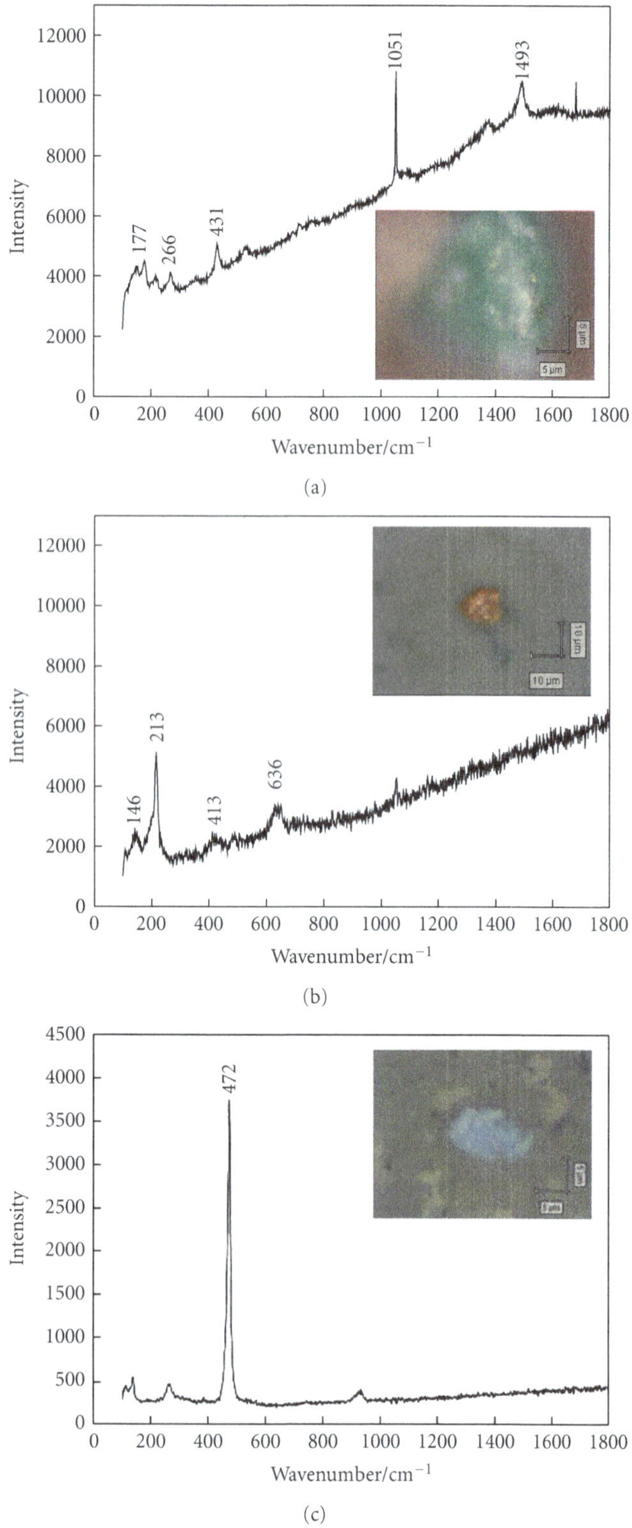

FIGURE 5: Raman spectrum of corrosion of sample 2: (a) green particle; (b) red particle; (c) blue particle.

Raman pattern of patina from two sampels which displayed vibration peaks at 1493 cm^{-1} (vs), 1051 cm^{-1} (vs), 431 cm^{-1} (vs), 266 cm^{-1} (m), and 177 cm^{-1} (s) due to malachite ($CuCO_3 \cdot Cu(OH)_2$ [8]. The red corrosive product demonstrates Raman vibration peaks at 636 cm^{-1} (s), 413 cm^{-1} (m), 213 cm^{-1} (vs), and 146 cm^{-1} (m) corresponding to cuprous oxide (Cu_2O) [9], as shown in Figures 5(b) and 6(b). Figure 5(c) shows Raman pattern of blue corrosive product from sample 2 with a vibration peak at 472 cm^{-1} (vs) due to copper sulphide (CuS) [10, 11].

4. Discussion

The analytical result of XRD indicates that two kinds of corrosive products existed in most collected test samples, which are Cu_2O and $PbCO_3$. Actually, the signal of patina was not detected by XRD, which was confirmed by Raman spectroscopy as malachite ($CuCO_3 \cdot Cu\ (OH)_2$).

Cuprite (Cu_2O) is one kind of common corrosive products of bronze wares that underground burial environment [12, 13] that would form a tough red layer on the surface of

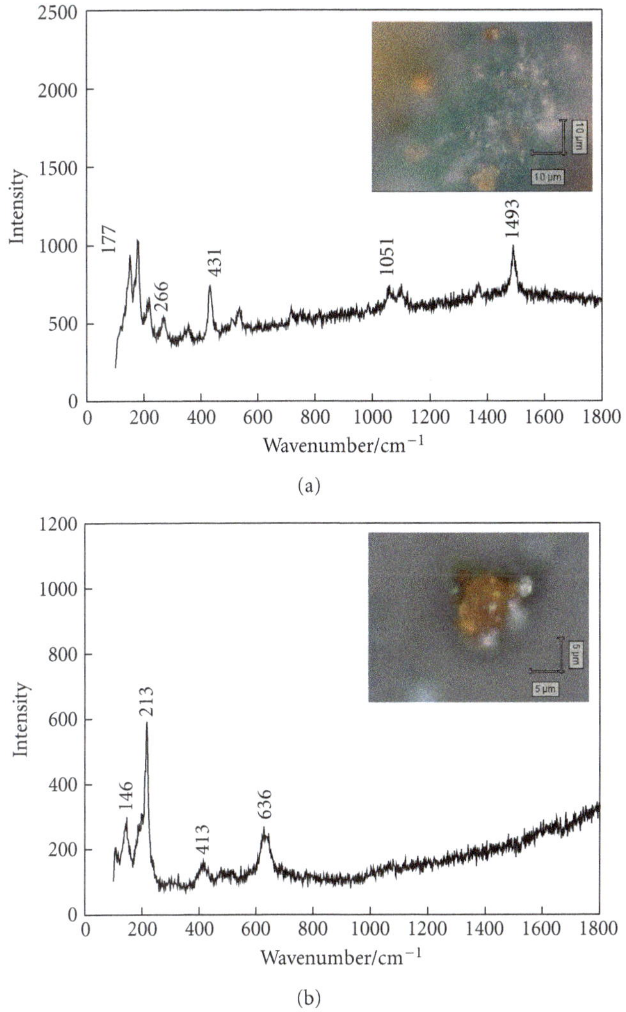

FIGURE 6: Raman spectrum of corrosion of sample 6: (a) green particle; (b) red particle.

bronze wares, which is regarded to delay corrosion of bronze alloy by keeping the harmful ion out [6, 14] and primary electrochemical reaction as follow.

Negative pole:

$$O_2 + 2H_2O + 4e \longrightarrow 4OH^-. \quad (1)$$

Positive pole:

$$Cu + H_2O \longrightarrow Cu_2O + 2H^+ + 2e. \quad (2)$$

Actually, transverse cracks observed in Cu_2O layer of cross section of sample 2 formed a free migrating pathway between the bronze alloy and the burial circumstance. The cracks in Cu_2O layer are likely to be a lot to produce along with the gradual corrosion of bronzes, its mechanism required further study.

Lead carbonate ($PbCO_3$) was confirmed by XRD that was a common lead corrosive phase in bronze wares [15]. Generally, the reaction of Pb to $PbCO_3$ will occur in burial environment as follow [16]:

$$Pb + O_2 \longrightarrow 2PbO, \quad (3)$$

$$PbO + CO_2 \longrightarrow PbCO_3. \quad (4)$$

By the result of SEM image and chemical element distribution, the sample 6 showed the migration out of Pb element form bronze alloy.

5. Conclusion

Photograph of cross section and component element data indicates that bronze substrate has been corroded severely, which must be consolidated before its restoration. The SEM image revealed transverse cracks in Cu_2O layer in sample 2 which had provided the pathway for element diffusion between the bronze alloy and the burial circumstance.

Acknowledgments

This work has been partially supported by the Sci. & Tech. Key Projects of Shaanxi Province of China under Grant no. 2009k01-43, by the Fundamental Research Funds for the Central Universities, by social science Fund of Shaanxi Normal University no. 09SYB05, by Scientific Research Project from Education Department of Shaanxi Provincial Government (2010JZ31), and by Cultural Relics Research Project of Chinese State Administration of Cultural Heritage (20090106).

References

[1] S. Wei-gang, "Xi, Li Yi and Li mountain park—concurrently discuss function of Li Yi of emperor Qin Shihuang mausoleum," *Archaeology and Cultural Relics*, no. 4, pp. 67–71, 2009.

[2] S. Wei-gang, *Study on Qin tombs unearthed from Xin feng*, M.S. thesis, Northweat University, 2009.

[3] R. F. Tylecote, "The effect of soil conditions on the long-term corrosion of buried tin-bronzes and copper," *Journal of Archaeological Science*, vol. 6, no. 4, pp. 345–368, 1979.

[4] L. Yu, Y. Si-xun, and Z. Xiao-mei, "Study on bronze corrosion from Jin State tombs of Zhou dynasty, Tian ma-Qu village," *Sciences of Conzervation and Archeaology*, vol. 2, pp. 9–18, 2000.

[5] L. Yan-ping, C. Xiao-lin, C. Yu-bin, and W. Zhi-qiang, "Bronze sample corrosion experiment in soil in archaeological site," *Archaeology and Cultural Relics*, no. 6, pp. 95–98, 2006.

[6] J. Pu-Jun, Q. Ying, H. Ya-Li, H. Si-Ping, and H. Wen-Hu, "Analysis of bronze corrosion product from No.1,2 tombs of Jiu lian dun tombs," *Jian Han Archaeology*, no. 1, pp. 112–119, 2009.

[7] P. Valério Pedro, R. J. C. Silva, A. M. Monge Soares et al., "Technological continuity in early iron age bronze metallurgy at the South-Western Iberian Peninsula—a sight from Castro dos Ratinhos," *Journal of Archaeological Science*, vol. 37, no. 8, pp. 1811–1819, 2010.

[8] L. I. McCann, K. Trentelman, T. Possley, and B. Golding, "Corrosion of ancient Chinese bronze money trees studied by

Raman microscopy," *Journal of Raman Spectroscopy*, vol. 30, no. 2, pp. 121–132, 1999.

[9] M. C. Bernard and S. Joiret, "Understanding corrosion of ancient metals for the conservation of cultural heritage," *Electrochimica Acta*, vol. 54, no. 22, pp. 5199–5205, 2009.

[10] S. Xu, Q. Wang, J. H. Cheng, Q. H. Meng, and Y. Jiao, "Preparation and characteristics of porous CuS microspheres consisted of polycrystalline nanoslices," *Powder Technology*, vol. 199, no. 2, pp. 139–143, 2010.

[11] P. J. Jin, Z. Q. Yao, M. L. Zhang, Y. H. Li, and H. P. Xing, "A pigment (CuS) identified by micro-raman spectroscopy on a chinese funerary lacquer ware of West Han Dynasty," *Journal of Raman Spectroscopy*, vol. 41, no. 2, pp. 222–225, 2010.

[12] F. J. R. de Oliveira, D. C. B. Lago, L. F. Senna, L. R. M. de Miranda, and E. D'Elia, "Study of patina formation on bronze specimens," *Materials Chemistry and Physics*, vol. 115, no. 2-3, pp. 761–770, 2009.

[13] W. Hui-Zhen, W. Guo-Feng, Z. Hong, and J. Pu-Jun, "Corrosion mechanism and conservation study of bronze ge of Shang dynasty," *Archaeology and Cultural Relics*, no. 3, pp. 93–96, 2001.

[14] J. J. Shim and J. G. Kim, "Copper corrosion in potable water distribution systems: influence of copper products on the corrosion behavior," *Materials Letters*, vol. 58, no. 14, pp. 2002–2006, 2004.

[15] J. La-jiang, "Science and technology archaeology study of bronze of early Qin dynasty," Northwest University, 2010.

[16] E. Rocca, C. Rapin, and F. Mirambet, "Inhibition treatment of the corrosion of lead artefacts in atmospheric conditions and by acetic acid vapour: use of sodium decanoate," *Corrosion Science*, vol. 46, no. 3, pp. 653–665, 2004.

Laser Surface Alloying of 316L Stainless Steel with Ru and Ni Mixtures

M. B. Lekala,[1] J. W. van der Merwe,[1] and S. L. Pityana[2]

[1] *School of Chemical and Metallurgical Engineering, University of the Witwatersrand and DST/NRF Centre of Excellence in Strong Materials, Johannesburg 2001, South Africa*
[2] *Council for Scientific and Industrial Research, National Laser Centre, Pretoria 0001, South Africa*

Correspondence should be addressed to J. W. van der Merwe, josias.vandermerwe@wits.ac.za

Academic Editor: Muthukannan Duraiselvam

The surfaces of AISI 316L stainless steel were laser alloyed with ruthenium powder and a mixture of ruthenium and nickel powders using a cw Nd:YAG laser set at fixed operating parameters. The microstructure, elemental composition, and corrosion characteristics of the alloyed zone were analyzed using optical and scanning electron microscopy (SEM), energy dispersive X-ray spectroscopy (EDX), and corrosion potential measurements. The depth of alloyed zone was measured using the AxioVision program and found to be approximately 1.8 mm for all the alloyed specimens. Hardness profile measurements through the surface-substrate interface showed a significant increase from 160 HV for the substrate to a maximum of 247 HV for the alloyed layer. The sample laser alloyed with 80 wt% Ni-20 wt% presented the most noble corrosion potential (E_{corr}) of −0.18 V and the lowest corrosion current density (i_{corr}).

1. Introduction

Minor additions of ruthenium to the bulk volume of steels resulted in a significant improvement of corrosion resistance in many reducing environments [1]. Ruthenium modified alloys possesses properties which render them candidate alloys to replacing the expensive nickel-based alloys which are currently used in more aggressive corrosion environments [2]. However, owing to the high-cost associated with ruthenium, bulk alloying is currently not a feasible means, although opportunities to explore the method exist. For instance, Streicher [3] observed a synergistic benefit when ruthenium and nickel were added together to steels. This observation offers an opportunity to reduce the amount of ruthenium per bulk volume added in the alloying process, yet presents significant improved corrosion resistance. Thus, minor additions of ruthenium together with nickel present an economically sound approach of modifying corrosion properties of alloys. Furthermore, since corrosion is a surface phenomenon, an equally cost-effective approach is to add these only on the surface, where protection is most required.

Laser surface modification techniques have been extensively studied for selective improvement of surfaces for wear, hardness, and corrosion [4–7]. The laser surface alloying technique is particularly applicable in cases where a change in the chemical composition and microstructure of the surface is required. The laser surface alloying technique enables external alloying elements to be added into the bulk material via a laser generated melt pool. Generally, the external alloy material is either preplaced on the desired surface of the substrate or fed into the melt pool. The alloy material reacts with the molten surface to create a new alloyed layer which exhibits unique properties, such as high corrosion resistance. Studies of corrosion resistance on laser surface alloyed of ferritic Fe-40Cr alloy with ruthenium (Ru) showed improvement when subjected to certain corrosive environments [8]. Despite these beneficial results, surface modification of alloys with ruthenium (Ru) to improve their corrosion properties has received little attention. In South Africa, there has been a renewed interest in investigating the effects of adding small amounts of Ru metals for improving corrosion resistance of stainless steels. It has been established that small additions of Ru in the steel shifts the corrosion potential to more noble values [1]. Streicher [3] pointed out that addition of Ni also has potential benefits.

Table 1: Composition of the powders used on sample.

Alloy name	Powder composition		Substrate material
	Nickel wt%	Ruthenium wt%	
Alloy 1	0	100	AISI 316
Alloy 2	80	20	AISI 316
Alloy 3	50	50	AISI 316

In this study, laser surface alloying of AISI 316 with Ru and with (Ru + Ni) mixed powders was investigated. The effect of the amounts of Ru added into the alloyed layer was studied by selecting a mixture with the following compositions (80 wt% Ni + 20 wt% Ru) and (50 wt% Ni + 50 wt% Ru). The objective(s) has been to keep the Ru content low, while maintaining superior corrosion resistance. The microstructure, chemical composition, hardness, and corrosion behaviour of the alloyed layers were analysed using SEM, EDX, hardness tester, and corrosion potentials.

2. Experimental Details

2.1. Laser Surface Alloying. The AISI 316 stainless steel was cut into $10 \times 5 \times 0.5$ cm rectangular plates. The surfaces of the plates were sandblasted and cleaned with acetone prior to laser surface alloying. Nickel and ruthenium were in the form of powders of commercial purity, 99.6 wt% and 99.9 wt%, respectively, were used to surface alloy AISI 316 stainless steel samples. The powders were mixed to specific Ru : Ni wt% proportions, and Table 1 shows the powder weight ratios used on each sample. The mixed powders were preplaced on the steel surface using a chemical binder. The thickness of the preplaced powder coatings could be controlled to approximately 1 mm.

The laser surface alloying was performed with a Rofin Sinar DY044 continuous wave Nd : YAG laser. A 600 μm optical fibre was used to guide the laser beam to the laser processing head. The processing head was mounted on a KUKA's articulated robot arm. The laser head was set at a fixed distance of 12 mm above the substrate. The laser process parameters were kept the same for all samples. The laser power, beam diameter, and scan speed were 4 kW, 4 mm, and 0.8 m/s, respectively. These parameters were chosen after a number of experimental tests done to establish acceptable alloys. Large coated areas were made by parallel tracks overlapped by 2 mm. The whole laser surface alloying operation was carried out in an inert argon environment to prevent oxidation.

2.2. Material Characterization. Metallographic specimens were prepared by cutting the samples transversally across the alloyed layer and mounting the pieces separately in a bakelite or/and lucite powder using a mounting press. The samples were etched electrolytically in 60 wt% nitric acid in distilled water at 1.5 V for 20 s. The microstructures and the elemental composition profile were evaluated using Zeiss Axiotech 25 HD microscope and JSM 5800 LV SEM with energy dispersive X-ray spectroscopy (EDS), respectively.

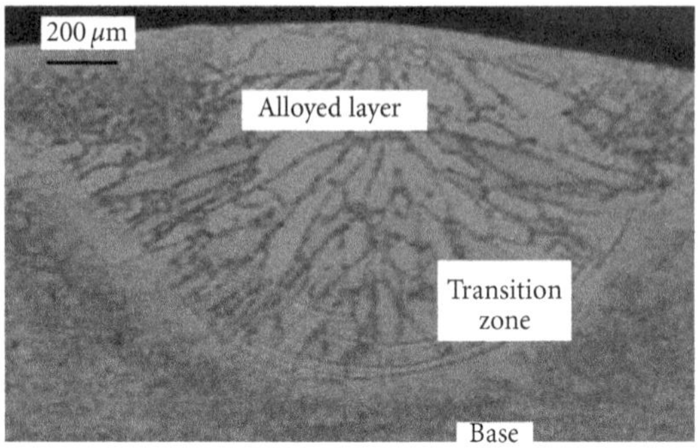

Figure 1: Typical cross section of the laser alloyed zone.

The hardness was determined using a Future-Tech FM-700 Vickers Micro-hardness testing instrument.

2.3. Electrochemical Tests. The corrosion tests were carried out in 80% sulphuric acid solution which was kept at 60°C using a thermostat-controlled bath. The corrosion performance of the laser alloyed surfaces was evaluated by means of electrochemical polarization measurements using an Autolab potentiostat, which utilizes platinum as the opposite electrode and a saturated silver-silver chloride electrode as the reference electrode. Potentiodynamic polarization curves were obtained for each alloy. A scanning rate of 0.1 mV/sec was used to conduct all the measurements.

3. Results and Discussion

3.1. Microstructure of the Laser Alloyed Layers. Figure 1 shows an optical micrograph of the cross sectional view of Alloy 1; sectioned perpendicular to the scanning direction. The width and depth of the cross section were approximately 4 mm and 1.8 mm, respectively. The alloyed zone was free of cracks and pores. The cross section showed three distinct regions: the alloyed layer, the transition zone and heat affected zone. The microstructure in the alloyed zone showed columnar grains which are attributed to rapid melting and directional solidification. Similar cross sections were obtained for the other alloys.

Figure 2 shows high magnification optical micrographs of alloys 1 and 2. The microstructure of alloy 1 shown in Figure 2(a), consisted of dendritic microstructures, while that of alloy 2 (Figure 2(b)) showed both dendrites and columnar grains of different orientations. This confirms that grain growth took place in preferential directions. The microstructures in Figure 2 are typical of weld beads which cooled under nonequilibrium conditions [9]. The ruthenium distribution was homogenous throughout the weld bead.

The average chemical composition of the laser alloyed layer obtained by EDX analysis is shown in Table 2. The elemental composition of the alloyed layer is consistent with the alloying material. The alloying material was melted and dissolved into the base material.

3.2. Microhardness. Microhardness measurements across the bead-substrate interface revealed a significant increase in

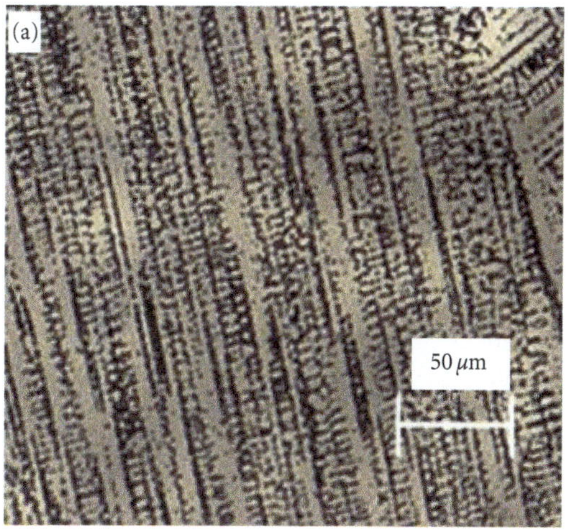

(a)

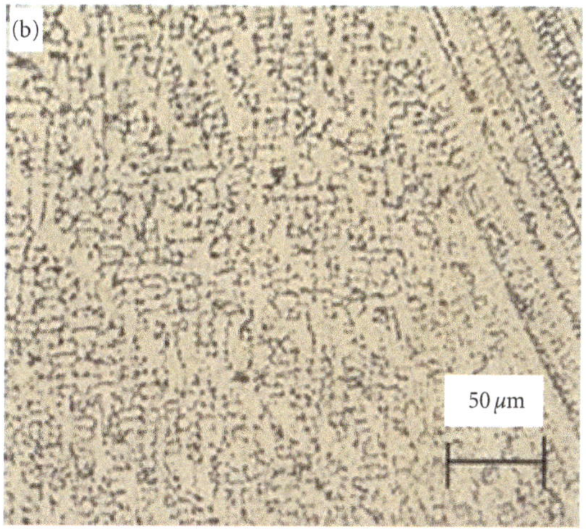

(b)

Figure 2: Optical microstructures of laser alloyed AISI 316 SS: (a) alloy 1 (9.6% Ru) and (b) alloy 2 (5.6% Ru).

Table 2: EDX elemental composition of the laser alloyed layer (wt%) with an error of 0.2 wt%.

Alloys	Al	Si	Cr	Mn	Fe	Ni	Mo	Ru
Alloy 1	0.1	0.4	16.2	1.6	61.2	9.1	1.8	9.6
Alloy 2	0.1	1.3	17.1	1.3	50.3	22.8	1.5	5.6
Alloy 3	0.1	1.2	14.7	1.3	53.5	16.9	1.0	9.5

hardness, varying from 158 HV for the AISI 316 to substrate 247 HV for the laser alloyed bead, for alloy 1, Figure 3. The increased hardness of alloys 1 and 2 can be attributed to their high Ru and Ni content, as well as the microstructural changes due to laser heating. The hardening effect of both ruthenium and nickel on alloys is well-known [10].

3.3. Potentiodynamic Polarization Behaviour. Potentiodynamic polarisation curves of the samples tested in 80% sulphuric acid at 60°C are shown in Figure 4. Although all samples exhibited passivation, there was a level of instability associated with the passive layer that formed on the bare

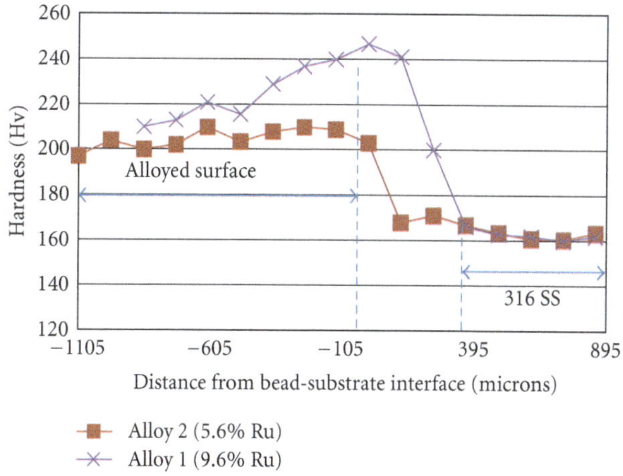

Figure 3: Variation of hardness with the distance from the bead/substrate interface in an AISI 316 stainless steel surface alloyed with Ru. The hardness measurements showed an error of approximately 3%.

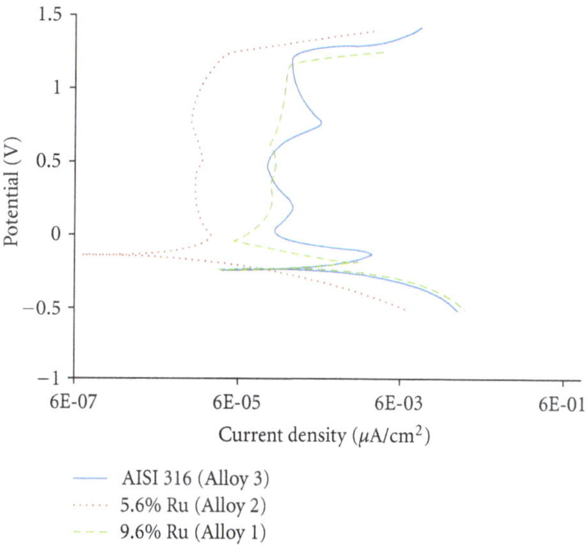

Figure 4: Comparison of potentiodynamic polarization curves for AISI 316 SS substrate, 5.6 wt% Ru steel and 9.6 wt% Ru steel.

substrate material. The bare substrate and the 9.6 wt% Ru surface corroded at similar corrosion potentials (E_{corr}) and they both passivated at comparable current densities, i_{pass}. The passive region of the 5.6 wt% Ru sample, indicated that anodic reaction was inhibited with the enhancement of polarization potential. In addition, it can be seen from Figure 4 that the 5.6 wt% Ru sample exhibited the noblest corrosion potential (E_{corr}) with a value of −0.18 V and the lowest corrosion current density (i_{corr}) with a value of approximately 5×10^{-5} A/cm^2. Conversely, the substrate showed the lowest E_{corr} with a value of −0.25 V and the highest i_{corr} with a value of 1×10^{-3} A/cm^2, which suggested that the 5.6 wt% Ru alloy exhibited the best anticorrosive properties. Despite its higher Ru content, alloy 1 showed less improvement on the corrosion properties of AISI 316 SS than alloy 2. The better corrosion properties of alloy 2 can be attributed to the higher nickel content. On the other hand, the presence of undissolved Ru particles in alloy 1 is a sign of poor alloying and most likely led to poor corrosion properties. The results show that the nature of the microstructure and other alloying elements such as Ni play a vital role on the influence of Ru on corrosion properties of steels. Higher nickel content enhanced the influence of Ru. This phenomenon has been observed by Steicher [3] and Higginson [11], although at lower concentrations of less than 0.5 wt% Ru.

4. Conclusions

Higher Ru content on the surface does not necessarily give better corrosion behavior. The effect of Ru on the corrosion behavior of the surface depended on the amount of Ni present. Higher nickel contents showed more effective improvements in corrosion resistance. Thus, laser surface alloying with Ru and Ni together present an economical way of using the two elements, because the Ru amount can be kept optimally low for maximum corrosion enhancement. The nature of the alloyed surface was greatly affected by the variation in the composition of the preplaced powder, thus showing that the laser surface alloying is system dependent. Further investigations into the surface alloying with ruthenium and nickel, particularly to identify their optimal composition for maximum corrosion improvements on various steel surface, is highly recommended. The laser surface alloying with Ru and nickel can be applied to various alloys. The surface alloying approach might be particularly suitable for thick engineering components or plates which require better corrosion properties, although the alloying should be sufficient to ensure that all the elements are in solution.

Acknowledgments

The authors would like to thank the CSIR-National Laser Centre for using its facilities. The Department of Science and Technology and the National Research Foundation, South Africa, are thanked for funding and support.

References

[1] J. H. Potgieter, A. M. Heyns, and W. Skinner, "Cathodic modification as a means of improving the corrosion resistance of alloys," *Journal of Applied Electrochemistry*, vol. 20, no. 5, pp. 711–715, 1990.

[2] P. A. Olubambi, J. H. Potgieter, and L. Cornish, "Corrosion behaviour of superferritic stainless steels cathodically modified with minor additions of ruthenium in sulphuric and hydrochloric acids," *Materials and Design*, vol. 30, no. 5, pp. 1451–1457, 2009.

[3] M. A. Streicher, "Development of pitting resistant Fe-Cr-Mo alloy," *Corrosion*, vol. 30, no. 3, pp. 77–91, 1974.

[4] L. A. B. Mabhali, S. L. Pityana, and N. Sacks, "Laser surface alloying of aluminum (AA1200) with Ni and SiC powders," *Materials and Manufacturing Processes*, vol. 25, no. 12, pp. 1397–1403, 2010.

[5] L. Mabhali, S. Pityana, and N. Sacks, "Laser alloying of Al with mixed Ni, Ti and SiC powders," *Journal of Laser Applications*, vol. 22, no. 4, pp. 121–126, 2010.

[6] S. Anandan, L. Pityana, and J. Dutta Majumdar, "Structure-property-correlation in laser surface alloyed AISI 304 stainless steel with WC + Ni + NiCr," *Materials Science and Engineering A*, vol. 536, pp. 159–169, 2012.

[7] J. D. Majumdar and I. Manna, "Laser surface alloying of AISI 304-stainless steel with molybdenum for improvement in pitting and erosion-corrosion resistance," *Materials Science and Engineering A*, vol. 267, no. 1, pp. 50–59, 1999.

[8] S. C. Tjong, J. S. Ku, and N. J. Ho, "Laser surface alloying of ferritic Fe-40Cr alloy with ruthenium," *Surface and Coatings Technology*, vol. 90, no. 3, pp. 203–209, 1997.

[9] R. W. K. Honeycombe and H. K. D. H. Bhadeshia, *Steels: Microstructure and Properties*, Butterworth-Heinemann, Oxford, UK, 2nd edition, 1995.

[10] C. A. Hampel, *Rare Metals Handbook*, New York, NY, USA, 1954.

[11] A. Higginson, *The passivation of Fe-Cr-Ru alloys in acidic solutions [Ph.D. thesis]*, University of Manchester, 1987.

Analysis of Short-Term Steel Corrosion Products Formed in Tropical Marine Environments of Panama

Juan A. Jaén,[1] Josefina Iglesias,[2] and Cecilio Hernández[2]

[1] *Departamento de Química Física, CITEN, Lab. No. 105, Edificio de Laboratorios Científicos-VIP, Universidad de Panamá, Panama, Panama*
[2] *Laboratorio de Análisis Industriales y Ciencias Ambientales, Universidad Tecnológica de Panamá, Panama, Panama*

Correspondence should be addressed to Juan A. Jaén, jjaen@ancon.up.ac.pa

Academic Editor: Flavio Deflorian

A low-carbon steel A-36 and two conventional weathering steels A-588 and COR-420 exposed at four atmospheric test stations located in (i) Tocumen, an urban site near the Pacific Ocean, (ii) Sherman-Open, (iii) Sherman-Coastal, and (iv) Sherman-Breakwater on the Caribbean coast of Panama. Kinetics of the short-term atmospheric corrosion process and the relationship with exposure time and environmental characteristics of each site were investigated. The atmospheric exposure conditions, particularly the time of wetness, deposition of chloride, and the washing effect of contaminants on the metal surface by rain are of upmost importance in determining the corrosion behaviour and composition of rust. The corrosion products were mainly identified using room temperature and low temperature (80 K) Mössbauer spectroscopy, FTIR, and X-ray powder diffraction. In all samples, γ-FeOOH and α-FeOOH were the main constituents. Maghemite (γ-Fe_2O_3), magnetite (Fe_3O_4), and Akaganeite (β-FeOOH) were also identified.

1. Introduction

The study of the corrosion products which form on steel is of upmost importance since it renders valuable information for the understanding of corrosion mechanisms. Different types of steel corrode at rates determined by the environment. Corrosivity in marine tropical regions is usually regarded as extreme due to the high temperatures, time-of-wetness, high atmospheric contaminants (i.e., chlorides and SO_2), and other factors such as rainfall and winds.

There are several recent reports [1–10] on the corrosion products formed on mild and weathering steels in different marine regions. In Panama, there have been some systematic studies of atmospheric corrosion of steels. The studies of Southwell et al. [11, 12], the cooperative efforts of the Ibero-American Map of Atmospheric Corrosiveness (MICAT) [13], and the Project Anticorrosive Protection of Metals in the Atmosphere (PATINA) [14] are noteworthy. We also have investigated phase composition and other characteristics of the corrosion products of steel exposed to the tropical climate of Panama [15–18]. In this paper, the updated corrosion results are reported for carbon and weathering steel coupons, after a short-term exposure at different locations in Panama. An attempt is also made to correlate with corrosion data and meteorological and pollution parameters. Corrosion products were characterized by means of Mössbauer spectroscopy. Powder X-ray diffraction and infrared spectroscopy were used as complementary techniques. Mössbauer spectroscopy is especially suitable for studying steel corrosion due to its capability of identifying and quantifying the frequently found amorphous or noncrystalline phases of rust [19–22].

2. Materials and Methods

The mild steel (A-36) and weathering steels (A-588 and CSN COR 420) used in this study have been described elsewhere [18]. Their chemical composition is presented in Table 1. Coupons of 150 mm × 100 mm were used for atmospheric exposure. After blasting, cleaning, and degreasing with acetone, they were exposed, on inert racks, for prolonged

Table 1: Chemical composition of the studied steels.

Type of steel	Composition (mass %)					
	C	Cu	Mn	P	Si	Cr
A-36	0.150 ± 0.004	<0.01	0.42	0.20	0.24	<0.01
A-588	0.163 ± 0.002	0.36	0.75	0.10	0.23	0.18
COR 420	0.099 ± 0.002	0.27	0.55	0.023	0.20	0.37

Table 2: Starting date of exposure to the atmosphere.

Test site	Type of steel		
	A-36	A-588	COR 420
Tocumen	01/30/07	01/30/07	07/24/07
Sherman-Open	02/02/07	02/02/07	07/23/07
Sherman-Coastal	02/02/07	02/02/07	07/23/07
Sherman-Breakwater	01/31/07	01/31/07	07/23/07

periods. Panama is a tropical country, very humid and warm, with temperatures averaging 27°C all year long. Panama effectively experiences two seasons: namely, wet season and dry season. During the dry season (from the middle of December through the end of April) there is little or no precipitation; however in the wet season (from May to the beginning December) there is frequent and heavy precipitations.

Four test sites were used for atmospheric exposure. One at an urban test site in Tocumen (T) (9°4′2″N 79°24′16″W), Panama city, Pacific coast, and three other test sites in the tropical marine environment of Fort Sherman, Caribbean coast, Fort Sherman-Open (SO) (9°21′43″N 79°57′16″W), Coastal (SC) (9°22′19″N 79°57′2″W), and Breakwater site (SB) (9°22′25″N 79°56′50″W). The Breakwater site is a splash zone, thus there are a high amount of air-borne salt particles. Taking advantage of existing facilities, the samples were exposed at 45° to the horizontal in the Tocumen site and at 30° in the Sherman sites, as has been done in other works [13, 18]. Vera et al. [23] demonstrated that the exposure angle has some influences in the corrosion rate and the morphology of the rust, but with no effect on rust composition. The program of exposure is given in Table 2. Exposure of steels A-36 and A-588 started during dry season, whereas exposure of steel COR 420 started during the rainy season. This schedule was used because of a delay in arrival of materials to the laboratory. Corrosion results going back three years and rust characterization samples resulting from one year of testing are presented.

The atmosphere at each site was characterized from meteorological and atmospheric pollution data [24]. The average annual environmental parameters and ISO classifications are shown in Table 3. After exposure, corrosion rates (corrosion penetration p) were calculated from weight losses according to standard methods [25].

The corrosion products were characterized by XDR using Cu (K_α) radiation with a Bruker D8 Advance diffractometer, equipped with a detector angle Vantek 2000 at the Materials Institute of Universidad Autónoma de México (UNAM). IR spectra were recorded using a Nicolet Avatar 360 FTIR spectrometer. The resolution of the spectra was 4 cm^{-1}. Mössbauer spectra (MS) were recorded using a conventional spectrometer of constant acceleration with a $^{57}Co(Rh)$ source of nominal activity of 10 mCi (370 MBq). A closed-cycle cryostat (CCS 850 Janis) was employed for low temperature measurements. Calibrations were done with a standard α-iron foil absorber at room temperature. The Mössbauer data was evaluated with the Recoil software (University of Ottawa, Canada) using Voigt base fitting.

3. Results and Discussion

In general, after 3 months of exposure, most of the samples had their surface totally covered by brownish-orange rust. Weathering steels were slightly darker than mild steel. Most of the specimens at the Tocumen, Sherman-Open, and Sherman-Coastal sites had a good appearance of rust on both sides (the skywards facing side and the side facing the floor), but the later showed irregularities, and a nonuniform distribution of the rust, besides tracks of the route of corrosion liquid.

The samples exposed to the atmosphere at Sherman-Breakwater site had thicker corrosion product layers of dark chocolate colour, completely covering the steel substrate. In fact, the steel was totally corroded by the edges and exfoliates with relative ease. Sections were easily detached, exposing very dark and adherent rust in the central part of the specimen, which we denominated internal rust.

3.1. Corrosivity of Exposure Stations. As can be seen in Figure 1, the rust layer had a parabolic growth trend at the Tocumen and Sherman-Open stations, but for the Breakwater station, a linear growth behaviour was noted. The situation is more complex in the Coastal station, in which linear behaviour was observed for steels A-36 and COR 420, while steel A-588 showed a parabolic behaviour. It is sometimes difficult to differentiate between parabolic and linear behaviours with short-time exposures.

The corrosion rates ($\mu m\, y^{-1}$) obtained in the different exposure stations from tests going back two years are given in Table 4. It can be ascertained that corrosion rates are moderate at the Tocumen, Sherman-Open, and Sherman-Coastal stations. SO_2 deposition is not an important parameter in the aggressiveness of the test sites, thus corrosion can be linked to the deposition rate of chlorides. The amount of Cl^- deposited changed slightly between the Tocumen, Sherman-Open, and Sherman-Coastal stations, with a tendency to decrease along with the decrease in distance from shore, as expected. In steels A-36 and A-588, there is a trend that shows corrosion rate increase going from the Tocumen site, to the Sherman-Open and Sherman-Coastal sites. The Sherman sites have similar levels of chloride deposition rates (few $mgCl^- \, m^{-2} d^{-1}$). In the case of Sherman-Breakwater, the incidence of chlorides is substantial, increasing the values of corrosion rates and corrosion to all steels. It is interesting to note that in the dry season there is a higher deposition rate of chlorides. Some authors have reported [26–28] that deposition of chlorides depends on several factors, which include wind direction, wind speeds, rainfall, and distance

Table 3: Corrosivity at test stations based on environmental data during one year of exposure.

Test site	Distance from seashore (m)	Cl^{-1} (mg·m²/d)	SO_2 (mg·m²/d)	Time of wetness %	Classification of the environment ISO 9223	Corrosion class (ISO)
Tocumen	5000	11.7	4.0	55.3	$S_1P_0\tau_4$	C_3
Fort Sherman-Open	600	22.9	3.8	66.3	$S_1P_0\tau_5$	C_4
Fort Sherman-Coastal	50	29.0	3.9	66.3	$S_1P_0\tau_5$	C_4
Fort Sherman-Breakwater	3	1490.0	4.3	66.3	$S_3P_0\tau_5$	$>C_5$

Table 4: Two year results of the corrosion rates ($\mu m\, y^{-1}$).

Exposure time (months)	Tocumen			Sherman-Open			Sherman-Coastal			Sherman-Breakwater		
	A-36	A-588	COR 420	A-36	A-588	COR 420	A-36	A-588	COR 420	A-36	A-588	COR 420
3	22.1	33.3	63.7	31.6	37.2	51.2	47.3	52.7	67.1	1011.4	801.0	177.9
6	41.8	38.8	48.0	33.9	44.5	55.0	42.5	42.3	67.1	963.0	664.2	358.6
9	38.1	39.1	34.1	36.8	37.4	29.3	37.0	36.0	44.6	771.0	504.3	456.6
12	38.5	35.2	38.3	42.2	36.6	39.8	48.9	40.9	42.6	718.8	469.3	473.1
24	24.4	37.2	19.6	26.7	22.2	17.5	40.2	29.9	—	654.6	561.8	523.6

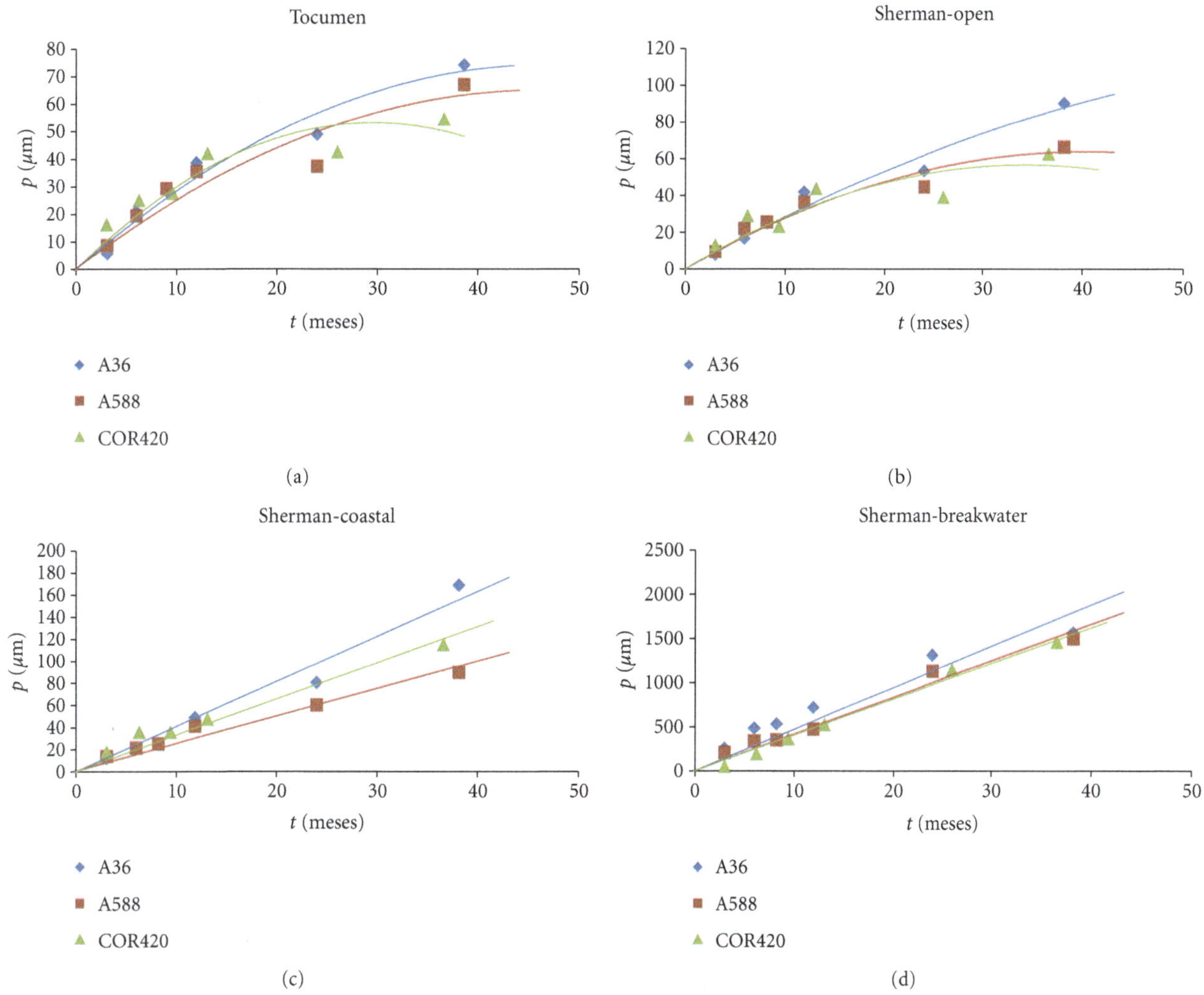

Figure 1: Penetration as a function of exposure time of the steels.

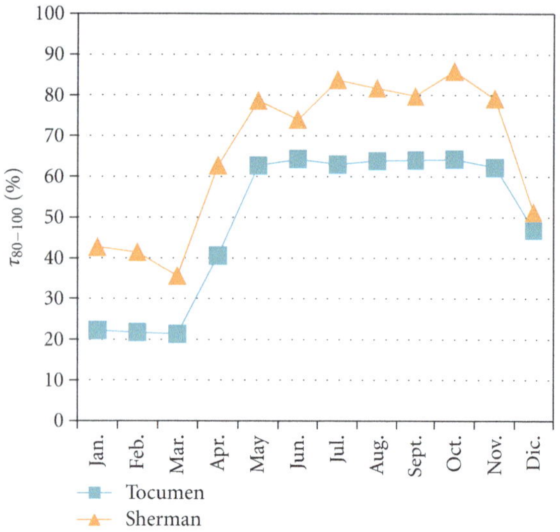

FIGURE 2: Time of wetness (RH 80%–100%) at the Tocumen in Sherman areas.

to shore. Deposition rates of chlorides in marine sites are usually in the range between 15 and 1500 $mg \cdot m^{-2} \cdot d^{-1}$ [26]. Our results from Breakwater station are much higher than usual, which is consistent with other places in the Caribbean [26, 27].

Even though it is not entirely clear, because of the outliers in the weight loss measurements, it seems that there is a transitional behaviour, particularly for steel COR 420; so that there is a different corrosion rate trend before and after a transition time (ca. 6 to 9 months). Before transition the corrosion rate is higher, afterwards there is a deceleration of corrosion. These types of changes in the kinetic of atmospheric corrosion have been reported for marine environments with high chloride content and high humidity [30]. The corrosion rates observed on weathering steel COR 420 at all test sites except Sherman-Breakwater are comparatively high in the initial phases of exposure, levelling off with time.

Exposure of the COR 420 steel began in the rainy season, while for steels A-36 and A-588 exposure started in the dry season. It is inferred that the conditions prevailing at the beginning of exposure determine the kinetics and the nature of corrosion products. Figure 2 shows that the time of wetness in the Sherman stations is larger at the Tocumen station. The greater period during which the metal surface of steel is covered by liquid films results in increased corrosion rates. On the other hand, the washing away of pollutants during the rainy season is the reason for the lower corrosion rate for COR 420 steel at the Sherman-Breakwater station. The deposition of pollutants combined with rainfall and wetting time is critical in corrosion rates. Data for the historical average annual rainfall in the Sherman area, shown in Figure 3, backs up this conclusion.

The rate of corrosion of a metal in a given environment, generally follows the Passano exponential equation $P = A \cdot t^n$ [31], where P represents the corroded amount expressed as penetration corroded (μm) or as weight loss, t is time in years, and A and n represent empirical constants. The value of the constant n is closely correlated to the progress of corrosion in subsequent years, and there may be three cases.

(a) If $n = 1$, the process occurs at constant rate, as there is a direct relationship between the exposure duration and penetration.

(b) If $n < 1$ would imply a decrease in corrosion rate.

(c) If $n > 1$ would imply an acceleration of the corrosion rate over time.

The value of n between 0.5 and 1.0 indicates that the corrosion products are not sufficiently protective. A value of less than 0.5 is obtained when the corrosion products quickly reach a parabolic growth, developing a barrier to the passage of aggressive agents, and considerably decreasing the corrosion rate.

The carbon steel A-36, the weathering steel A-588, and COR 420 follow the bilogarithmic relation at all sites of exposure, the results are presented in Table 5. It is interesting to note that the values of the intercepts are very similar between the Tocumen and Sherman-Open stations, with few significant differences, if the standard error in the measurement is considered. The values of n, determined as the slope of the bilogarithmic relation, indicate that the rust of carbon steel is not protective. Moreover, the weathering steels produce more protective rusts; the COR 420 steel being the steel which gives added protection. Therefore, systematically lower corrosion rates are observed in these steels after two years, when it begins to show the protective properties of weathering steel. The highest corrosion rates recorded at the Sherman-Breakwater station correlated with the highest category of air-borne chloride. The "n" values greater than 0.5 were observed at this exposure site revealing an acceleration of the diffusion process as a result of the rust detachment by erosion, flaking, cracking, and so forth.

3.2. Characterization of Corrosion Products. The IR analyses of the rust from samples at the different test sites were carried out. Figure 4 shows typical FTIR spectra of the corrosion products of the carbon and weathering steels exposed for three months, nine months, and one year. The characteristics bands at 1022 cm^{-1} and 746 cm^{-1} typical of lepidocrocite (γ-FeOOH) and at 885 cm^{-1} and 795 cm^{-1} typical of goethite (α-FeOOH) are observed on the rust layer formed on steels exposed at the Tocumen, Sherman-Open, and Sherman-Coastal stations. These are common phases found in steel corrosion products. In addition to these bands, there is one band at 1120 cm^{-1}, which Mendoza and Corvo [27] suggest may be due to the formation of a complex between the Fe and SO_2, but do not rule out the possibility that the band corresponds with feroxyhyte absorption (δ'-FeOOH) and/or lepidocrocite (γ-FeOOH). Given the low levels of SO_2 we have observed, it seems more likely that this band corresponds to one of the iron oxyhydroxides. In some cases intensity higher than expected [32] is observed, indicating the contribution of amorphous feroxyhyte (δ'-FeOOH), which has reported a broad and strong Fe-OH bending at 1110 cm^{-1}. The band at 1635 cm^{-1} corresponds to the

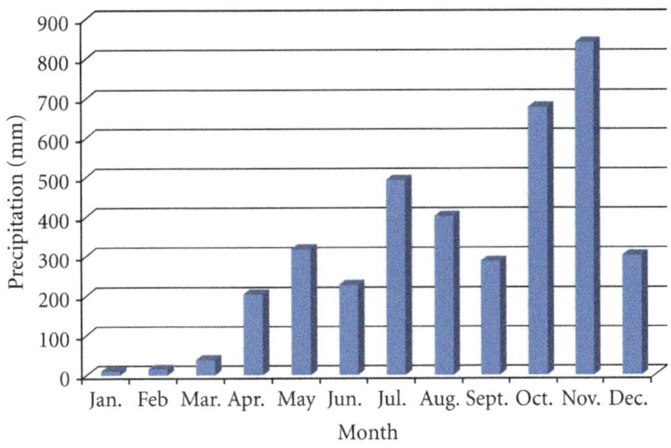

FIGURE 3: Data of the historical average annual rainfall in the Sherman area [29].

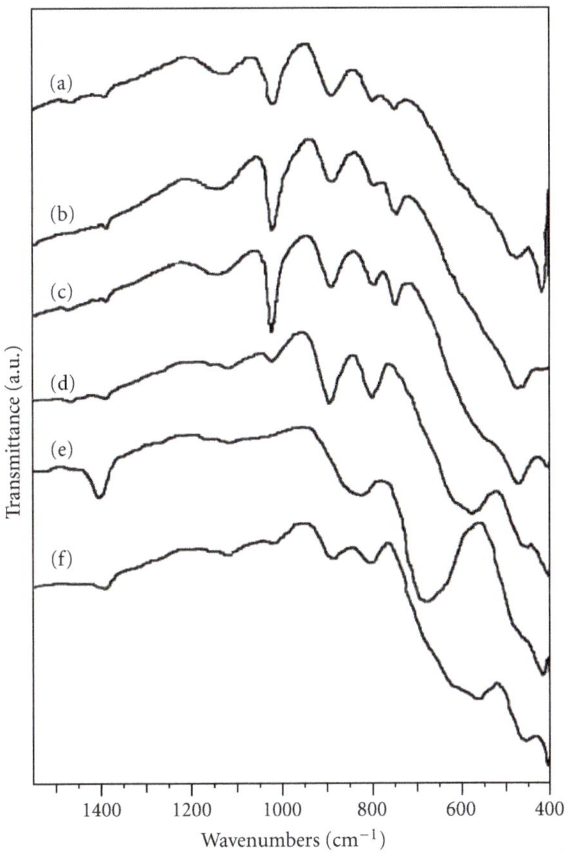

FIGURE 4: Fourier transformed infrared spectra of rusts formed at (a) Tocumen site A-588 for 6 months, (b) Sherman-Coastal site A-36 for 3 months, (c) Sherman-Open site COR 420 for 3 months, (d) Sherman-Breakwater site COR 420 for 1 year, and internal rusts formed at (e) Sherman-Breakwater rust A-588 for 1 year (f) Sherman-Breakwater rust COR 420 for 1 year.

TABLE 5: Summary of the results of the log-log plot of corrosion versus time.

log-log parameter	Tocumen			Sherman-Open			Sherman-Coastal			Sherman-Breakwater		
	A-36	A-588	COR420	A-36	A-588	COR420	A-36	A-588	COR420	A-36	A-588	COR420
A	30.04	29.06	32.6	32.59	30.92	31.90	42.23	36.53	40.67	726.40	556.50	395.70
n	0.92	0.73	0.47	0.92	0.71	0.55	0.95	0.76	0.56	0.72	0.82	1.37

bending vibration mode of H_2O, from which is inferred that these rust products contain molecular water adsorbed. The FTIR spectra obtained from the COR 420 rust, in addition to the bands of lepidocrocite and goethite, show that a shoulder band is located at 558 cm^{-1}, which is attributed to the maghemite.

The spectra for the A-36, A-588, and COR 420 steels exposed at Sherman-Breakwater site, also showed the typical bands of lepidocrocite and goethite. The lepidocrocite appear with low intensities in the FTIR spectra of the A-36 and A-588 steels, but with significant intensities in the sample of COR 420 steel. In the COR 420 sample the band is not only very intense, but also quite sharp, thus this phase should be well-crystallized in this rust. In addition, all samples clearly show a band at 578 cm^{-1} (see Figure 4(d)) approximately corresponding to magnetite (Fe_3O_4). The relative amounts of goethite and magnetite appear to increase with exposure duration at the expense of lepidocrocite phase.

In the product formed in the internal parts of panels exposed at Sherman-Breakwater (i.e., rust removed from the flaked parts corresponding to the interior of the panels) of steels A-36 and A-588, akaganeite (β-FeOOH) could be identified by the strong and broad bands at approximately 830 cm^{-1} and 667 cm^{-1} (see e.g., Figure 4(e)). In the COR 420 steel absorptions at 884 cm^{-1} and 799 cm^{-1}, corresponding to the goethite (α-FeOOH), plus an intense band 577 cm^{-1}, typical of magnetite (Fe_3O_4), could clearly be observed, as shown in Figure 4(f).

The formation of akaganeite is also observed in the rust in the internal flaking parts of the A-36 and A-588 steels. Curiously, akaganeite was not clearly detected in the COR 420 steels by FTIR, but instead magnetite was observed. We assume that this is due to the rain washing of chloride contaminants during the exposure in the rainy season. Considering the predominant phases in the internal parts of the COR 420 steels specimens were composed of goethite and magnetite, it follows that the washing of these samples in the early stages of exposure limits the possibilities of reaching the critical levels of chlorides required for the formation of akaganeite, as proposed by Santana Rodríguez et al. [33]. The magnetite is formed by the reducing conditions prevailing inside the flaked steel, where oxygen diffusion is limited. The formation of magnetite and akaganeite is most probably favoured in the presence of Cl^-, by the transformation of lepidocrocite and ferric species of FeOOH.

In order to establish whether or not there were differences in the corrosion products between the skyward and downward facing sides, we also undertook a comparative analysis using infrared FT-IR spectroscopy. There were no significant differences in the nature of the compounds in the rust, but there were differences in the relative amounts of phases in some cases. In previous work [34], which studied the corrosion products of carbon steel A-36 exposed to the weather in the tropical marine environment of Fort Sherman on the Caribbean coast of Panama, and compared the results with those obtained in an urban site in Panama city, with the help of X-ray diffraction, it was concluded that basically there are differences in the characteristics of particle size between the skyward facing and downward facing sides. No

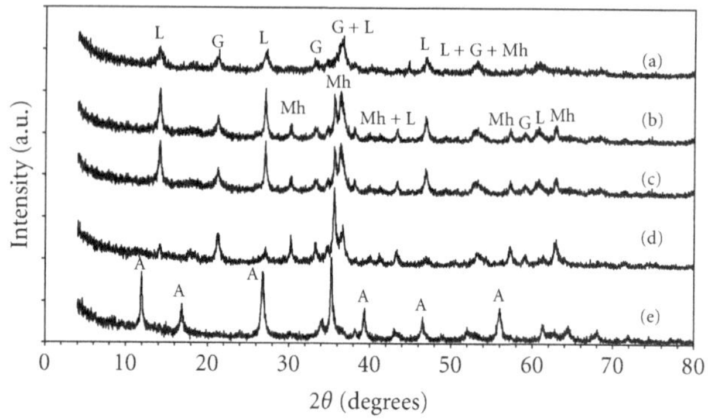

Figure 5: XRD patterns of some characteristic samples. (a) Tocumen site A-588 for 3 months, (b) Sherman-Open site A-588 for 3 months, (c) Sherman-Breakwater site A-588 for 1 year, (d) Sherman-Breakwater site COR 420 for 1 year, and internal rusts formed at (e) Sherman-Breakwater rust A-588 for 1 year. A, G, L, and Mh stand for akaganeite, goethite, lepidocrocite, and maghemite, respectively.

differences in the nature of the corrosion products on both sides were reported working with carbon steel A-36 at the station in Panama city.

X-ray powder diffraction pattern for all samples was recorded to confirm the identification of the phases present due to corrosion. The X-ray diffraction patterns corroborated with the previous FTIR analysis, although the relative intensities of the diffraction peaks detected were changed from coupon to coupon. However, the predominance of lepidocrocite or goethite based only on X-ray diffraction results was difficult to determine, since XRD only allows the analysis of crystalline constituents. Some phases like magnetite (Fe_3O_4) or maghemite (γ-Fe_2O_3) could not be differentiated by XRD measurements due to their similar crystalline structure. This oxide was always found in considerable proportions in the rust from weathering steels COR 420, particularly in Sherman-Breakwater samples. The presence of akaganeite, as expected for saline environments with high amounts of chlorides, was detected in a few samples by the presence intensities at 2θ angles of 11.8° and 55.7°. A discussion of the implications of these outcomes was presented previously [18] and will be extended later using Mössbauer results. Representative X-ray diffraction patterns are shown in Figure 5. Characteristics peaks associated with goethite (G), lepidocrocite (L), maghemite (Mh), and akaganeite (A) are shown.

Mössbauer analysis of the phase composition of the corrosion products of mild steel (A-36) and two weathering steels (A-588 and COR 420) formed after 3-months exposure to the tropical marine atmosphere of Panama has been reported elsewhere [18]. The results show that amorphous or crystallized iron oxyhydroxides goethite α-FeOOH and lepidocrocite γ-FeOOH are early corrosion products. Maghemite γ-Fe_2O_3 and magnetite Fe_3O_4 were also identified in the most aggressive conditions. The formation of akaganeite β-FeOOH was observed when chlorides were occluded within the rust.

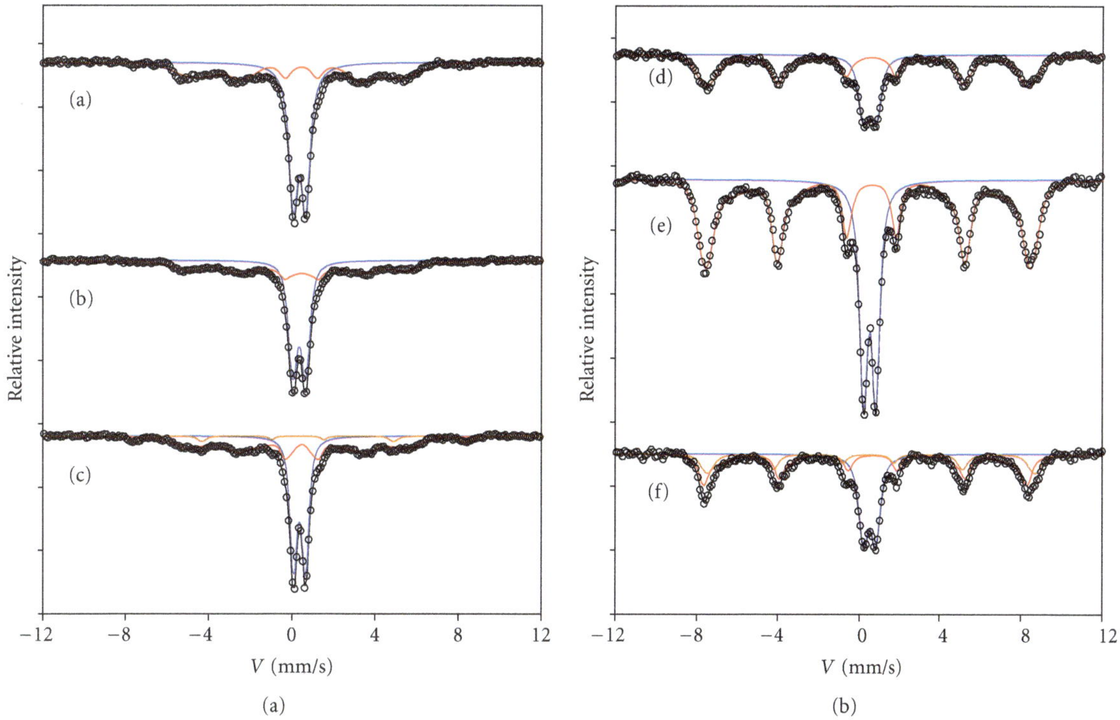

FIGURE 6: Transmission Mössbauer spectra of rust of mild steel A-36 and weathering steels formed after one year exposure at Tocumen site at 295 K: (a) A-36, (b) A-588, and (c) COR 420, and at 80 K: (d) A-36, (e) A-588, and (f) COR 420.

Figures 6 and 7 show the Mössbauer spectra of skyward corrosion products of the carbon steel and weathering steels coupons after 1 year of exposure at the Tocumen and Sherman-Breakwater sites. Table 6 summarizes the Mössbauer results as relative abundance determined basically from the 80 K spectra; but in the case of α-FeOOH, both room temperature and 80 K data were used because of the temperature dependence of the Mössbauer spectrum profile of this phase. α-FeOOH particles of >15 nm are magnetic at 298 K, while superparamagnetic α-FeOOH of <15 nm displays a doublet at room temperature and a magnetic sextet at 80 K [22]. Mössbauer spectrum of superparamagnetic α-FeOOH with particles <8 nm remains a doublet at 80 K [34, 35].

The fitting parameters indicate that the corrosion products of most samples contain γ-FeOOH, magnetic α-FeOOH, and superparamagnetic α-FeOOH, several also have γ-Fe_2O_3. The fitting of magnetic α-FeOOH at 298 K showed collapsing magnetic profile due to a broad distribution with the average of the hyperfine magnetic field ranging from 25.0–32.2 T, showing the poor crystallinity of this phase, particularly in the samples from Tocumen, Sherman-Open, and Sherman-Coastal. It is noteworthy that, at a given exposure duration, the relative concentration of superparamagnetic α-FeOOH is mostly larger for carbon steel A-36 and weathering A-588, compared to weathering steel COR 420. Rust formed at Sherman-Coastal station also contains magnetic γ-Fe_2O_3, phase which has also been observed in similar previous studies [15–17]. On one hand, superparamagnetic α-FeOOH has been shown to be important for the formation of the protective layer on weathering steels [4, 22]; on the other hand, Oh et al. [35] have linked the formation of magnetic γ-Fe_2O_3 and large particles of α-FeOOH to the increase of corrosion rates of different steels. Therefore, there is a lower corrosion rate for steel A-36 and A-588 exposed to mild corrosion conditions of Tocumen and Sherman-Open sites, compared with those exposed in Sherman-Coastal, which is attributed to superparamagnetic α-FeOOH. It is possible that γ-Fe_2O_3 enhances the corrosion rate by forming nonprotective patina. The presence of large particle of γ-Fe_2O_3 in all steel COR 420 coupons can be caused by the high humidity exposure existed during the rainy season. It is known that periods of wet-dry cycles, with long wet and shor-dry periods, form large fractions of γ-Fe_2O_3 [22]. Other factors like composition of the steel and the morphology should be examined.

Disregarding the type of steel, the Mössbauer results show that magnetic γ-Fe_2O_3 is always formed in large quantities at the Sherman-Breakwater site. This indicates that the harsh environmental condition of contamination by chlorides promotes the formation of γ-Fe_2O_3. The washing effect on the corrosion rate is noticed during the rainy season, affecting the composition of corrosion products due to the natural elimination of chloride contamination from the oxide layer. This may explain why there was a relatively low amount of γ-Fe_2O_3 in the COR 420 corrosion products of steel exposed at the Sherman-Breakwater site, and lower corrosion rate observed for the A-36 and A-588 steels.

Large amounts of magnetite and akaganeite are produced in the inner layer corrosion products from steels

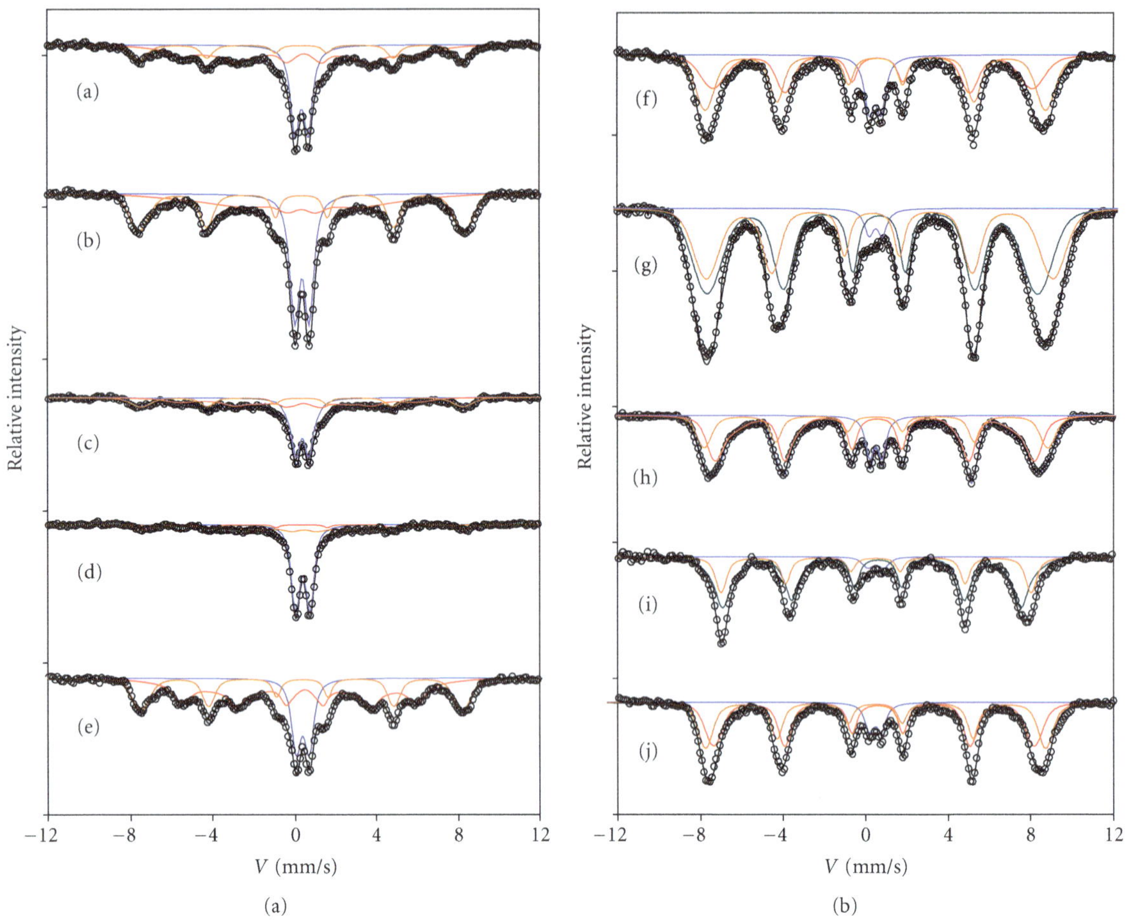

FIGURE 7: Transmission Mössbauer spectra of rust of mild steel A-36 and weathering steels formed after one year exposure at Sherman-Breakwater site at 295 K: (a) A-36, (b) internal rusts A-36, (c) A-588, (d) internal rusts A-588, and (e) COR 420, and at 80 K: (f) A-36, (g) internal rusts A-36, (h) A-588, (i) internal rusts A-588, and (j) COR 420.

exfoliated. This is due to the high chloride contamination that accumulates in the inner parts of the coupons, where salt is occluded. As we have already pointed out, β-FeOOH as corrosion product is typical of coastal areas that have high concentrations of chloride due to marine aerosols. It is presumed that β-FeOOH is formed instead of goethite and its presence may be related to the Fe_3O_4 in the rust, leading to high corrosion rates.

The differences in composition between the samples could be related to the conditions prevailing at the beginning of the exposure, giving different rates of corrosion, affecting the nature of corrosion products, and also to the composition of steels. The current results suggest that under severe chloride contamination, such as in the Sherman- Breakwater site, corrosion is controlled primarily by this pollutant and time of wetness, not the products of corrosion.

Protection ability is associated with corrosion products, alloying elements, exposure environment, and pollutants. In the initial stages of atmospheric corrosion, the formation of γ-FeOOH is predominant, but it further converts to goethite, in concordance with observations after long-term exposures [35, 36]. Actually, there might be favourable conditions for the formation of other corrosion products, such as maghemite, as previously pointed out. After a year, in the wet conditions of the atmosphere of Panama, the layer structure in rust is still loose, not providing protection. The corrosion rate data suggest that after three year's exposure, the dense corrosion-protective patina of fine rust particles in the weathering steels barely starts forming. Thus, the proposed protective ability index (PAI) of the rust layer [37] and the classification of the corrosion rate of the nonprotective rust layer [38] may be used with caution as these criteria were defined from long-term exposures. In addition, as indicated by Cook [22], it is magnetic goethite, that is, large particle goethite >15 nm exhibiting magnetic sextet at 298 K in the Mössbauer spectrum, that is the component that can be identified in the XRD patterns. As we obtained the fractions of the components by Mössbauer, for the rust to be a protective coating, the suggested value of more than two in weathering steel should be used for the ratio (α_m/γ^*), where α_m and γ^* are the mass ratio of crystalline α-FeOOH, and the total of γ-FeOOH, β-FeOOH and spinel-type iron oxide, respectively. This ratio in the skyward samples is less than 2 in all samples, but the ratio (β-FeOOH + spinel-type iron oxide)/(β-FeOOH + γ-FeOOH + spinel-type iron oxide), was less than 0.5 in samples exposed

TABLE 6: Mössbauer relative concentrations (%) for corrosion products after exposure at tests sites for six months and one year.

Exposure time	Type of steel	Exposure site	α-FeOOH (m)[1]	α-FeOOH (s)[2]	α-FeOOH Total	γ-FeOOH	γ-Fe_2O_3 /Fe_3O_4	β-FeOOH
6 months	A-36	Tocumen	23.6	41.6	65.2	34.8	—	—
		Sherman-Open	36.7	35.6	72.3	27.6	—	—
		Sherman-Coastal	25.3	41.1	66.4	33.6	—	—
		Sherman-Breakwater	38.8	44.7	83.5	16.5	—	—
	A-588	Tocumen	38.0	27.7	65.7	34.3	—	—
		Sherman-Open	48.4	17.4	65.8	34.2	—	—
		Sherman-Coastal	43.3	17.2	60.5	39.5	—	—
		Sherman-Breakwater	51.0	28.8	79.8	9.9	10.3	—
	COR 420	Tocumen	44.2	15.6	59.8	32.1	8.1	—
		Sherman-Open	30.4	12.4	42.8	37.1	20.1	—
		Sherman-Coastal	31.6	7.0	38.6	35.2	26.1	—
		Sherman-Breakwater	—	21.9	20.5	12.2	65.9	—
		Sherman-Breakwater I*	40.0	7.0	47.0	10.9	19.0	23.0
1 year	A-36	Tocumen	43.8	23.0	66.8	33.2	—	—
		Sherman-Open	31.3	22.5	53.8	39.5	6.7	—
		Sherman-Coastal	46.7	7.5	54.2	37.9	7.9	—
		Sherman-Breakwater	20.7	19.3	40	13.6	47.0	—
		Sherman-Breakwater I*	—	—	—	3.4	40.7	55.9
	A-588	Tocumen	48.3	15.0	63.3	36.7	—	—
		Sherman-Open	37.8	16.0	53.8	39.5	6.7	—
		Sherman-Coastal	23.6	20.8	44.4	40.6	15.0	—
		Sherman-Breakwater	33.5	25.8	59.3	11.6	29.1	—
		Sherman-Breakwater I*	—	—	—	—	29.0	71.0
	COR 420	Tocumen	32.1	8.9	41.0	37.0	22.0	—
		Sherman-Open	21.9	11.6	33.5	37.5	29.0	—
		Sherman-Coastal	28.1	14.9	43.0	39.3	17.7	—
		Sherman-Breakwater	38.9	11.5	50.4	8.6	41.0	—

* Corrosion product from the inner layer of flaked simple.
[1] α-FeOOH (m): magnetic goethite (>15 nm) exhibiting magnetic sextet at 298 K.
[2] α-FeOOH (s): superparamagnetic goethite (<15 nm) exhibiting superparamagnetic doublet at 298 K and magnetic sextet at 80 K.

to mild corrosion conditions of the Tocumen, Sherman-Open, and Sherman-Coastal sites, whereas it was more than 0.5 in Sherman-Breakwater. Hence, the rust obtained after one year exposure under mild conditions should be nonprotective but inactive, whereas the rust from Sherman-Breakwater is active, showing high corrosion rates.

4. Conclusions

(i) The atmospheric corrosion behaviour of samples of carbon steel A-36 weathering steels A-588, and COR 420 exposed to the humid tropical weather at different sites in Panama proceeded according to the well-known bilogarithmic law for weight loss. This process could be linked to the high chloride content and high relative humidity. The difference in corrosion of steels at the various sites with mild conditions was small, with a variation of A from 30 to 42 and n from 0.5 to 1.0 (using corrosion depth in mm, and years as time). The difference was large for the Sherman-Breakwater station, a highly contaminated site with chlorides, with A values varying from 396 to 726 and n values from 0.7 to 1.4. The corrosion rate observed on weathering steel COR 420 is high in the initial phases of exposure (rainy season) levelling off with time. There is a transitional behaviour at the Sherman-Breakwater site for weathering steel COR-420, with different corrosion rates before and after 6 to 9 months.

(ii) There is a strong correlation between compositions of rust formed in dependence of the atmospheric exposure conditions, particularly the time of wetness, deposition of chloride, and the washing effect of contaminants on the metal surface by rain. The dominant phases after short-term exposure were amorphous or crystalline oxyhydroxides such as lepidocrocite (γ-FeOOH) and goethite (α-FeOOH). Maghemite (γ-Fe_2O_3) and magnetite (Fe_3O_4) were also found. Akaganeite (β-FeOOH) was identified as prominent component in the most aggressive conditions, with high chlorine amounts, and occluded within the rust.

(iii) The rust obtained after one year exposure is nonprotective, inactive in the mild conditions of the Tocumen, Sherman-Open, and Sherman-Coastal sites, but active in the rust from the Sherman-Breakwater site.

Acknowledgment

The authors are thankful to Secretaría Nacional de Ciencia, Tecnología e Innovación (Panamá), for financial help through the Project FID06-227, SENACYT, 2006.

References

[1] Q. C. Zhang, J. S. Wu, J. J. Wang, W. L. Zheng, J. B. Chen, and A. B. Li, "Corrosion behavior of weathering steel in marine atmosphere," *Materials Chemistry and Physics*, vol. 77, pp. 603–608, 2002.

[2] R. A. Antunes, R. A. CI, and D. L. Araújo de Faria, "Characterization of corrosion products formed on steels in the first months of atmospheric exposure," *Materials Research*, vol. 6, no. 3, pp. 403–408, 2003.

[3] K. Asami and M. Kikuchi, "In-depth distribution of rusts on a plain carbon steel and weathering steels exposed to coastal-industrial atmosphere for 17 years," *Corrosion Science*, vol. 45, no. 11, pp. 2671–2688, 2003.

[4] D. C. Cook, "Application of Mössbauer spectroscopy to the study of corrosion," *Hyperfine Interactions*, vol. 153, no. 1–4, pp. 61–82, 2004.

[5] T. Kamimura, S. Hara, H. Miyuki, M. Yamashita, and H. Uchida, "Composition and protective ability of rust layer formed on weathering steel exposed to various environments," *Corrosion Science*, vol. 48, no. 9, pp. 2799–2812, 2006.

[6] W. Han, G. Yu, Z. Wang, and J. Wang, "Characterisation of initial atmospheric corrosion carbon steels by field exposure and laboratory simulation," *Corrosion Science*, vol. 49, no. 7, pp. 2920–2935, 2007.

[7] Y. Li, "Mössbauer effect of rust layer formed on steel in different marine corrosion zones," *Corrosion*, vol. 63, no. 9, pp. 850–856, 2007.

[8] Y. C. Sica, E. D. Kenny, K. F. Portella, and D. F. Campos Filho, "Atmospheric corrosion performance of carbon steel, galvanized steel, aluminum and copper in the North Brazilian coast," *Journal of the Brazilian Chemical Society*, vol. 18, no. 1, pp. 153–166, 2007.

[9] F. Corvo, T. Perez, L. R. Dzib et al., "Outdoor-indoor corrosion of metals in tropical coastal atmospheres," *Corrosion Science*, vol. 50, no. 1, pp. 220–230, 2008.

[10] J. G. Castaño, C. A. Botero, A. H. Restrepo, E. A. Agudelo, E. Correa, and F. Echeverría, "Atmospheric corrosion of carbon steel in Colombia," *Corrosion Science*, vol. 52, no. 1, pp. 216–223, 2010.

[11] A. L. Alexander, B. W. Forgenson, H. W. Mumdt, C. R. Southwell, and S. J. Thompson, "Corrosion of metals in tropical environments, part-1 test methods used and result obtained for pure metals and a structural steel," NRL Report 4929, Washington, DC, USA, 1957.

[12] C. R. Southwell, B. W. Forgeson, and A. L. Alexander, "Corrosion of metals in tropical environments, part-2 atmospheric corrosion of ten structural steels," NRL Report 5002, Washington, DC, USA, 1957.

[13] M. Morcillo, E. Almeida, B. Rosales, J. Uruchurtu, and M. Marrocos, Corrosión y Protección de Metales en las Atmósferas de Iberoamérica. Parte I: Mapas de Iberoamérica de Corrosividad Atmosférica (Red Temática MICAT, XV.1/CYTED) Madrid: CYTED, 1999.

[14] M. Morcillo, E. Almeida, F. Fragata, and Z. Panossian, Corrosión y Protección de Metales en las Atmósferas de Iberoamérica. Parte II: Protección Anticorrosiva de Metales en las Atmósferas de Iberoamérica (Red Temática PATINA, XV.D/CYTED) Madrid: CYTED, 2002.

[15] J. A. Jaén and B. Fernández, "Mössbauer spectroscopy study of steel corrosion in a tropical marine atmosphere," *Electrochimica Acta*, vol. 34, no. 6, pp. 885–886, 1989.

[16] J. A. Jaén, M. S. De Villalaz, L. De Araque, and A. De Bósquez, "Kinetics and structural studies of the atmospheric corrosion of carbon steels in Panama," *Hyperfine Interactions*, vol. 110, no. 1-2, pp. 93–99, 1997.

[17] J. A. Jaén, M. S. de Villalaz, L. de Araque, C. Hernández, and A. de Bósquez, "Study of the corrosion products formed on carbon steels in the tropical atmosphere of Panama," *Revista de Metalurgia*, pp. 32–37, 2003.

[18] J. A. Jaén, A. Muñóz, J. Justavino, and C. Hernández, "Characterization of initial atmospheric corrosion of conventional weathering steels and a mild steel in a tropical atmosphere," *Hyperfine Interactions*, vol. 192, no. 1–3, pp. 51–59, 2009.

[19] H. Leidheiser Jr. and S. Musić, "The atmospheric corrosion of iron as studied by Mössbauer spectroscopy," *Corrosion Science*, vol. 22, no. 12, pp. 1089–1096, 1982.

[20] A. K. Singh, T. Ericsson, L. Häggström, and J. Gullman, "Mössbauer and x-ray diffraction phase analysis of rusts from atmospheric test sites with different environments in Sweden," *Corrosion Science*, vol. 25, no. 10, pp. 931–945, 1985.

[21] N. R. Furet, C. Haces, F. Corvo, C. Díaz, and J. Gomez, "Corrosion rate determination using57Fe Mössbauer spectra of corrosion products of steel," *Hyperfine Interactions*, vol. 57, no. 1–4, pp. 1833–1838, 1990.

[22] D. C. Cook, "Spectroscopic identification of protective and non-protective corrosion coatings on steel structures in marine environments," *Corrosion Science*, vol. 47, no. 10, pp. 2550–2570, 2005.

[23] R. Vera, B. M. Rosales, and C. Tapia, "Effect of the exposure angle in the corrosion rate of plain carbon steel in a marine atmosphere," *Corrosion Science*, vol. 45, no. 2, pp. 321–337, 2003.

[24] ISO 9223, *Corrosion of Metal and Alloys—Classification of Corrosivity of Atmospheres*, International Standards Organization, Geneva, Switzerland, 1992.

[25] ISO 9226, *Corrosion of Metals and Alloys. Method for Determination of Corrosion Rate of Standard Specimens for the Evaluation of Corrosivity*, International Standards Organization, Geneva, Switzerland, 1991.

[26] F. Corvo, C. Haces, N. Bethancourt et al., "Atmospheric corrosivity in the Caribbean area," *Corrosion Science*, vol. 39, no. 5, pp. 823–833, 1997.

[27] A. R. Mendoza and F. Corvo, "Outdoor and indoor atmospheric corrosion of carbon steel," *Corrosion Science*, vol. 41, no. 1, pp. 75–86, 1999.

[28] M. Morcillo, B. Chico, L. Mariaca, and E. Otero, "Salinity in marine atmospheric corrosion: its dependence on the wind

regime existing in the site," *Corrosion Science*, vol. 42, no. 1, pp. 91–104, 2000.

[29] ETESA—Empresa de Transmisión Eléctrica, S.A. Hidrometeorología de ETESA, 2011, http://www.hidromet.com.pa.

[30] Y. Ma, Y. Li, and F. Wang, "The atmospheric corrosion kinetics of low carbon steel in a tropical marine environment," *Corrosion Science*, vol. 52, no. 5, pp. 1796–1800, 2010.

[31] R. F. Passano, "The Harmony of Outdoors Weatherfng Steels," in *Proceedings on the Outdoor Wheathering of Metals and Metalic Coating*, Philadelphia, Pa, USA, 1934.

[32] A. Raman, B. Kuban, and A. Razvan, "The application of infrared spectroscopy to the study of atmospheric rust systems-I. Standard spectra and illustrative applications to identify rust phases in natural atmospheric corrosion products," *Corrosion Science*, vol. 32, no. 12, pp. 1295–1306, 1991.

[33] J. J. Santana Rodríguez, F. J. Santana Hernández, and J. E. González González, "XRD and SEM studies of the layer of corrosion products for carbon steel in various different environments in the province of Las Palmas (The Canary Islands, Spain)," *Corrosion Science*, vol. 44, no. 11, pp. 2425–2438, 2002.

[34] J. A. Jaén Osorio, De Gracia de Araque Ld. A Study of the Corrosion Products on Carbon Steel Exposed in the Caribbean Coast of Panama. In Congreso Latinoamericano de Corrosión LATINCORR, 2006/9 Congreso Iberoamericano de Corrosión y Protección/6 Congreso de Corrosión NACE Latinoamericano, Fortaleza, Brazil, 2006.

[35] S. J. Oh, D. C. Cook, S. J. Kwon, and H. E. Townsend, "Studying the atmospheric corrosion behavior of weathering steels at a mild marine environment," *Hyperfine Interactions*, vol. 4, pp. 49–54, 1999.

[36] S. J. Oh, D. C. Cook, and H. E. Townsend, "Atmospheric corrosion of different steels in marine, rural and industrial environments," *Corrosion Science*, vol. 41, no. 9, pp. 1687–1702, 1999.

[37] M. Yamashita, H. Miyuki, Y. Matsuda, H. Nagano, and T. Misawa, "The long term growth of the protective rust layer formed on weathering steel by atmospheric corrosion during a quarter of a century," *Corrosion Science*, vol. 36, no. 2, pp. 283–299, 1994.

[38] S. Hara, T. Kamimura, H. Miyuki, and M. Yamashita, "Taxonomy for protective ability of rust layer using its composition formed on weathering steel bridge," *Corrosion Science*, vol. 49, no. 3, pp. 1131–1142, 2007.

6

Influence of Simulated Acid Rain Corrosion on the Uniaxial Tensile Mechanical Properties of Concrete

Ying-zi Zhang,[1] Ying-fang Fan,[2] and Hong-nan Li[3]

[1] *Department of Civil Engineering, Harbin Institute of Technology, Weihai, Shandong 264209, China*
[2] *Institute of Road and Bridge Engineering, Dalian Maritime University, Dalian, Liaoning 116026, China*
[3] *State Key Laboratory of Coastal and Offshore Engineering, Dalian University of Technology, Dalian, Liaoning 116023, China*

Correspondence should be addressed to Ying-fang Fan, fanyf72@yahoo.com.cn

Academic Editor: Facundo Almeraya

An experimental study on the uniaxial tensile property of concrete exposed to the acid rain environment was carried out. Acid rain with pH level of 1.0 was deposed by the mixture of sulfate and nitric acid solution in the laboratory. Dumbbell-shaped concrete specimens were immersed in the simulated acid rain completely. After being exposed to the deposed mixture for a certain period, uniaxial tensile test was performed on the concrete specimens. The results indicate that elastic modulus, tensile strength, and peak strain have a slight increase at the initial corrosion stage, and with the extension of corrosion process, elastic modulus and tensile strength decrease gradually, while the peak strain still increases. It is found that the compressive strength is more sensitive than the tensile strength in aggressive environment. Based on the experimental results, an equation was proposed to describe the ascending branch of the stress-strain curve of the concrete corroded by acid rain.

1. Introduction

Acid rain has become an issue of widespread concern in Asia, and it was first recognized as a potential environmental pollution problem in the late 1970s and early 1980s. It was also an increasing problem in China during the last decades [1–4]; one third of China's land mass was affected by acid rain. The acid rain is composed of acid substances, such as H_2SO_4, HCl, and HNO_3, which readily react with $Ca(OH)_2$ and form $CaSO_4 \cdot 2H_2O$, $Ca(NO_3)_2 \cdot 4H_2O$, and $CaCl_2 \cdot 6H_2O$. Because the solubility of these products is higher than that of $CaCO_3$, which is formed at the surface by the reaction of $Ca(OH)_2$ with CO_2 (carbonation), they are dissolved in rainwater and penetrate into the inner pores of concrete. After the evaporation of rainwater, these salts redeposit and give rise to stress in the concrete structure, resulting in cracking, mass loss, and eventually structural failure. In the early 1950s, US Environmental Protection Agency reported that the cost of repairing or replacing the concrete structure every year due to acid rain in the United States was more than \$5 billion dollars. In Britain, the economic loss due to acid rain was 0.15% of the GNP. Guo [5] in 2001 estimated that the direct material loss due to acid rain only in Guangzhou was 1.2 billion Chinese Yuan. Although some ways have been used to reduce acid rain, the acid rain region in China even continue to expand, therefore, the property of the corroded concrete have become to be a critical issue and attracted a lot of researchers worldwide [6–12]. However, due to the complicated mechanism, there remains only very limited available literature on the tensile property of concrete corroded by acid rain. In this paper, acid rain was simulated in the laboratory, a series of C30 dumbbell-shaped specimen and the accompanying specimens were exposed to different simulated aggressive environments for different periods. A series of tests, including weighing, compressive tests, and tensile tests were conducted at different conditioned stages to determine mass loss, compressive strength, and tensile properties.

2. Experimental Details

2.1. Materials and Specimen Preparation. C30 concrete was applied to cast the specimen. The dosage of the concrete is shown in Table 1. The specimens for tension test were

TABLE 1: Mix proportion of the concrete.

Cement/kg	Sand/kg	Pebble/kg	Water/kg	Sand ratio/%
450	541	1200	200	31

dumbbell-shaped, as shown in Figure 1 and the accompanying specimens were cubic, 100 mm in length, which were used to measure the compressive strength.

2.2. Accelerate Corrosion Technique. Concrete is subjected to a variety of exposure conditions, such as frequency of drying and wetting cycles, different temperatures, superficial carbonation, and drying shrinkage. All the conditions will affect the mode of attack. Therefore, it is difficult to exactly duplicate the field attack condition in the laboratory. Though the field corrosion test can better simulate the actual corrosion process, it always took much more time to fulfill the corrosion. Therefore, the accelerate corrosion test is applied in this paper. Wang et al. [13] pointed out that the acid rain is due to sulfuric-acid in most parts of China. Therefore, only sulfuric acid type acid rain is simulated in this study, which was achieved by mixing sulfate and nitric acid solutions with pH value of 1.0, and the specimens were divided into several groups in the different corrosion time, the corrosion conditions are listed in Table 2 in detail.

2.3. Test Device and Procedure. Liu [14] used site and accelerate tests to study the concrete corrosivity, it was found that the results showed a good coincidence in two different exposure conditions. Obviously, the submerging method is more suitable for cement concrete, therefore, it is adopted for accelerated test in this study. After being cured in a fog room for 28 days, the specimens were naturally cured in the laboratory for 120 days, and then they are placed in the anticorrosion bucket containing simulated acid salutation (Figure 2). Since pH values of the solutions were changing as the conditioning continued, the acidity of solution was recorded by PB-10 Sartorius Acidometer every day, acid was added to the solution periodically and mixed well with the solution so that a constant pH value was guaranteed. After being corroded for different exposure periods, a group of 3 specimens will be picked out to be dried for about 5 days, followed by visual inspection, weighting, compressive tests, tensile tests, to determine mass loss, compressive strength, tensile strength, and elastic modulus, respectively. The compressive test was carried out on an electrohydraulic servo-compressive testing system (YAW-YAW2000A, Figure 3) with a capacity of 2000 kN. Following present Chinese code for concrete compressive testing (GB/T50081-2002) [15], the loading rate was 0.3 MPa/s. The tensile tests were carried out on the MTS810 electro-hydraulic dynamic-static universal testing machine. The MTS810 machine is a programmable displacement controlled loading system and can produce tensile and compressive loads at a maximum nominal rate up to 150 mm/s. Its force capacity is 100 kN and stiffness is 2.6×10^8 N/s. The measurement system consists of DPM-8H strain amplifier, INV306D intelligent signal processor, and DASP software. The highest frequency of MTS810 can

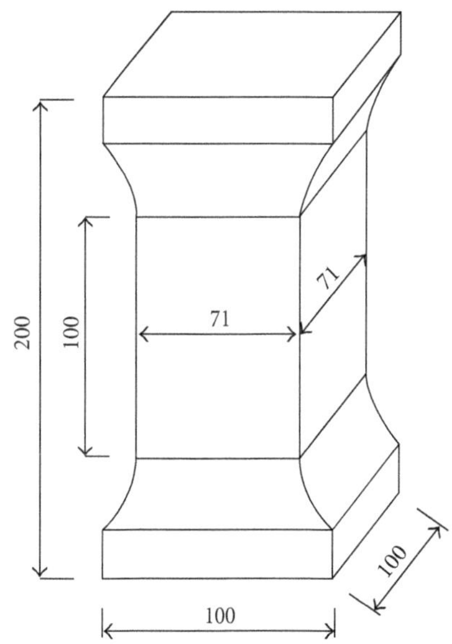

FIGURE 1: Dimensions of dumbbell-shaped specimen (unit: mm).

reach 10^4 Hz. The machine can offer different strain rates, strain rate of 10^{-5}/s is applied in this test. Foil strain gauges, 50 mm in length and 5 mm in width, were used to measure strain histories during tensile process. Two-crossed strain gauges were glued onto the middle of each side of specimen to measure the longitudinal and lateral strains respectively (Figure 4). In addition, the linear variable differential transformer (LVDT) was fixed on the specimens to measure the displacement after the appearance of cracks. The load device and instrument arrangement are shown in Figure 5.

3. Results and Discussion

Uniaxial tensile tests were carried out on the dumbbell-shaped specimens. For multiple reasons, such as the randomicity, diversity, and complexity of distribution of concrete, and the limited capacity of the testing system, testing equipments, testing techniques, and so on, the center of each cross-section tensile strength of specimen is not the same location, therefore, no method could make the test specimen achieve the perfect axial tensile state, and these may be influence the accuracy of the test results. The 25 tensile specimens were achieved ideal results.

3.1. Visual Observations. During the test, specimen were picked out at the assignment time for visual inspection. From Figure 6, it is found that the concrete surface became unsmooth with the increase of the corrosion time, up to 33 days, the surface became to be honeycomb, large amounts of coarse aggregates can be seen, and concrete at the specimens became to be powdery, and the surface colour is similiar to what was observed by Xie [12] and Fan et al. [16].

3.2. Mass Change. During the test, specimens were periodically picked out from the simulated acid rain for visual

Table 2: Experimental conditions.

Specimen C30	Designation	Exposing condition	Number of specimen	Immersion time (days)
Dumbbell-shaped specimens	$H_2SO_4 + HNO_3$ (molar ratio is 9: 1)	pH 1.0	4 × 10	0, 5, 11, 15, 18, 21, 24, 27, 30, 33
Cube specimens	$H_2SO_4 + HNO_3$ (molar ratio is 9: 1)	pH 1.0	3 × 10	0, 5, 11, 15, 18, 21, 24, 27, 30, 33

Figure 2: Concrete specimens exposed to the similated acid rain.

Figure 4: Arrangements of strain gages of on the tensile specimen.

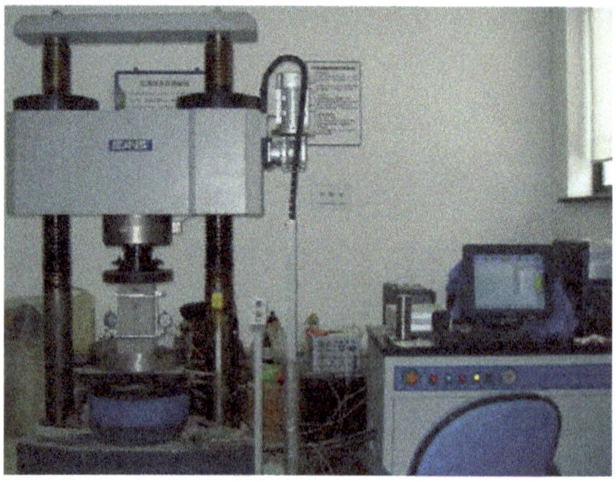

Figure 3: Electrohydraulic servo testing system.

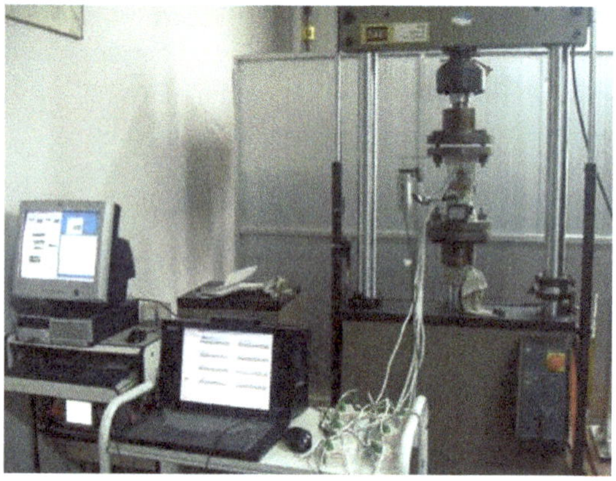

Figure 5: MTS810 Electrohydraulic dynamic-static universal testing machine.

inspection. It was observed that the surface of the concrete changed became very rough, honeycomb voids were formed, and coarse aggregates were exposed, this caused the specimen's mass change. Mass change versus corrosion time for the specimens is shown in Figure 7. It can be seen that mass of cube specimens and dumbbell-shaped specimens all first increase and then decrease with the increasing of corrosion duration, which is similar to the results obtained by the literature [16].

3.3. Tensile Strength and Compressive Strength. During the tensile test, it was found that a number of coarse aggregates were broken in the short corrosion time (such as day 5 and day 11), the fracture surface of the specimen mainly passes through the aggregate and the aggregate-mortar interfaces, the fracture of the specimen became flattened with the increasing corrosion duration, it can be inferred that dense degree between the aggregate and mortar had a slight increase, therefore, the tensile strength became larger than that of the no-corrosion specimen (see Table 2), however, the fracture surface failure modes were mainly coarse aggregates bond failure for severe deterioration (day 33), typical modes of which are depicted in Figure 8. Strength of concrete exposed to acid solution for different periods were listed in Table 3 in detail.

To quantify the deterioration in tensile and compressive strength of corroded concrete, the strength degradation

FIGURE 6: Surface description of the concrete specimens corroded for different period.

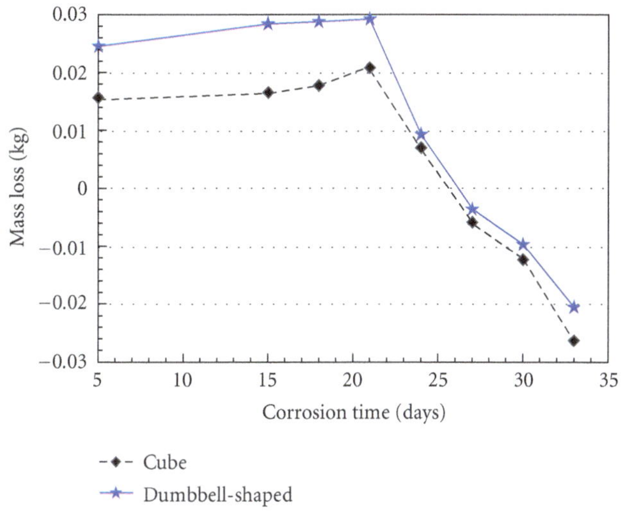

FIGURE 7: Relation between mass loss and corrosion duration.

ratio D_{fc} and D_{ft} are introduced. The degradation ratio are expressed as follows:

$$D_{fc} = \left(1 - \frac{f_{ct}}{f_{c0}}\right) \times 100\%,$$
$$D_{ft} = \left(1 - \frac{f_{ctt}}{f_{ct0}}\right) \times 100\%, \quad (1)$$

where D_{fc} and D_{ft} are the degradation ratio of corroded concrete, f_{c0} is the tensile strength of uncorroded specimen, f_{ct} is the tensile strength of specimen exposed to the acid solution for t days, f_{ct0} is defined as the compressive strength of uncorroded specimen, and f_{ctt} is the compressive strength of specimen exposed to the acid solution for t days. Based on the testing results, relation between strength loss and immersion time is shown in Figure 9. From Figure 9, it can be seen that the strength D_{fc} and D_{ft} decreases and then increases as conditioning continue, but their critical time is different. The critical value of the compressive strength degradation ratio is larger than that of the tensile strength degradation ratio, and D_{fc} showed greater decline than D_{ft}, this indicates that the compressive strength is more sensitive to aggressive environment than the tensile strength, but this conclusion needs to be further validated due to limited number of specimens in this paper.

3.4. Elastic Modulus. The modulus of elasticity is calculated a secant modulus in this paper, a secant modulus is calculated from the origin to a defined point on the stress-strain curve, usually within 40% to 60% of the sample's ultimate strength [17]. In this study, the elastic modulus was measured 50% the specimen's ultimate strength. Form the experiment result, the elastic modulus at different corrosion period is reported in Table 4. As can be seen from Table 4, elastic modulus exist some random fluctuation during the test.

To quantify the deterioration in elastic modulus of corroded concrete, the degradation ratio D_{Ect} is introduced. The degradation ratio are expressed as follows:

$$D_{Ect} = \left(1 - \frac{E_{ct}}{E_{c0}}\right) \times 100\%, \quad (2)$$

where D_{Ect} is the degradation ratio of corroded concrete; E_{c0} is the elastic modulus of uncorroded specimen; E_{ct} is the elastic modulus of specimen exposed to the acid solution for t days. Relation between elastic modulus loss and corrosion duration is shown in Figure 10. It can be seen that elastic modulus of the specimens are unstable during the tensile test, the same tendency can be observed on D_{Ec} for specimens exposed to the three solutions in literature [16]. There may be several reasons: (a) some construction variances during casting; (b) the surface of the specimen became unflattered with the increasing corrosion duration, this influenced the strain gauge paste and date acquisition; (c) There are personal errors, system error, and random error and so on; (d) the limited number of the specimens. However, the D_{Ect} has a slight decrease at initial stage (day 5) and then fluctuates increasingly when the conditioning continues, it means that the elastic modulus of the tensile specimens decreases when the increasing of corrosion duration.

The elastic modulus of corroded specimens is calculated by (5) proposed in the literature [18], which depended mainly on the fresh elastic modulus, poisson ratio, and void volume before and after corrosion. As can be seen from Table 4, few model calculation results have good agreement with the experimental results. The following reasons exist: (1) some construction variances during casting and conditioning process; (2) the personal error during the test; (3) there is only one control specimen during the

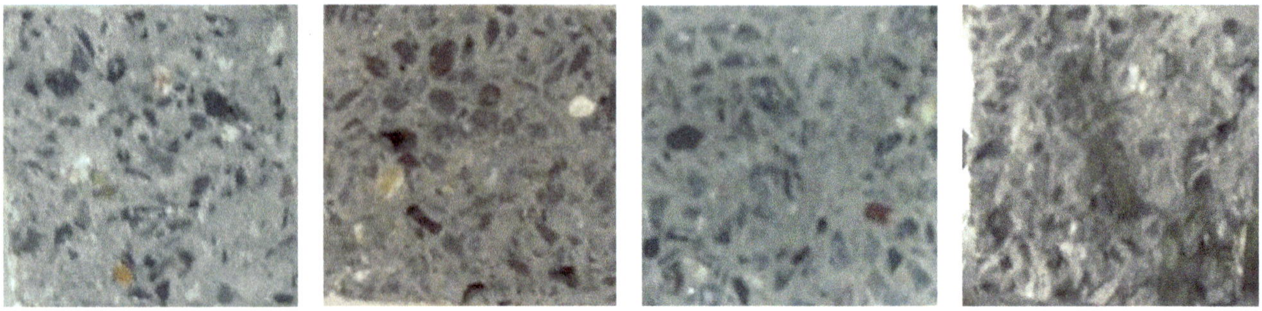

FIGURE 8: Typical fracture area of concrete in different corrosion period.

TABLE 3: Strength of concrete exposed to acid solution for different periods.

Immersion time/days	0	5	11	15	18	21	24	27	30	33
Tensile strength/MPa	2.738	2.915	2.938	2.946	2.837	2.704	2.698	2.702	2.676	1.727
Compressive strength/MPa	42.746	44.240	44.758	35.835	35.475	32.106	34.342	34.329	34.309	30.847

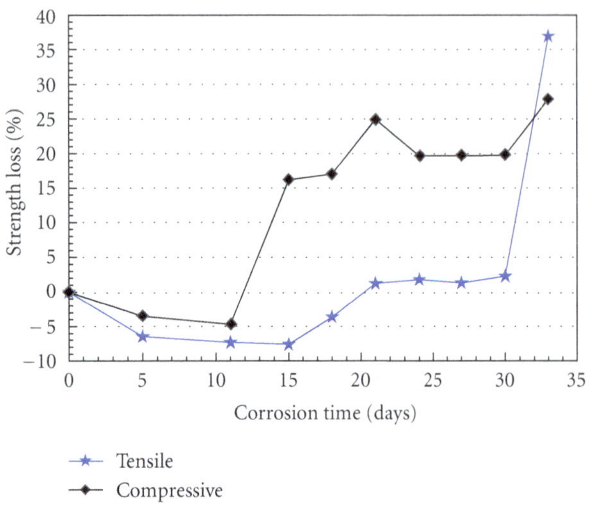

FIGURE 9: Relation between strength and corrosion time.

different corrosion time, however, the elastic modulus of concrete may change at different time; (4) specimen number is not enough; (5) there is a little volume expansion in this high simulated acid solutions.

3.5. Peak Strain. Peak strain is defined as the strain when the tensile stress reaches the maximum stress during the test in this paper. To quantify the deterioration in peak strain of corroded concrete, and the degradation ratio $D_{\varepsilon c}$ is introduced in the following:

$$D_{\varepsilon c} = \left(1 - \frac{\varepsilon_{\mathrm{ct}}}{\varepsilon_{\mathrm{c0}}}\right) \times 100\%, \tag{3}$$

where $D_{\varepsilon c}$ denotes the degradation ratio of corroded concrete, $\varepsilon_{\mathrm{c0}}$ is the strain of uncorroded specimen, and $\varepsilon_{\mathrm{ct}}$ is the strain of specimen exposed to the acid solution for t days. Relation between peak strain loss and corrosion duration is shown in Figure 11. It is shown that the peak strain of the specimens exposed to acid solutions will decrease with exposure days linearly (correlation coefficient is 0.916). It is obvious that the degree of corrosion damage increases with exposure time.

3.6. Stress-Strain Relation. During the tensile tests, we found that once the peak strain is reached, a visible crack will appear and specimens will fail subsequently in a very short time. Many reasons, such as some construction variances during casting, conditioning process, and impossibility of the perfect axial tensile state, lead to obtain the descending part of the stress-strain curve is difficult, therefore, only the ascending branch of the stress-strain curve will be discussed in this study. Typical stress-strain curves of concrete corroded by acid solutions were recorded in the experiment, which are plotted, respectively, in Figure 12.

From Figure 12, it can be seen that the changing trends in tension strain-stress relation for concrete corroded with different concentration are similar. At the initial stage (not exceeding 11 days), the 3 major mechanical property indexes, uniaxial tensile strength, peak strain, and elastic modulus increase gradually, with the increasing of corrosion duration, the uniaxial tensile strength and elastic modulus decrease, but the peak strain continues to increase, which shows that the damage increases with the corrosion duration.

In order to describe the ascending branch of stress-strain curves of tensile concrete specimen in different corrosion periods, a formula is presented which is based on the current Chinese concrete structure design code (GB/T 50010—2002) [15].

Firstly, the dimensionless form of the stress and the strain are obtained

$$x = \frac{\varepsilon}{\varepsilon_t}, \qquad y = \frac{\sigma}{f_t}, \tag{4}$$

where f_t is the uniaxial tensile strength of concrete, ε_t is the peak strain of the concrete specimen.

TABLE 4: Experimental data and model calculation results.

Immersion time/days	0	5	11	15	18	21	24	27	30	33
Test results/10^3 MPa	34.4	41.6	35.7	28.9	22.7	20.0	26.0	39.3	22.6	40.9
Model calculation results/10^3 MPa	34.4	35.0	35.6	35.1	35.0	35.1	34.6	34.3	34.2	33.9
Error/%	0	18.9	0.28	17.7	35.2	43.1	24.9	14.6	33.9	20.6

Note: error = 100% × |testresults − calculatedresults|.

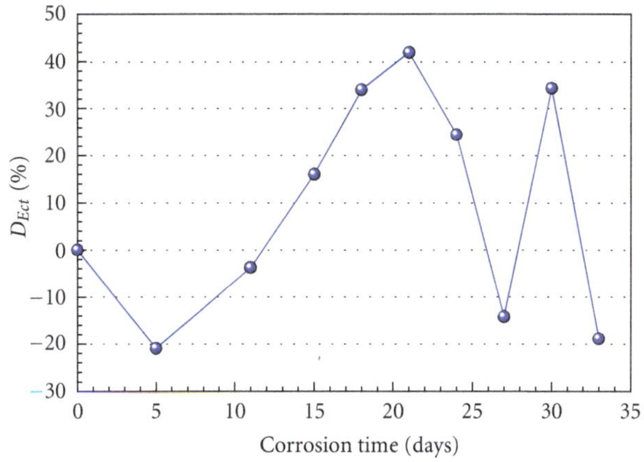

FIGURE 10: Relation between elastic modulus loss and corrosion duration.

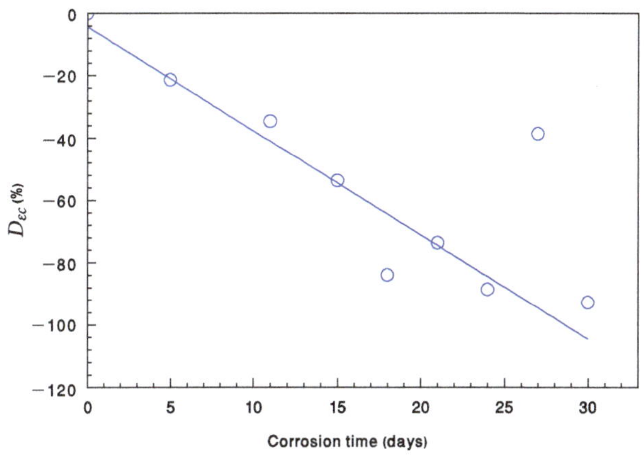

FIGURE 11: Relation between deterioration ratio of peak stain and corrosion duration.

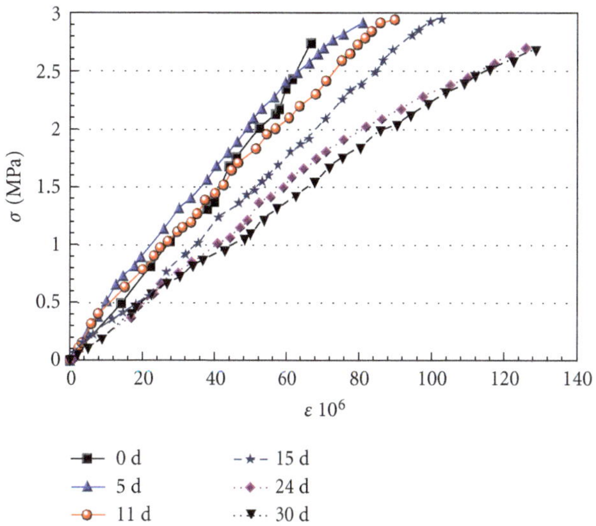

FIGURE 12: Ascending branch of stress-strain curves of concrete specimen corroded for different periods.

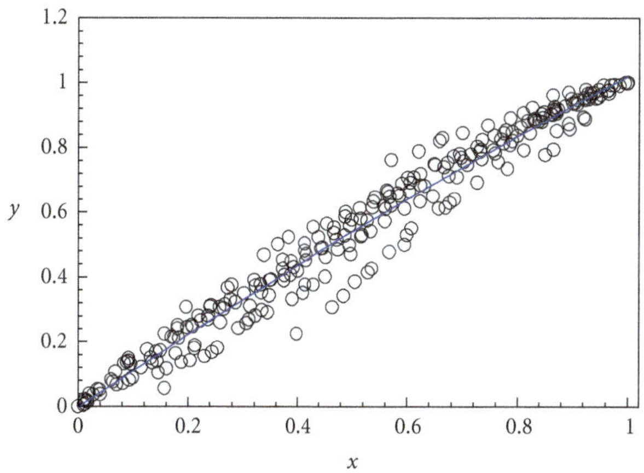

FIGURE 13: Ascending branch of stress-strain curves of corroded concrete specimen.

Therefore, the equation, which describes the ascending branch of stress-strain curves ($x \leq 1$) of tensile concrete specimen, can be written

$$y = Ax + Bx^m, \tag{5}$$

where A and B are polynomial coefficients, m is exponent of polynomial, which all can be obtained from the experiment. By fitting to the test data in this study, (6) can be obtained, where the correlation coefficient $R^2 = 0.97$. From Figure 13, it is indicated that the polynomial function can better describe the ascending branch of tensile stress-strain curves of concrete corroded by the acid rain,

$$y = 1.15x - 0.13x^2. \tag{6}$$

4. Conclusion

In order to investigate the tensile properties of the concrete corroded by the acid rain, laboratory accelerating tests were performed on C30 specimen, acid solution pH 1.0 was used as the corrosive medium herein. Based on the testing results, following conclusions can be drawn

(1) At the initial stage, the tensile strength of deteriorated specimens increases slightly, and then decreased

gradually, and the compressive strength is more sensitive to aggressive environment than the tensile strength.

(2) When the immersing period does not exceed 11 days, the 3 major mechanical property indexes, uniaxial tensile strength, peak strain and elastic modulus increase gradually, however, when the corrosion time exceeds 15 days, with the increasing of corrosion duration, the uniaxial tensile strength and elastic modulus decrease, while the peak strain still increases which shows that the damage increases with the corrosion duration.

(3) A polynomial model, which can describe the ascending branch of stress-strain curves of tensile concrete specimen corroded by acid rain in different periods, is proposed.

Acknowledgments

This paper is funded by the National Natural Science Foundation of China (Grant no. 50708010 and Grant no. 51178069), National Basic Research Program of China (no. 2011CB013605), Liaoning Provincial Funded project (Grant no. 20092149), and supported by the Fundamental Research Funds for the Central Universities.

References

[1] T. Larssen, H. M. Seip, A. Semb et al., "Acid deposition and its effects in China: an overview," *Environmental Science & Policy*, vol. 2, no. 1, pp. 9–24, 1999.

[2] X. M. Zhang, F. H. Chai, S. L. Wang, X. Z. Sun, and M. Han, "Research progress of acid precipitation in China," *Research of Environmental Sciences*, vol. 23, no. 5, pp. 527–531, 2010 (Chinese).

[3] Q. Hou, Y. X. Zhao, and X. B. Xu, "Spatial-temporal distribution of acid rain in china during 2005," *Advance in Climate Change Research*, vol. 2, no. 5, pp. 242–245, 2006 (Chinese).

[4] Q. Hou and Y. X. Zhao, "An analysis on characteristics of regional acid rain over China in 2007," *Advance in Climate Change Research*, vol. 5, no. 1, pp. 7–11, 2009 (Chinese).

[5] J. Guo, *Economic Loss Estimation on the Material Damage Caused by the Acid Deposition*, Beijing University of Chemical Technology, BeiJing, China, 2004.

[6] H. Okochi, H. Kameda, S. I. Hasegawa, N. Saito, K. Kubota, and M. Igawa, "Deterioration of concrete structures by acid deposition—an assessment of the role of rainwater on deterioration by laboratory and field exposure experiments using mortar specimens," *Atmospheric Environment*, vol. 34, no. 18, pp. 2937–2945, 2000.

[7] K. Attiogbe and S. H. Rizkalla, "Response of concrete to sulfuric acid attack," *ACI Materials Journal*, vol. 85, no. 6, pp. 481–488, 1988.

[8] L. H. Kong and G. James, "Concrete deterioration due to acid precipitation," *ACI Materials Journal*, vol. 84, no. 2, pp. 110–116, 1987.

[9] N. I. Fattuhi and B. P. Hughes, "The performance of cement paste and concrete subjected to sulphuric acid attack," *Cement and Concrete Research*, vol. 18, no. 4, pp. 545–553, 1988.

[10] X. B. Hu, "Analysis for simulation test on acid rain attacking concrete," *Journal of the Chinese Ceramic Society*, vol. 36, supplement 1, pp. 147–152, 2008 (Chinese).

[11] Y. Z. Zhang, Y. F. Fan, J. L. Liu, and X. X. Cao, "Experimental study on compressive performance of concrete C40 in simulated acid environment," *Journal of Building Materials*, vol. 13, no. 1, pp. 105–110, 2010.

[12] S. Xie, L. Qi, and D. Zhou, "Investigation of the effects of acid rain on the deterioration of cement concrete using accelerated tests established in laboratory," *Atmospheric Environment*, vol. 38, no. 27, pp. 4457–4466, 2004.

[13] T. J. Wang, L. S. Jin, Z. K. Li, and K. S. Lam, "A modeling study on acid rain and recommended emission control strategies in China," *Atmospheric Environment*, vol. 34, no. 26, pp. 4467–4477, 2000.

[14] H. L. Liu, D. Zhou, and D. S. Xie, "A study on influences of acid rain in Southwest China on concrete," *Journal of Harbin Institute of Technology*, vol. 29, no. 6, pp. 101–104, 1997.

[15] Code for Communications Ministry of P.R.China, *Code for Design of Concrete Structures (GB/T 50010—2002)*, China Architecture and Building Press, Beijing, China, 2002.

[16] Y. F. Fan, Z. Q. Hu, Y. Z. Zhang, and J. L. Liu, "Deterioration of compressive property of concrete under simulated acid rain environment," *Construction and Building Materials*, vol. 24, no. 10, pp. 1975–1983, 2010.

[17] D. M. Yan and G. Lin, "Influence of initial static stress on the dynamic properties of concrete," *Cement and Concrete Composites*, vol. 30, no. 4, pp. 327–333, 2008.

[18] Y. Z. Zhang, Y. H. Zhao, and Y. F. Fan, "A theoretical model for assessing elastic modulus of concrete corroded by acid rain," *Engineering Mechanics*, vol. 28, no. 2, pp. 175–180, 2011 (Chinese).

7

Effect of Minor Alloying Elements on the Corrosion Behavior of Fe40Al in NaCl-KCl Molten Salts

G. Salinas,[1] J. G. Gonzalez-Rodriguez,[1] J. Porcayo-Calderon,[2] V. M. Salinas-Bravo,[2] and M. A. Espinoza-Medina[3]

[1] *CIICAP, Universidad Autónoma del Estado de Morelos, Avendia Universidad 1001, Colonia Chamilpa, 62209 Cuernavaca, MOR, Mexico*
[2] *IIE, Gerencia de Procesos Térmicos, Avenida Reforma 120, 62490 Temixco, MOR, Mexico*
[3] *Facultad de Ingeniera Mecanica, Universidad Michoacana de San Nicolás de Hidalgo, 58000 Morelia, MICH, Mexico*

Correspondence should be addressed to J. G. Gonzalez-Rodriguez, ggonzalez@uaem.mx

Academic Editor: Vesna Mišković-Stanković

The hot corrosion behavior of Fe40Al intermetallic alloyed with Ag, Cu, Li, and Ni (1–5 at.%) in NaCl-KCl (1 : 1 M) at 670°C, typical of waste gasification environments, has been evaluated by using polarization curves and weight loss techniques and compared with a 304-type stainless steel. Both gravimetric and electrochemical techniques showed that all different Fe40Al-base alloys have a much higher corrosion resistance than that for stainless steel. Among the different Fe40Al-based alloys, the corrosion rate was very similar among each other, but it was evident that the addition of Li decreased their corrosion rate whereas all the other elements increased it. Results have been explained in terms of the formation and stability of an external, protective Al_2O_3 layer.

1. Introduction

Sodium and potassium impurities present in the form of chloride or sulfates are very corrosive constituents under certain combustion conditions such as waste incinerators and biomass-fired boilers [1, 2]. Early failure of the thermal components frequently occurs due to the complex reactions between the metallic materials and the hostile combustion environment. Problems with process equipment resulting from fireside corrosion have been frequently encountered in incinerators. The major problem is the complex nature of the feed (waste) as well as corrosive impurities which form low-melting point compounds with heavy and alkali metal chlorides which prevent the formation of protective oxide scales and then causes an accelerated degradation of metallic elements [1]. In particular, under reducing conditions such as those typical of the operation of waste gasification plants or even under localized reducing conditions, which frequently arise in the case of incorrect operation of waste incineration systems, it is difficult to form protective oxide scales such as Cr_2O_3, SiO_2, and Al_2O_3 on the surface of structural materials. Thus, the corrosion attack can be further enhanced under reducing atmospheres in the presence of salt deposits [2].

The effect of individual KCl, NaCl, and their mixtures with heavy metal chlorides or sulfates on the corrosion behavior of a series of alloy systems has been studied in detail so far [2–9]. It is generally realized that Cr is not as effective element for corrosion resistance of Fe-based and Ni-based alloys due to chloride attack. In contrast, alumina-(Al_2O_3-) forming alloys exhibit promising candidate materials considering the better high temperature resistance of alumina over chromia and their properties such as lower cost, low density, high strength, and good wear resistance [10–13].

In order to obtain more information on the corrosion mechanisms taking place, electrochemical techniques are more appropriated than gravimetric methods. Since a molten salt acts as an electrolyte, electrochemical techniques such as potentiodynamic polarization curves and linear polarization resistance (LPR) measurements can be used just

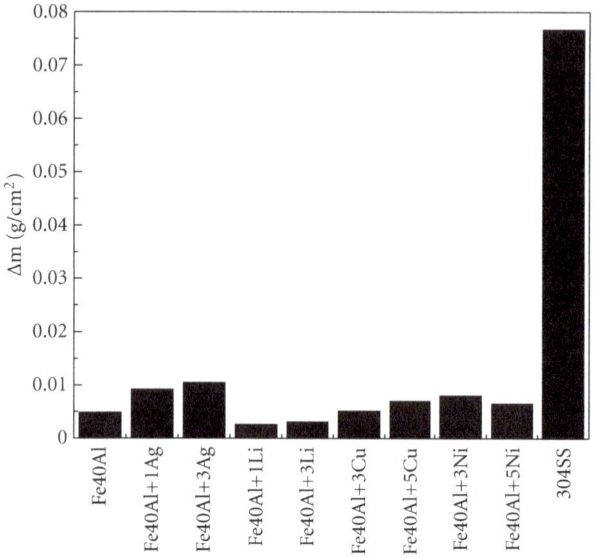

FIGURE 1: Effect of Ag, Li, Cu, and Ni on the weight loss for Fe40Al alloy in NaCl-KCl (1 : 1 molar ratio) deposit in static air at 670°C.

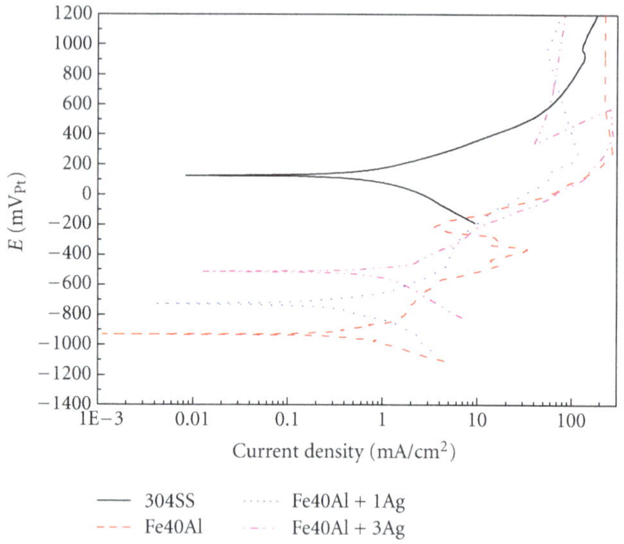

FIGURE 2: Effect of Ag on the polarization curve for Fe40Al alloy in NaCl-KCl (1 : 1 molar ratio) salt in static air at 670°C.

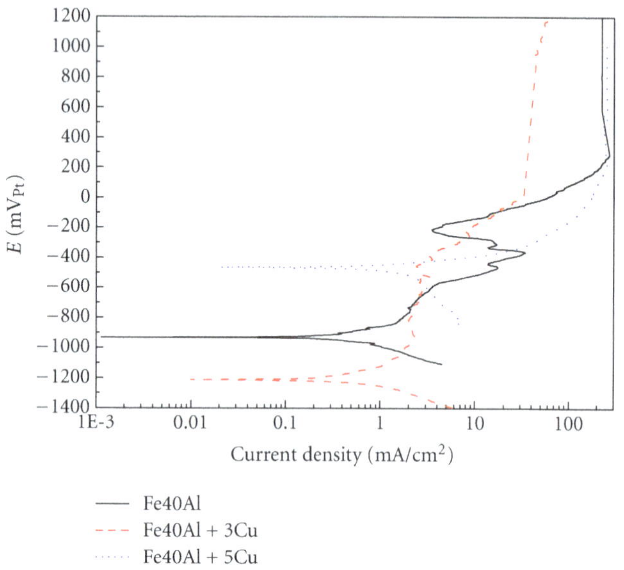

FIGURE 3: Effect of Cu on the polarization curve for Fe40Al alloy in NaCl-KCl (1 : 1 molar ratio) salt in static air at 670°C.

like Ducati et al. [14] showed in one of the earliest works on this topic. Zeng et al. used electrochemical impedance spectroscopy (EIS) studies for the corrosion of two-phase Cu-15Al alloy in the eutectic $(Li, K)_2CO_3$ at 650°C [15] and for Ni_3Al alloy in $(Li, Na, K)_2SO_4$ at 650°C [16]. They found that the complex plane of the electrochemical impedance suggested that the fast corrosion rate of these alloys was due to the formation of a nonprotective oxide and that the corrosion process was controlled by the diffusion of oxidants. Shirvani et al. [17] used polarization curves to study the corrosion performance of the slurry Si-modified aluminide coating on the nickel base super alloy In-738LC exposed to Na_2SO_4—20 wt.% NaCl melt at 750°C. Zhu et al. [18] used Tafel extrapolation, linear polarization resistance, chronopotentiometry, and EIS to determine corrosion rates of iron and iron-based alloys in molten carbonate melts for both cathode and anode in molten carbonate fuel cell (MCFC) environments. Cuevas-Arteaga and coworkers [19–21] have successfully used electrochemical techniques to study the corrosion of materials used in sulfate+vanadate environments and obtained information about both the type and mechanisms of corrosion on these environments. Thus, this work deals with an electrochemical study of the effect of adding alloying element such as Cu, Li, Ni, and Ag to a Fe40Al intermetallic alloy corroded in NaCl-KCl (1 : 1 M) at 670°C. For comparison, traditional weight loss experiments have been performed too.

2. Experimental Procedure

Cast ingots of binary Fe-40 at.% Al and ternary Fe40Al-X (X = 1 and 3 at.% for Cr or Li, 3 and 5 at.% for Ni or Ag) alloys were fabricated using a high-frequency vacuum induction furnace. Chemical elements of high purity (99.9%) were placed in an alumina crucible placed inside a graphite crucible in order to be induction melted under vacuum. The molten Fe-40Al, Fe40Al-X alloys were poured into a rectangular cooper mould. The produced ingots were cut by a diamond wheel cutter in small rectangular parallelepiped pieces of surface area ranging from 1.5 to 2.5 cm². Potentiodynamic polarization curves and weight loss measurements were used. Details of the experimental setup for the electrochemical cell used in this work are given elsewhere [22]. The body of the cell was a 15 mL silica crucible. The most important elements were a reference and auxiliary electrodes, made of a 0.5 mm diameter platinum (Pt) wire inside a mullite tube and filled with commercial refractory, ceramic cement. The amount of salt in each run was 0.5 g for an exposed area of 1.0 cm² under static

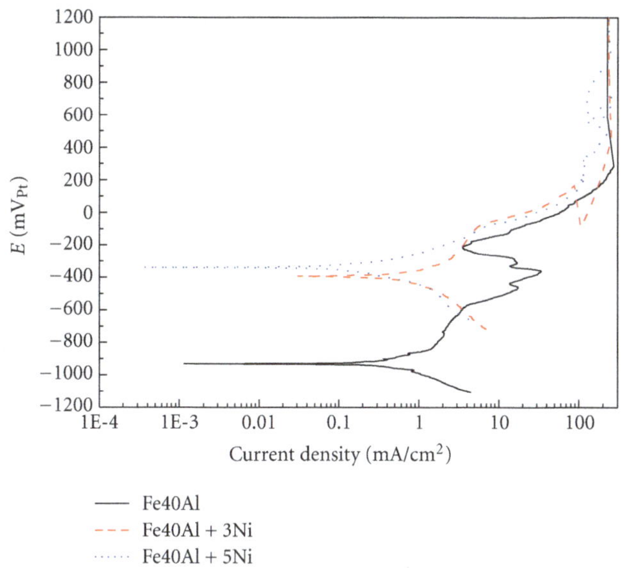

FIGURE 4: Effect of Ni on the polarization curve for Fe40Al alloy in NaCl-KCl (1 : 1 molar ratio) salt in static air at 670°C.

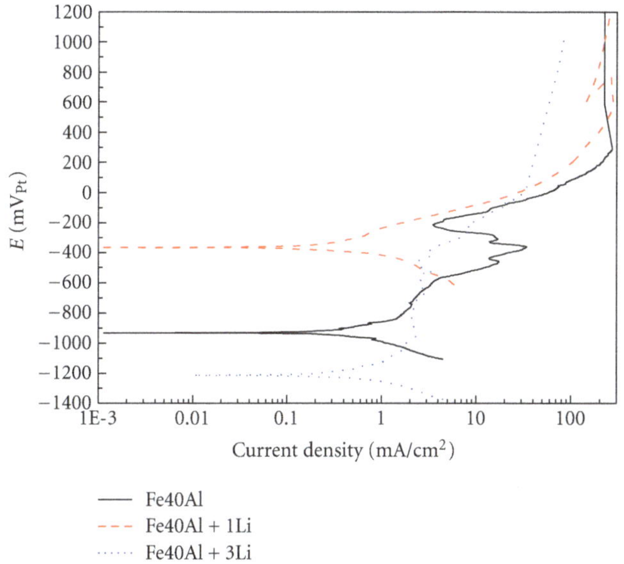

FIGURE 5: Effect of Li on the polarization curve for Fe40Al alloy in NaCl-KCl (1 : 1 molar ratio) salt in static air at 670°C.

conditions. The electrical contact was made by welding an 80 wt.% Cr-20 Ni wire to the specimen. Polarization curves were obtained by polarizing the specimens from −500 to +1500 mV_{Pt} with respect to the free corrosion potential value, E_{corr}, at a scanning rate of 1.0 mV/s. Corrosion current density values, I_{corr}, were calculated by using the Tafel extrapolation method, which is valid for the sweep rates used here according to [21] and taking an extrapolation zone of ±250 mV around the E_{corr} value once it was stable. The working salt consisted of 500 mg/cm^2 of an eutectic mixture of NaCl-KCl, 1 : 1 M, analytical grade for each test. The testing temperature was 670°C in static air condition.

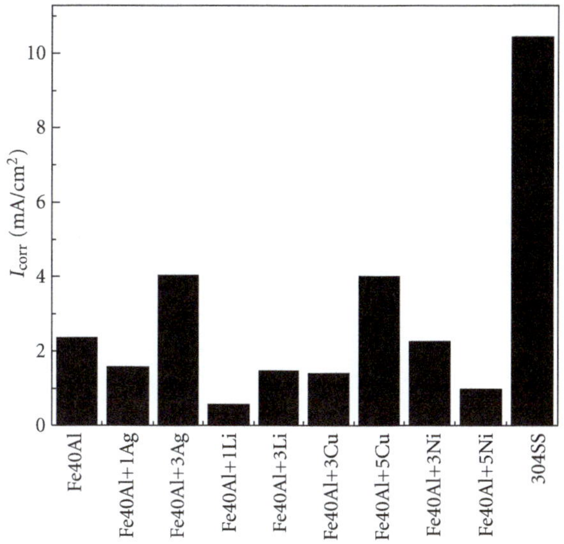

FIGURE 6: Effect of Ag, Li, Cu, and Ni on the I_{corr} values for Fe40Al alloy in NaCl-KCl (1 : 1 molar ratio) deposit in static air at 670°C.

Prior the tests, the specimen surfaces were prepared by the standard technique of grinding with SiC from 240 to 600 grit emery paper, washed with water, and degreased with acetone. Weight loss tests were carried out in electric furnaces in a static air during 100 hours. All materials were machined into specimens with dimensions of about 10 mm × 15 mm × 1.5 mm, then grounded to 600 grade emery paper, and cleaned in a supersonic bath of acetone. After the corrosion tests, the corrosion rate was measured as weight loss. Three specimens of each condition test were decaled and chemically cleaned according to ASTM G1 81 standard. For comparison, the same tests were done with a 304-type stainless steel. One of each heat was mounted in bakelite in cross-section and polished to analyze the subsurface corrosive attack using a scanning electron microscopy (SEM) aided with energy dispersive spectroscopy (EDX) to carry out microchemical analysis.

3. Results and Discussion

Figure 1 shows that the mass loss for the different alloys alters being exposed to an eutectic NaCl-KCl mixture at 670°C during 100 hours. It can be seen that 304-type stainless steel suffered the greatest corrosion rate, more than one order of magnitude higher than that for all the Fe40-based alloys. Li et al. [2] obtained very similar results with an Fe45Al alloy in NaCl-KCl mixture at 670°C during 100 hours. Among the different Fe40Al-based alloys, the addition of either 1 or 3% Li decreased the corrosion rate of unalloyed Fe40Al alloy, whereas the addition of Ag, Ag or Ni increased it; however, this difference is marginal.

The effect of 1 and 3% Ag on the polarization curves for Fe40Ag base alloy, together with that one for 304-type stainless steel, is shown on Figure 2. It can be seen that unalloyed Fe40Al-based alloy displays an active-passive behavior, with an E_{corr} value close to −927 mV and an I_{corr}

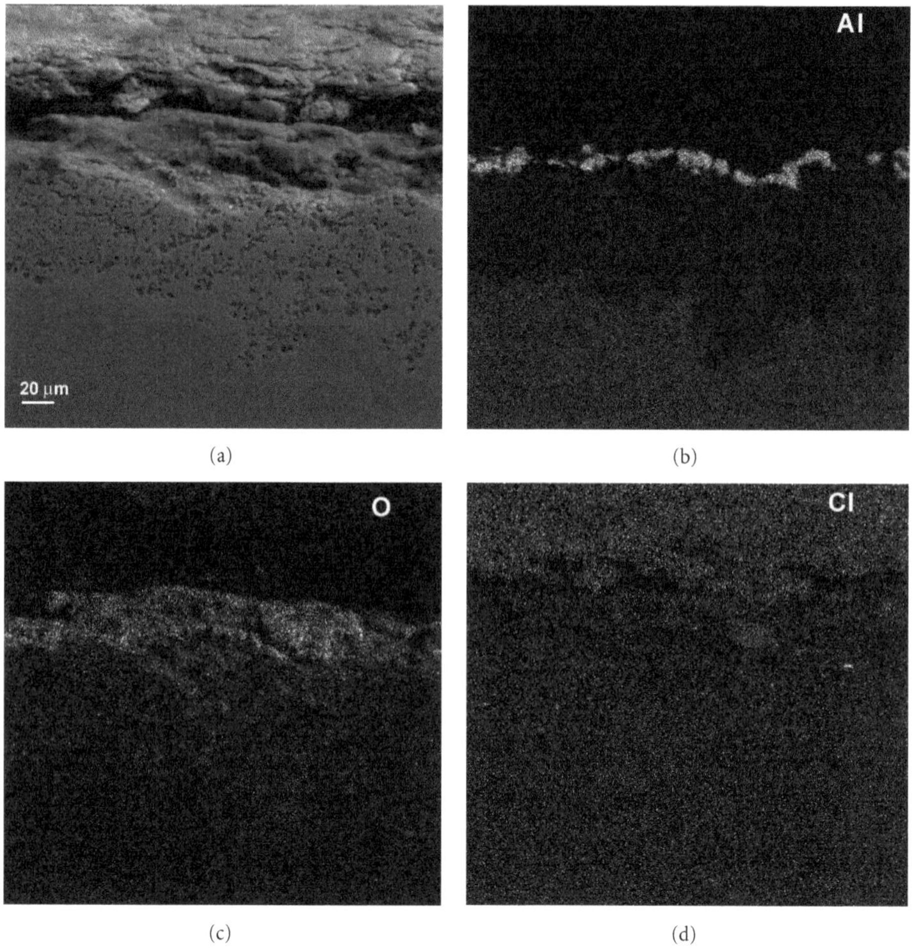

Figure 7: Micrograph of Fe40Al alloy corroded in NaCl-KCl at 670°C together with the X-ray mappings of Al, O, and Cl.

value of 0.15 mA/cm^2. As the applied potential is more anodic, the anodic current density value increases until it reaches a more or less constant value, that is, a passive region, between −800 and −600 mV, where a sudden increase in the current density value appears. Some anodic peaks are observed at −500 and −400 mV, respectively. A second, narrow passive region can be observed around −200 mV, after which the anodic current density increases once again until it reaches an anodic limit current density. This might be due to the presence of an external layer which acts as a barrier against the ingress of the aggressive species. When Ag was added into the alloy, the E_{corr} value is made more anodic and the I_{corr} value is increased. The presence of a passive region in the Ag-containing alloys is not very clear; however, for both alloys, around −450 mV the anodic current density value showed a more or less constant value, although this was not so evident for the alloy containing 3% Ag. The presence of an anodic limit current density, lower than that observed for the Fe40Al base alloy, is observed in both cases at a potential value close to 200 mV. The corrosion current density value for 304-type stainless steel was higher than that for the three Fe40Al-based alloys for at least one order of magnitude, and there was no evidence of a passive region. According to the mass loss data (Figure 1) 304-type stainless steel had the highest corrosion rate too.

The addition of 3% Cu did not affect very much the I_{corr} value as the addition of 5% Cu did, as can be seen on Figure 3. The free corrosion potential became more active with the addition of 3% Cu, but it became much nobler when 5% Cu was added. The passive region in the alloy containing 3% Cu was much wider and more stable than that shown by the unalloyed base alloy, starting around −1100 mV and finishing around −300 mV, where a breakdown potential is obtained. There was no evidence of apassive region for the alloy containing 5% Cu, only anodic dissolution, with the highest anodic current density value. Similarly, the addition of either 3 or 5% Ni shifted the E_{corr} value towards the noble direction, whereas the I_{corr} value was slightly decreased only with the addition of 5% Ni (Figure 4). The presence of a passive region was found only with the addition of 3% Ni between −300 and 0 mV, where the breakdown potential was reached. After this point, an anodic limit current density value is observed in both cases, with a value slightly lower than that found for unalloyed Fe40Al-base alloy.

Finally, when Li was added (Figure 5), the E_{corr} value became nobler with the addition of 1% but more active with

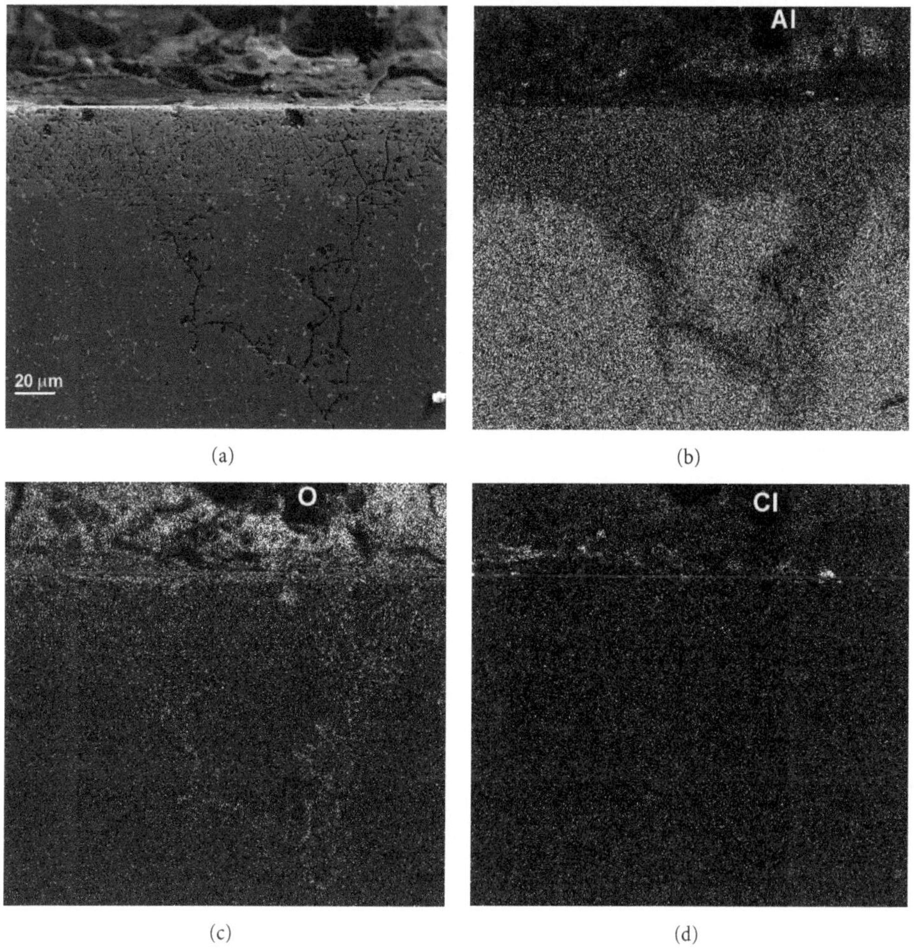

FIGURE 8: Micrograph of Fe40Al + 3Ag alloy corroded in NaCl-KCl at 670°C together with the X-ray mappings of Al, O, and Cl.

3% Li. In addition, the corrosion rate, expressed in terms of the corrosion current density values, was decreased in both cases, just as evidenced by the weight loss results in Figure 1. A much wider and stable passive region than that found on the unalloyed Fe40Al based-alloy is found between −1000 and −300 mV. Unlike this, no passive region can be seen with the addition of 1% Li, but only anodic dissolution until an anodic limit current density is reached. Thus, it can be seen that different alloying elements have a different effect on the electrochemical behavior of Fe40Al alloy in molten NaCl-KCl. Table 1 gives a summary of the different electrochemical parameters obtained from the polarization curves. Figure 6 gives the corrosion current density values, I_{corr}, for all the Fe40Al-based alloys, where it can be seen that the highest value corresponds to the 304-type stainless steel, whereas the lowest value corresponds to the Li-containing alloys. Similar results were obtained with the weight loss test (Figure 1), thus, both gravimetric and electrochemical techniques gave similar results: all the different Fe40Al-based alloys have a better corrosion performance than that for 304-type stainless steel in the NaCl-KCl mixture at 670°C.

Figure 7 shows a micrograph of corroded unalloyed Fe40Al-based alloy together with the X-ray mappings of Al, O, and Cl, where it can be seen the presence of both Al and O, maybe representing the formation of what seems to be a protective Al_2O_3 oxide, which is the responsible for the passive behavior observed in the polarization curve for this alloy (Figure 2); the absence of Al and O in some points indicates that the external Al_2O_3 oxide is porous and it has been dissolved by aggressive molten salt. Li et al. [2], corroding an Fe45Al alloy in NaCl-KCl salt at

TABLE 1: Electrochemical parameters obtained from polarization curves.

Alloy	β_a (mV/dec)	β_c (mV/dec)	I_{corr} (mA/cm^2)	E_{corr} (mV)
FeAl-base	311	223	1.5	−927
FeAl-1Ag	293	398	2.3	−730
FeAl-3Ag	269	432	4.0	−510
FeAl-1Li	159	65	0.5	−363
FeAl-3Li	346	134	1.4	−1212
FeAl-3Cu	169	180	1.4	−1212
FeAl-5Cu	78	191	4	−460
FeAl-3Ni	159	207	2.2	−395
FeAl-5Ni	177	222	0.9	−340
304SS	243	311	10.4	125

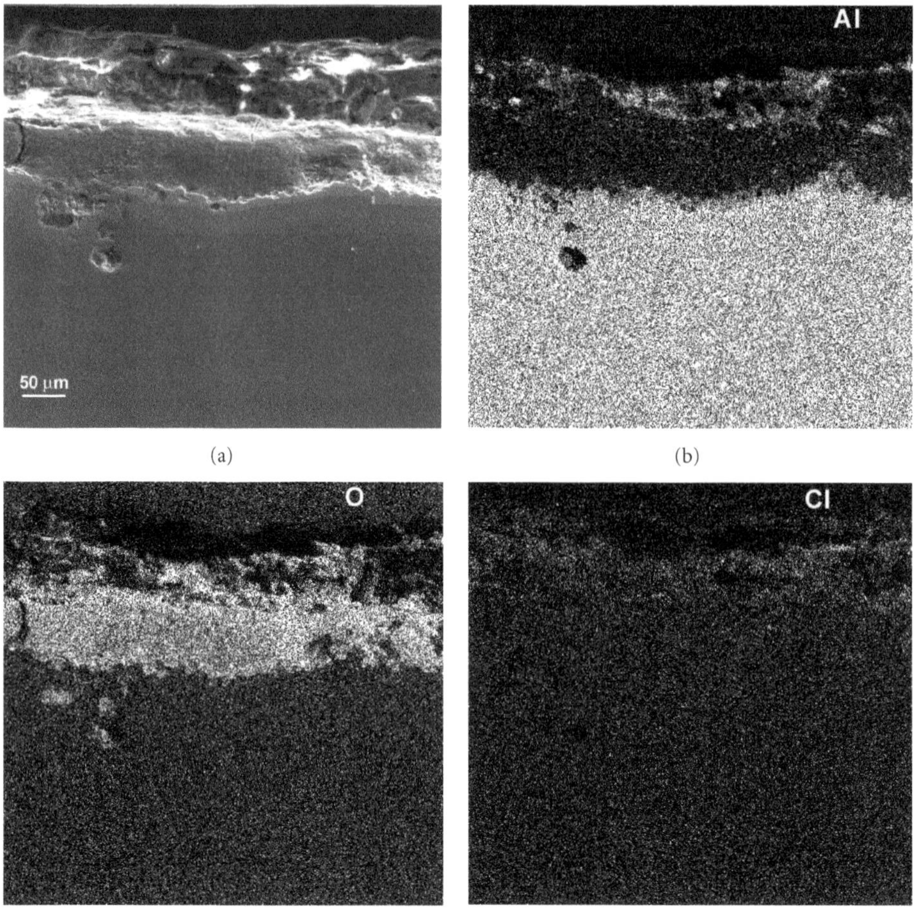

(a) (b) (c) (d)

FIGURE 9: Micrograph of Fe40Al + 3Li alloy corroded in NaCl-KCl at 670°C together with the X-ray mappings of Al, O, and Cl.

670°C during 48 hours, found Al_2O_3 as the main compound formed on the external layer. For the alloy containing 3 Ag, Figure 8, there is evidence of Al, O, and Cl, indicating the formation of an external protective Al_2O_3 oxide dissolved by the Cl-containing salt; however, there is evidence of internal oxidation along the grain boundaries. The formation of an external Al_2O_3 oxide is more evident for the alloy containing 3Li (Figure 9), where it can be seen that aggressive ions such as Cl^- practically remain outside the protective Al_2O_3 oxide, but, eventually, the molten salt dissolves this oxide and dissolves the underlying alloy. Both chlorine and aluminum seem to form a compound on the external salt, which, according to Li et al. [2], could be liquid $AlCl_3$, which melts at 192°C. An Al-depleted and an O-rich zone can be observed in all alloys. These two alloys showed an active-passive behavior in their polarization curves (Figures 2 and 5), which is due to the presence of a very adherent, protective Al_2O_3 layer.

Thus, it can be seen that the corrosion protection of an alloy against salt melt attack depends on the chemical stability of both the kind of metal and their compounds such as oxides and chlorides. In fact, a breakdown of the protective oxide readily occurs by dissolution into the melt, and the degradation rate can be especially fast if the oxide has a high solubility in the melt. Since from a thermodynamic point of view, Al is the most reactive element among all the different investigated elements here; Al chloride or oxide forms preferentially over the other elements. At the melt-substrate interface, the oxygen potential is low while the chlorine potential is relatively high, and, consequently, liquid aluminum chloride forms there and transports outwards, leaving an aluminum depleted layer as observed in Figures 7–9. Near the melt-atmosphere interface,the transition from aluminum chloride to its oxide occurs, which onlyrelatively low oxygen pressure is needed because of the very stablenature of aluminum oxide. Therefore, the preferential removal of aluminum from the matrix is enhanced, basically in the form of highly volatile liquid chloride. This process seems to be beneficial to help establish a protective alumina, Al_2O_3, scale due to a fast aluminum supply. The role of the different alloying elements is to help to make more stable the alumina layer on top of the alloy and increase the alloy corrosion rate in the melt.

4. Conclusions

A study of the effect of alloying elements such as Ag, Li, Cu, and Ni on the corrosion resistance of Fe40Al intermetallic alloy beneath a NaCl-KCl mixture in air at 670°C has been carried out by using both weight loss and potentiodynamic polarization curves. Both techniques have shown that the corrosion rate of the different Fe40Al-based alloys is much lower than that for 304-type stainless steel for nearly one order of magnitude. Addition of Li decreased the corrosion rate of unalloyed Fe40Al-base alloy, whereas the addition of Ag, Ni, or Cu increased it. However, the corrosion rate for the different Fe40Al-based alloys was within the same order of magnitude. Results have been explained in terms of the formation and stability of an external, protective alumina, Al_2O_3, layer.

References

[1] C. S. Ni, L. Y. Lu, C. L. Zeng, and Y. Niu, "Electrochemical impedance studies of the initial-stage corrosion of 310S stainless steel beneath thin film of molten $(0.62Li,0.38K)_2CO_3$ at 650°C," *Corrosion Science*, vol. 53, no. 3, pp. 1018–1024, 2011.

[2] Y. S. Li, M. Spiegel, and S. Shimada, "Corrosion behaviour of various model alloys with NaCl-KCl coating," *Materials Chemistry and Physics*, vol. 93, no. 1, pp. 217–223, 2005.

[3] Nobuo Otsuka, "Effects of fuel impurities on the fireside corrosion of boiler tubes in advanced power generating systems—a thermodynamic calculation of deposit chemistry," *Corrosion Science*, vol. 44, no. 2, pp. 265–283, 2002.

[4] Y. Shinata and Y. Nishi, "NaCl-induced accelerated oxidation of chromium," *Oxidation of Metals*, vol. 26, no. 3-4, pp. 201–212, 1986.

[5] N. Hiramatsu, Y. Uematsu, T. Tanaka, and M. Kinugasa, "Effects of alloying elements on NaCl-induced hot corrosion of stainless steels," *Materials Science and Engineering A*, vol. 120-121, no. 1, pp. 319–328, 1989.

[6] Y. S. Li, M. Sanchez-Pasten, and M. Spiegel, "High temperature interaction of pure Cr with KCl," *Materials Science Forum*, vol. 461-464, pp. 1047–1054, 2004.

[7] F. Wang and Y. Shu, "Influence of Cr content on the corrosion of Fe-Cr alloys: the synergistic effect of NaCl and water vapor," *Oxidation of Metals*, vol. 59, no. 3-4, pp. 201–214, 2003.

[8] C. J. Wang and Y. C. Chang, "NaCl-induced hot corrosion of Fe-Mn-Al-C alloys," *Materials Chemistry and Physics*, vol. 76, no. 2, pp. 151–161, 2002.

[9] Y. S. Li, Y. Niu, and W. T. Wu, "Accelerated corrosion of pure Fe, Ni, Cr and several Fe-based alloys induced by $ZnCl_2$–KCl at 450°C in oxidizing environment," *Materials Science and Engineering A*, vol. 345, no. 3, pp. 964–970, 2003.

[10] F. H. Stott and G. C. Wood, "Internal oxidation," *Materials Science and Technology*, vol. 4, no. 12, pp. 1072–1078, 1988.

[11] G. Han and W. D. Cho, "High-temperature corrosion of Fe_3Al in 1% Cl_2Ar," *Oxidation of Metals*, vol. 58, no. 3-4, pp. 391–413, 2002.

[12] F. Lang and Z. Yu, "Corrosion behavior of Fe–40Al sheet in N_2–$11.2O_2$–$7.5CO_2$atmospheres with various SO_2 contents at 1273 K," *Intermetallics*, vol. 11, no. 2, pp. 135–141, 2003.

[13] Y. Kawahara, "High temperature corrosion mechanisms and effect of alloying elements for materials used in waste incineration environment," *Corrosion Science*, vol. 44, no. 2, pp. 223–232, 2002.

[14] U. Ducati, G. L. Coccia, and G. Caironi, "Electrochemical evaluation of hot corrosion resistance of metallic materials," *Materials Chemistry and Physics*, vol. 8, no. 2, pp. 135–145, 1983.

[15] C. L. Zeng, P. Y. Guo, and W. T. Wu, "Electrochemical impedance spectra for the corrosion of two-phase Cu-15Al alloy in eutectic $(Li, K)_2CO_3$ at 650°C in air," *Electrochimica Acta*, vol. 49, no. 9-10, pp. 1445–1450, 2004.

[16] C. L. Zeng, W. Wang, and W. T. Wu, "Electrochemical impedance models for molten salt corrosion," *Corrosion Science*, vol. 43, no. 4, pp. 787–801, 2001.

[17] K. Shirvani, M. Saremi, A. Nishikata, and T. Tsuru, "Electrochemical study on hot corrosion of Si-modified aluminide coated In–738LC in Na_2SO_4–20 wt.% NaCl melt at 750°C," *Corrosion Science*, vol. 45, no. 5, pp. 1011–1021, 2003.

[18] B. Zhu, G. Lindbergh, and D. Simonsson, "Electrochemical impedance studies of the initial-stage corrosion of 310 S stainless steel beneath thin film of molten $(0.62Li, 0.38K)_2CO_3$ at 650°C," *Corrosion Science*, vol. 41, no. 9, pp. 1497–1513, 1999.

[19] C. Cuevas-Arteaga, J. Uruchurtu-Chavarín, J. Porcayo-Calderon, G. Izquierdo-Montalvo, and J. Gonzalez, "Study of molten salt corrosion of HK–40 m alloy applying linear polarization resistance. And conventional weight loss techniques," *Corrosion Science*, vol. 46, no. 11, pp. 2663–2679, 2004.

[20] C. Cuevas-Arteaga, J. Uruchurtu-Chavarín, J. G. González, G. Izquierdo-Montalvo, J. Porcayo-Calderón, and U. Cano, "Corrosion evaluation of Alloy 800 in sulfate/vanadate molten salts," *Corrosion*, vol. 60, no. 4, pp. 520–527, 2004.

[21] C. Cuevas-Arteaga, "Corrosion study of HK–40 m alloy exposed to molten sulfate/vanadate mixtures using the electrochemical noise technique," *Corrosion Science*, vol. 50, no. 3, pp. 650–663, 2008.

[22] J. G. Gonzalez-Rodriguez, O. L. Arenas, J. Porcayo-Calderon, V. M. Salinas-Bravo, M. Casales, and A. Martinez-Villafañe, "An electrochemical study of the effect of B on the corrosion of atomized Fe40Al intermetallics in molten Na_2SO_4," *High Temperature Materials and Processes*, vol. 25, no. 2, pp. 293–301, 2006.

8

Atomistic Modeling of Corrosion Events at the Interface between a Metal and Its Environment

Christopher D. Taylor

Materials Technology - Metallurgy (MST-6), Materials Science and Technology, Los Alamos National Laboratory, Los Alamos, NM 87545, USA

Correspondence should be addressed to Christopher D. Taylor, cdtaylor@lanl.gov

Academic Editor: Osama Alyousif

Atomistic simulation is a powerful tool for probing the structure and properties of materials and the nature of chemical reactions. Corrosion is a complex process that involves chemical reactions occurring at the interface between a material and its environment and is, therefore, highly suited to study by atomistic modeling techniques. In this paper, the complex nature of corrosion processes and mechanisms is briefly reviewed. Various atomistic methods for exploring corrosion mechanisms are then described, and recent applications in the literature surveyed. Several instances of the application of atomistic modeling to corrosion science are then reviewed in detail, including studies of the metal-water interface, the reaction of water on electrified metallic interfaces, the dissolution of metal atoms from metallic surfaces, and the role of competitive adsorption in controlling the chemical nature and structure of a metallic surface. Some perspectives are then given concerning the future of atomistic modeling in the field of corrosion science.

1. Introduction to Corrosion Mechanisms

The following statement was made by Pletnev in regards to the corrosion of iron by chlorides in acidic media.

> *It is a common point of view that the chemical nature and structure of the surface of a metal, which is in contact with an electrolyte, are decisive in the kinetics of electrochemical reactions that proceed on this surface [1].*

This "common point of view" is manifested in the great body of work in corrosion science that proposes mechanisms that explain how corrosion reactions occur (see, for instance, the monograph edited by Marcus) [2]. Corrosion in aqueous environments proceeds via an electrochemical mechanism, in which the coupled anodic and cathodic reactions take place at unique sites within the material/environment interface. The reactions themselves involve transfer of electrons or ions—often both—across the electrochemical double layer [3]. For this reason, the mechanisms via which corrosion proceeds can be strongly influenced by perturbations in the surface and interfacial environment.

In many cases, the chemical reactions that together constitute a proposed mechanism or reaction scheme should be considered as placeholders: representative entities that subsume a host of microkinetic processes such as mass transport, surface adsorption and desorption, and bond-making/bond-breaking chemical reactions. Pltenev's "chemical nature and structure" of the materials/environment interface are, in fact, rather complex quantities. Figure 1 contains an illustration that presents, schematically, some of the important elements of this interface and the processes relevant to corrosion [4].

Important classes of variables that impact the process and nature of corrosion in a material include the temperature, local environment, and the stress on the material. Environmental factors may include flow rates, pH, and speciation [5, 6]. The nature of the electrolyte is also significant: different corrosion mechanisms may be favored when the electrolyte is changed [7]. The conductivity of the electrolyte also affects the Ohmic drop in potential between the cathodic

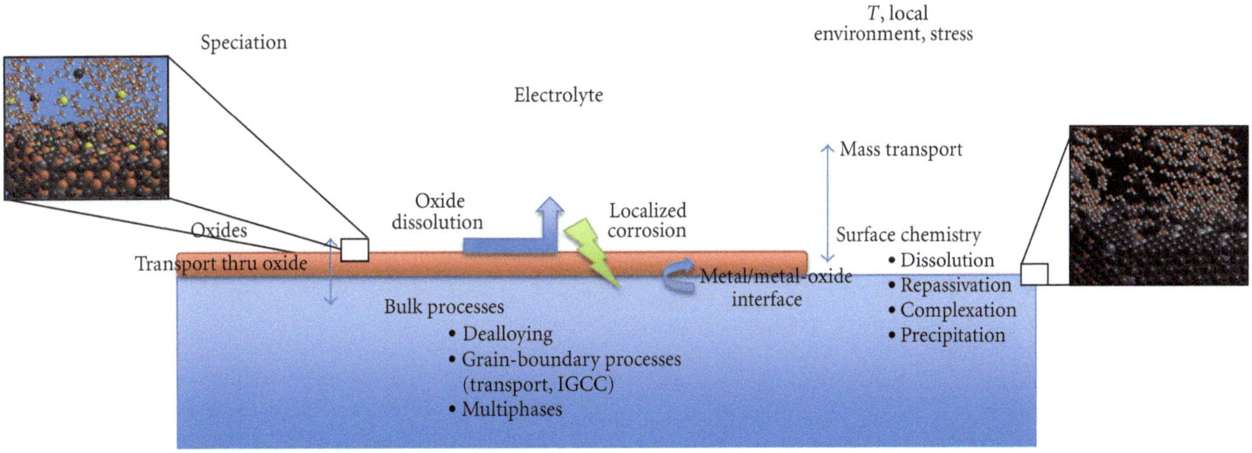

FIGURE 1: Schematic of the materials/environment interface. Features relevant to corrosion are labeled. Reproduced from [4].

and anodic reaction sites in situations such as crevice corrosion and pitting [8]. Mass transport processes at the interface often define the kinetics of corrosion reactions. These processes may occur across a hydrodynamic boundary layer, through an oxide film, or involve migration of species within the metallic component [9–13]. Each of these rates may be corrosion limiting. The microstructure of the metal itself can have important consequences for the process of corrosion. Corrosion may be intergranular, in which case corrosion occurs preferentially at the grain boundaries [14–17]. Crystallographic pitting has been observed in some cases, where the corrosion morphology reflects the crystallographic order of the material substrate [18, 19]. Galvanic coupling can occur between precipitates in the material and the bulk matrix [20], and stress-corrosion cracking pathways may be controlled by microstructural effects and phase distributions [21]. Grain boundary engineering has, in some cases, been employed to optimize corrosion resistance of certain materials [17, 22, 23].

While not comprehensive, the above list serves to illustrate that corrosion rate is a highly nonlinear function of multiple materials and environmental variables. This nonlinearity is, ultimately, the reason why, for some scientists and engineers, corrosion science presents such a rich and complex field of study.

2. Atomistic Models

Scientists working in the fields of surface science and solid-state physics have made it possible, in recent years, to directly simulate interfacial reactions with atomistic resolution using electronic structure methods or interatomic potentials, finding particular applications in the design and modeling of heterogeneous catalysts [24–26]. Whereas heterogeneous catalysis might be described as the deployment of a material to manipulate its environment, corrosion is the inverse process: the modification and depletion of a material by its environment. The deterioration of a solid-state material as a consequence of its interactions with a fluid phase, however, involves processes that are typically slower and involve higher-energy transition states than the catalysis of a molecular transformation by a substrate [27]. Although it is not, at the present time, possible to include all of the elements contained within Figure 1 within some massively parallel computational simulation of corrosion (although for some particular mechanisms and materials this has been attempted) [28, 29], it is possible to deconstruct certain elements deemed to be critical within a given application. For instance, in cases of localized corrosion or corrosion cracking, one may focus on the chemistry occurring at a bare-metal/solution interface [30–32]. In cases where oxide growth is relevant, one can focus on ensembles that consist of defect states in the oxide material [33, 34]. It is also possible to look at dissolution processes from oxide surfaces or to interrogate questions regarding bulk or surface thermodynamic stability [4, 35, 36]. The field of computational materials science is currently growing such that, by coupling continuum kinetic or fracture mechanics models with atomistic scale codes or phase-field theory, it may soon be possible to realistically simulate increasingly comprehensive portions of Figure 1 [28, 29, 37]. The list of examples and references provided in this section should by no means be considered comprehensive, but it is an attempt to capture recent efforts, as well as historical modeling studies that are relevant to scientific studies of corrosion phenomena conducted in the present research group.

Atomistic modeling techniques can, at present, be grouped into four main categories: molecular modeling, static band-structure calculations, molecular dynamics, and Monte Carlo methods [38–40]. Molecular modeling encompasses the field of (generally) small-molecule calculations [40]. Molecular orbital theory is used to solve the electronic structure problem, which in turn allows access to molecular properties, such as HOMO-LUMO (highest occupied and lowest unoccupied molecular orbitals, resp.) gap, bond lengths and bond angles, and energetics of reactant, intermediate, transition, and product states. Extensions of molecular orbital theory allow, for example, the modeling of cationic states of metals when dissolved in solution phase [41, 42], the study of metallic nanoparticles [43], investigations of bonding of water and oxygen to metal atoms and clusters [44, 45], and the evaluation of molecular properties of chemical inhibitors [46, 47].

Static band-structure calculations are drawn more from the fields of density functional theory and solid-state physics [48]. These calculations were originally developed to evaluate the properties of solids, in a similar way to how molecular orbital theory calculations are used to evaluate the properties of molecules. From a reaction point of view, the electronic and atomistic properties of reactant, product, intermediate, and transition states can also be evaluated using these methods, allowing insights to be gained regarding rate-limiting steps, for example, in mass-transport processes such as solid-state diffusion [49–51] or phase transformations [52]. These techniques can be extended to look at surfaces as well, as has been applied heavily in the field of heterogeneous catalysis [24] and electrocatalysis [25], and, more recently, to problems in corrosion [35, 53–57]. This technique, as well as molecular modeling, is limited primarily by the computational cost of performing such calculations. Ensembles containing small numbers of atoms, either isolated as a cluster in space or in cells with periodic boundary conditions, must therefore be carefully constructed to imitate the system one is trying to model, and the appropriate caveats should be given regarding artifacts that may be induced by these idealized conditions. Furthermore, the inability to readily incorporate dynamic effects means that reactions which require significant thermal reorganization—such as the formation of solvation shells around ions during a dissolution event—may be difficult to capture reliably. Overall, however, these methods can be powerfully applied to interrogate detailed, atomistically resolved reaction mechanisms and mass transport processes relevant to corrosion. Their dependence upon electronic structure evaluation makes them the most flexible and reliable, in terms of fundamental, deterministic physics but, at the same time, makes them exceedingly expensive in terms of computational cost and user time.

Molecular dynamics simulations can be used to simulate reaction events, particularly those involving high energy or thermally stimulated processes. While simulations using electronic structure techniques to evaluate the forces on atoms and their thermal trajectories can be performed, the cost of such methods is usually far too high to enable useful time scales to be reached within the scope of the simulation. Therefore, modelers employing molecular dynamics will typically apply interatomic potential techniques. These techniques employ simple algorithms for evaluating forces on atoms as a function of their relative positions in space, for example, the embedded atom method [58–60]. Based on this technique, one can simulate the evolution of corrosion morphologies, diffusion processes in materials, the behavior of defects in materials, such as grain boundaries, and the rates at which oxidation or other processes occur. Examples include the oxidation of aluminum and zirconium metals [28, 34], the structure of the metal/water interface [61–69], and the evolution of radiation damage effects in nanolayered materials [70].

Kinetic Monte Carlo simulations go beyond dynamic effects but utilize atomistically determined information for the rate constants of fundamental atomistic and molecular transformations to perform stochastic updates to the state of a system, which, in this context, may be considered its atomistic configuration [71, 72]. This method, therefore, is not limited to dynamic time scales but, instead, depends on the accuracy of the rate information passed to the model from lower length scale models. These models have been applied to look at atomic and molecular arrangements of adsorbates on metal surfaces [73] and growth of molecular films [74], as well as to bimetallic dissolution problems [75–77]. Monte Carlo models can also be applied to search for equilibrium structures, such as crystallographic pitting morphologies or defect clusters in materials such as metals and oxides [78, 79].

In the following, several applications of atomistic modeling to understand the details of various corrosion reactions are reviewed, drawing primarily upon studies performed within our research group. The paper concludes with a perspective on the state of contemporary modeling in corrosion research and some statements concerning our expectations for the future of this field.

3. Electrochemistry of Water on Metallic Surfaces

The structure of water molecules on metallic surfaces has been studied by various experimental and theoretical methods [80, 81]. One detail that has emerged from these studies is that liquid water tends to be more structured at the interface, due to the necessity for water molecules to arrange in such a way to preserve hydrogen bonding characteristics, despite the discontinuity presented by the interface [63, 81–83]. Additionally, metal-water interactions may impose their own preferences/constraints on the interfacial structure of water [84, 85]. A common model that has been adopted by modelers and experimentalists alike is the "bilayer" [86]. In this structure, water molecules adsorb at atop sites on the metal surface via donation of electron density from the lone pair on oxygen to the frontier orbitals or "dangling bonds" of the surface metal atoms. A second layer of water molecules arranges on the surface in such a way as to form hexagonal overlayer hydrogen bonded to the water molecules directly bound to the surface. Molecular dynamics simulations have suggested that such networks are rather stable on many metallic surfaces, and this general picture has been supported by neutron diffraction experiments [63, 82, 83].

In a work we recently performed to examine the role of ordering of water molecules in crack tips, motivated by the observation of fast transport of ions to the crack tip during stress-corrosion cracking [87], we utilized the *ab initio* molecular dynamics technique to examine the emergence of water structure on the Ni(001) surface [88]. A short molecular dynamics simulation was performed using density functional theory and the Hellman-Feynman theorem to determine forces on the atoms [38]. The code VASP was used to perform these simulations [89–93]. The simulation was short in the sense that only a few picoseconds of time could be simulated, whereas estimates of 100 ps for water equilibration in supercell molecular dynamics simulations have been given in the literature [94]. The simulations are performed using three-dimensionally periodic supercells. This geometry allows the band structure of the exposed Ni(100) surface to be modeled realistically, as well as for bulk water to be imitated in the periodic ensemble. The periodicity in the direction normal to the slab means that the simulation is actually considering an array of nickel slabs, spaced about 2 nm apart. Similar choices are often made when simulating surface chemistry, as the slabs are usually spaced far enough apart to avoid direct interactions. However, in the case where the environment may possess some ordering, we may expect that the interaction between slabs could become more significant. In this way, the simulation we are performing is somewhat imitative of the situation in which there is a tight 2 nm width crack front.

Over the course of the molecular dynamics simulation, the position of the oxygen atoms was monitored to look for evidence of structure, as opposed to homogenization, across the entire volume of the fracture space. The probability distribution for oxygen atoms obtained from the molecular dynamics simulation is reproduced here in Figure 2. Similarly to the previous findings cited above, water molecules show some ordering in the region close to the metal. This ordering can be discerned by the presence of pronounced peaks in the distribution function of oxygen close to the interface. Since the slab spacing is small in this simulation (2 nm, or enough for 5 water layers), the ordering appears to persist throughout the entire region of the simulation cell. This result implies that a 2 nm crack is sufficiently narrow to confine the water molecules in such a way as to enforce certain "packings" of the water molecules. From a surface point of view, the results suggest that some ordering of water molecules may be expected to occur for a distance of up to 1 nm from the metallic surface. This ordering of water molecules will have implications for the diffusion of ions and molecules along surfaces, such as the newly formed surfaces that grow during a cracking event, or penetration along internal cracks in a substance.

In addition to organizing into "ice-like" configurations, water molecules have the opportunity to react on metal surfaces to form hydrided surfaces, oxidized surfaces, or hydroxylated surfaces [95]. The nature of this reaction is strongly dependent on the identity of the metal, as well as the electrochemical potential. For instance, as the electrochemical potential is made cathodic, the Fermi level of the metal is raised, which makes electron transfer to the protons in water to become more favorable, leading to the formation of atomic hydrogen via the mechanism:

$$H_2O + e^- \longrightarrow H_{(ads)} + OH^-. \quad (1)$$

On the other hand, making the surface more anodic depletes electrons from the metal, lowering the Fermi level

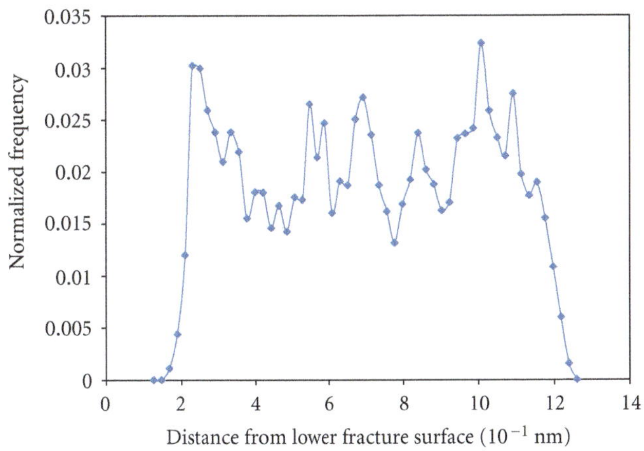

FIGURE 2: The ordering of water molecules during molecular dynamics simulation of the region in between the two Ni(100) fracture surfaces exposed to water. The normalized frequency refers to the frequency with which oxygen may be found at a given distance from the electrode surface, normalized such that the integrated frequency goes to unity.

and thus making the metal a good electron acceptor. Hence, such reactions as

$$H_2O \longrightarrow OH + H^+ + e^-, \quad (2)$$

$$H_2O \longrightarrow O + 2H^+ + 2e^-, \quad (3)$$

become energetically favorable.

In previous work, we developed a methodology for simulating the electrochemical double layer in density functional theory models of interfaces [96]. Using this method we were able to predict the potentials at which hydrogen, water, hydroxide, and oxygen are stable species on metallic surfaces, such as Ni(111) [30, 31]. These predictions are represented in Figure 3, which plots the free energy for each of the possible surface phases (OH, O, H, and H_2O) that may be produced when water reacts with a metallic surface. The free energies are obtained by computing the energy of slab models for the surface that consist of several layers of Ni(111) planes, exposed to a supercell of water molecules, in addition to the adsorbed surface phase. The energies are corrected for the electrochemical potential using the method cited above [96]. The results correlated well with experimental determinations for the onset of oxidation of the metallic surfaces [97].

We also used the same technique to directly evaluate the nature of bonding between copper and hydroxide [98]. We found that, even when significant changes in the electrochemical potential were made, the overall binding energy did not change significantly, and this was attributed to the fact that, when the Fermi energy of the metal was raised or lowered in response to changes in the applied potential, the orbitals of the bonding hydroxide species tended to follow. The adsorption geometry, on the other hand, did change significantly, with rotation of the OH dipole moving in such a way as to counteract the electric field across the double layer. The binding of water, on the other hand, does

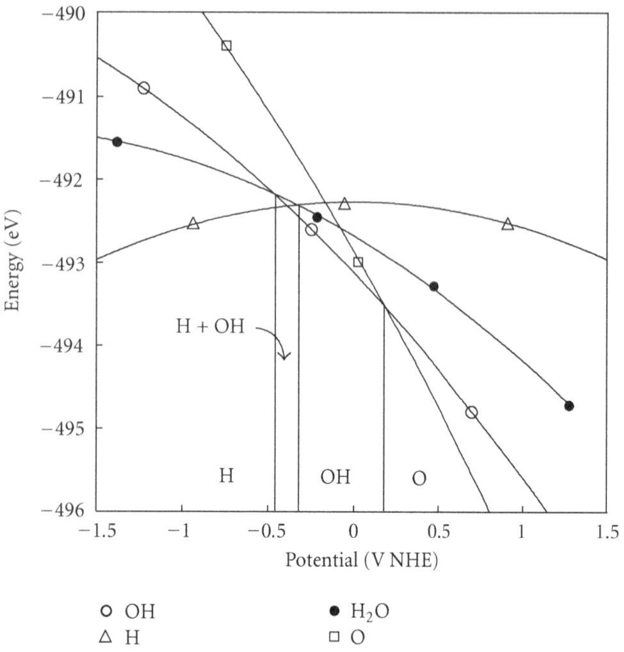

FIGURE 3: Phase diagram for surface hydride, molecular water, hydroxy and oxy species on Ni(111) at pH 0. Reproduced with permission from [30].

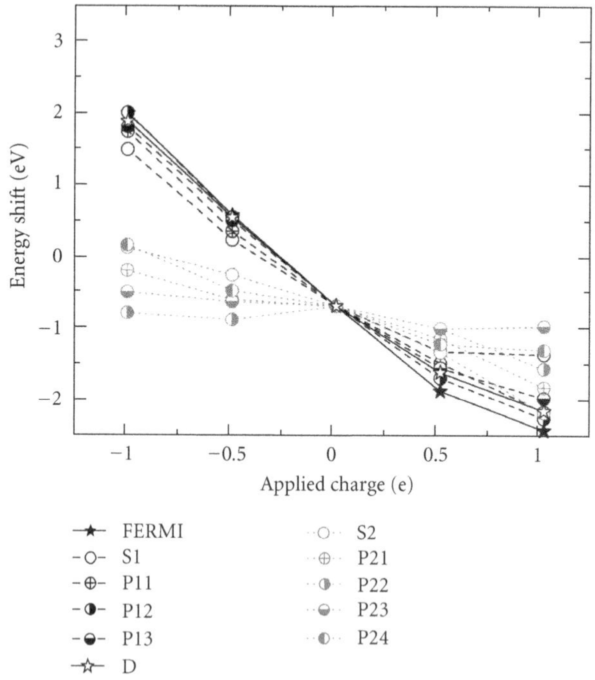

FIGURE 4: The shifts in energy of the Fermi level, copper d-band center (D), s and p bands of the oxygen of adsorbed OH (black symbols, dashed lines, S1, P11, P12, P13), and s and p bands of the oxygen of first-layer water molecules (gray symbols, dotted lines, S2, P21, P22, P23, P24) relative to the zero-charge values. Reproduced with permission from [98].

change more significantly, because the orbitals on water molecules are engaged much more weakly with the frontier orbitals on the metal surface; the metal-oxygen bond in the chemisorption of water is such that the highest occupied molecular orbitals on the water molecule are beyond the range of the electrochemical double layer. This relationship between the electronic structure of the metal (FERMI and D in Figure 4) and the hydroxyl molecular orbitals (S1 and P11, P12, P13 bands in Figure 4) and the water molecular orbitals (S2, P21–P24 in Figure 4) is shown as a function of the applied surface charge in Figure 4. It can be seen that the energy levels of the hydroxyl, which is bonded to the surface, closely track the Fermi level of the metal, whereas the energy levels associated with the water molecule are only weakly correlated.

In the same study we tracked the change in the OH geometry as the potential was made more anodic. During this process, the proton of the adsorbed OH begins to transfer to the second layer of H_2O close to the interface, tending towards the formation of H_3O^+ and an adsorbed O species (an important step in surface oxidation by H_2O). One consequence of this OH-H_2O adduct formation is an increase in the capacitance of the system due to the delocalization of positive charge across the first "inner-layer" of the solution environment. These simulations represent prepassive chemistry: events that will occur prior to, or during, the formation of passive oxide layers on nickel and copper metal.

In a related study, a series of geometry optimizations was performed for a model Cu(111)-H_2O interface, with the potentials made successively more anodic at each step [99]. The model was, therefore, set up in an equivalent way to a cyclic voltammetry experiment, although kinetic processes were not explicitly taken into account. When the resulting geometries were plotted, it could be seen that, given a threshold surface charge density, phase transitions on the surface could be expected to emerge. These began with the simple water dissociation reactions, described above, but ultimately led to destabilization of the metallic copper surface and place exchange between the electron-rich oxygen species and the electron-depleted copper atoms. This series of phase transformations is plotted in Figure 5.

The stepwise charging of the model system and subsequent relaxation of the atomic structure prompted the following structural changes (Figure 5).

(a) Chemisorption of H_2O to the atop site of a Cu atom:

$$H_2O_{(aq)} \longrightarrow H_2O_{(atop)}. \quad (4)$$

(b) Migration of H_2O from the atop site to a bridge site:

$$H_2O_{(atop)} \longrightarrow H_2O_{(bridge)}. \quad (5)$$

(c) Deprotonation of the bridging H_2O to form OH_{ads} + $H^+_{aq} + e^-$:

$$H_2O_{(bridge)} \longrightarrow HO_{(bridge)} + H^+_{(aq)} + e^-. \quad (6)$$

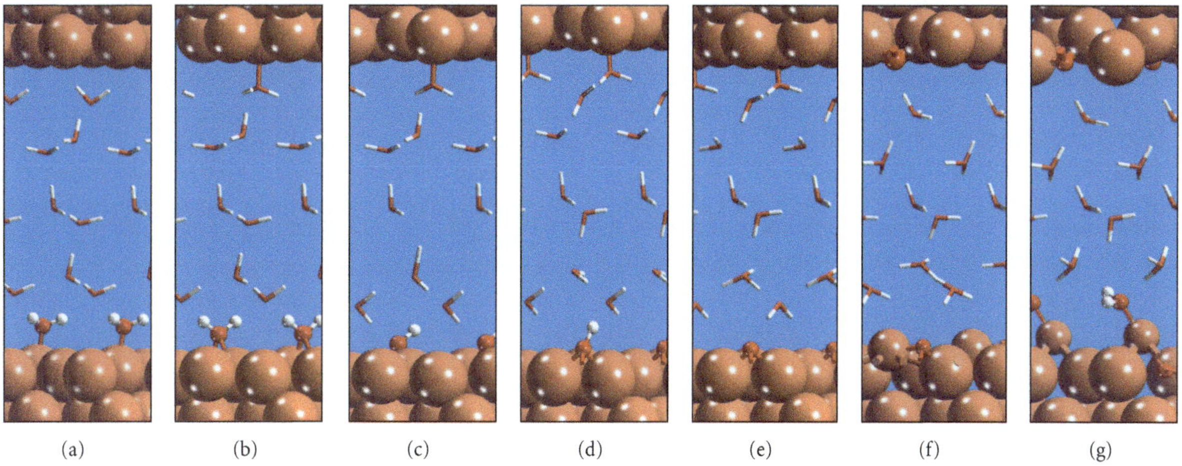

FIGURE 5: Charge-induced reaction sequence for the activation and migration of H_2O (a)–(c), OH (c)–(e) and O (e)–(g) on Cu(111), followed by oxidation of Cu (f)-(g) effected by adsorbed oxygen [99].

(d) Migration of the bridging OH to the threefold hollow site, consistent with the emergence of a second type of OH resonance in the surface enhanced resonance spectrum recorded by Niaura [100]:

$$HO_{(bridge)} \longrightarrow HO_{(3\text{-fold})}. \quad (7)$$

(e) Deprotonation of OH_{ads} to form $O_{ads} + H^+{}_{aq} + e^-$:

$$HO_{(3\text{-fold})} \longrightarrow O_{(3\text{-fold})} + H^+{}_{(aq)} + e^-. \quad (8)$$

(f) Surface reconstruction by eruption of a Cu adatom from the surface layer. Adsorbed oxygen acts as a tether:

$$Cu_{(metal)} + O_{(3\text{-fold})} \longrightarrow Cu_{(bridge)}\text{-}O_{bridge}. \quad (9)$$

(g) Complete displacement of the copper adatom by oxygen, resulting in a Cu_2O-like adlayer, in which O is adsorbed at the vacancy site left by the Cu atom, and this partially adsorbed Cu atom is now coordinated to H_2O, according to the reaction:

$$H_2O_{(l)} + Cu_{(bridge)}\text{-}O_{(bridge)} \longrightarrow H_2O\text{-}Cu_{(ads)} + O_{(ads)} + \text{Vacancy}. \quad (10)$$

We have since generalized this work to a larger range of transition metals using a simplified version of the technique that does not require polarization of the interfacial cells but simply shifts the free energy of electrochemical reactions by an energy term $nF\Delta U$ [32]. This method works most simply when the transfers of an electron and proton occur simultaneously, as in the reaction schemes presented above, as these energies can be related to the energy of H_2 for a reference potential of the standard hydrogen electrode. The utility of this method to capture the leading terms for the free energy changes of electrochemical reactions was demonstrated in a joint paper between our research group and that of Rossmeisl and Nørskov [101].

4. Dissolution of Metal Atoms

Having demonstrated the capability to model the electrochemical reactions involving water on bare metal surfaces, we moved on to consider what is perhaps the quintessential reaction in any aqueous corrosion scheme—the dissolution event itself. This reaction is typically expressed by the chemical equation:

$$M_{(s)} \longrightarrow M^{z+}_{(aq)} + ze^-. \quad (11)$$

The above chemical equation highlights a tension between the forces of metal-metal bonding and the formation of bonds with species present in the aqueous solution. An analysis of the complexity contained within this tension was recently made through a series of papers published by Gileadi [27, 102, 103]. To understand how theoretically challenging this concept is, consider that molecular water is bound to transition and noble-metal close-packed surfaces by energies no more than a few tenths of an eV [104], whereas the surface cohesive energies are an order of magnitude greater [105]. The reason that such dissolution events do occur, on a regular and (from a corrosion mitigation perspective) inconveniently frequent basis, is that electric fields at the interface (of order 0.1 V/Å) provide a gradient sufficiently strong enough to stabilize the formation and extraction of that metal atom in its ionic form [102].

One way to interrogate this process via atomistic simulation is to begin by simply considering the role of metal-metal bonding in controlling corrosion morphologies. This issue has been addressed previously to some extent via kinetic Monte Carlo simulations of the dealloying of binary alloy systems and of crystallographic pitting [75, 76, 106]. We recently published work on Fe-Tc systems, showing how the degree of local coordination and local composition affect metal atom cohesive energies [4]. It was shown in that work that the number of nearest neighbors alone was insufficient, however, to capture the strength of these interactions, but that the organization of these neighbors was of great significance. We performed a similar study on nanoparticles, close-packed and stepped surfaces of copper, and showed that generally surface cohesive energies correlated linearly with the number of nearest neighbors, but significant noise existed in this correlation such that individual topological features must indeed be taken into account [105].

Because it is fundamentally challenging to directly model the interplay between metal-metal cohesion and metal-solvent attraction using typical materials dynamics simulation techniques (such as embedded atom potentials) [107–110], we decided to build a model system that could explore aspects of the dissolution reaction using highly controlled, density functional theory calculations [111]. The model we built to study this system consisted of a Cu nanoparticle, with a single Cu adatom adsorbed on the surface in contact with a hemispherical "nanodroplet" close to a dozen water molecules (see Figure 6(a)). The metal particle was designed to be large enough to provide a fair representation of metal-metal bonding and electronic band structure, and the solution model is sufficient to model one to two layers of an emerging ion's solvation sheath. The configuration of the solvent molecules was obtained by performing a quantum mechanical molecular dynamics simulation at 300 K and then cooling the system to an annealed configuration, which was governed by the hydrogen bonding between water molecules and their interactions with the metal particle. We then performed a series of geometry optimizations, similar to what had been performed for water molecules on Cu(111) surfaces in the study mentioned above, but rather than varying the system charge state, we varied the height of the metal atom above the particle surface stepwise.

To understand some of the mechanistic details of the study we performed the simulation under three separate sets of conditions (no water, water, and water plus an applied electric field, thus simulating a double layer). We plotted the charge states of the adatom as a function of the height above the surface (or, equivalently, as a function of the dissolution process), as well as the associated energies (Figures 6(b) and 6(c), resp.).

In the absence of solvent, the energy of the adatom appears to rise in a quasi-harmonic way, which will at some point vary from harmonicity and tend asymptotically to the value for the surface-binding energy of the adatom. There is some weak charging of the surface adatom. When the solvent molecules are present, the charge on the adsorbed copper atom is significantly more cationic. The Bader method yields a charge of +0.36 for the copper adatom in the solvent phase, even before dissolution occurs. Thus water molecules appear to "pre-charge" adsorbed metal atoms on the surface, as they form a partial solvation shell around the adatom. As the bond is broken (due to raising of the surface metal atom), the energy increases, as does the charge. Furthermore, the copper atom begins to form more and more bonds with the surrounding water molecules, eventually obtaining a rudimentary solvation shell.

The early part of the energy-distance curve, for distances less than 1 Å, is not so different from the case of the vacuum or "no solvent." It appears that *the energy in the early stages of dissolution is mostly concerned with the strong effect of breaking bonds with the metal surface.* In the latter part of the curve, for distances greater than 1 Å, the formation of bonds with the solvent environment alleviates the high energies incurred when the metal-surface bond is broken, in agreement with the "make-before-break" argument as posed in the manuscript by Gileadi [102]. The development of what appears to be a +1 charge on the Cu ion may be an indicator that the +1 state develops first, and then transition to +2 takes place as an outer-layer, electron-transfer event. The application of the electric field further lowers the energy trajectory of the dissolving ion.

5. Surface Chemistry Control of Corrosion Processes

Although water chemistry and bare-metal dissolution remain important focus areas for the development of our understanding of corrosion mechanisms, in many cases corrosion takes place in more complex media. For instance, in the oil and gas industry, it is known that numerous corrosion processes take place simultaneously, and, depending on the local conditions, one or other of these processes will be dominant at a given location [112]. Documented corrosion mechanisms that occur during crude oil refining operations include ammonium-chloride corrosion, ammonium bisulfide corrosion, high-temperature sulfide corrosion, cyanide accelerated corrosion, or stress corrosion cracking that may be brought on by hydrogen, chloride, sulfide, or carbonate [112].

Recently, we embarked upon a study of ammonium chloride corrosion mechanisms, focusing on the surface adsorption processes associated with this reaction scheme [113]. Ammonium chloride corrosion occurs when conditions in the process stream are such that the ammonia and hydrogen chloride in the vapor phase crystallize as hygroscopic NH_4Cl particulates and interact with condensed water vapor to form highly saturated chloride solutions attended by aggressive corrosion behavior [112, 114, 115]. It was demonstrated by Shimbarevich and Tseitlin that increasing the chloride content for a given NH_3 concentration can cause enhanced corrosion, and it was suggested that this occurs due to adsorption of chloride ions on the surface and the subsequent enhancement of anodic dissolution process [116]. At the same time it was shown that increasing the NH_3 concentration made the solution more basic, thus favoring OH adsorption on the surface, as opposed to Cl adsorption and hence hastening passivation [117].

In order to begin to understand the significance of surface processes to the corrosion mechanism occurring when mild steel is exposed to ammonium chloride and to attempt to evaluate these mechanisms that are driven by consideration of surface processes, we applied the techniques we had developed for water-metal interfaces to a more complex mixture of reactive species. In particular, we modeled the competitive adsorption of NH_x, OH_x, H, and Cl species on an exposed Fe(110) facet as a function of the local electrochemical potential. The relevant phase diagram, computed

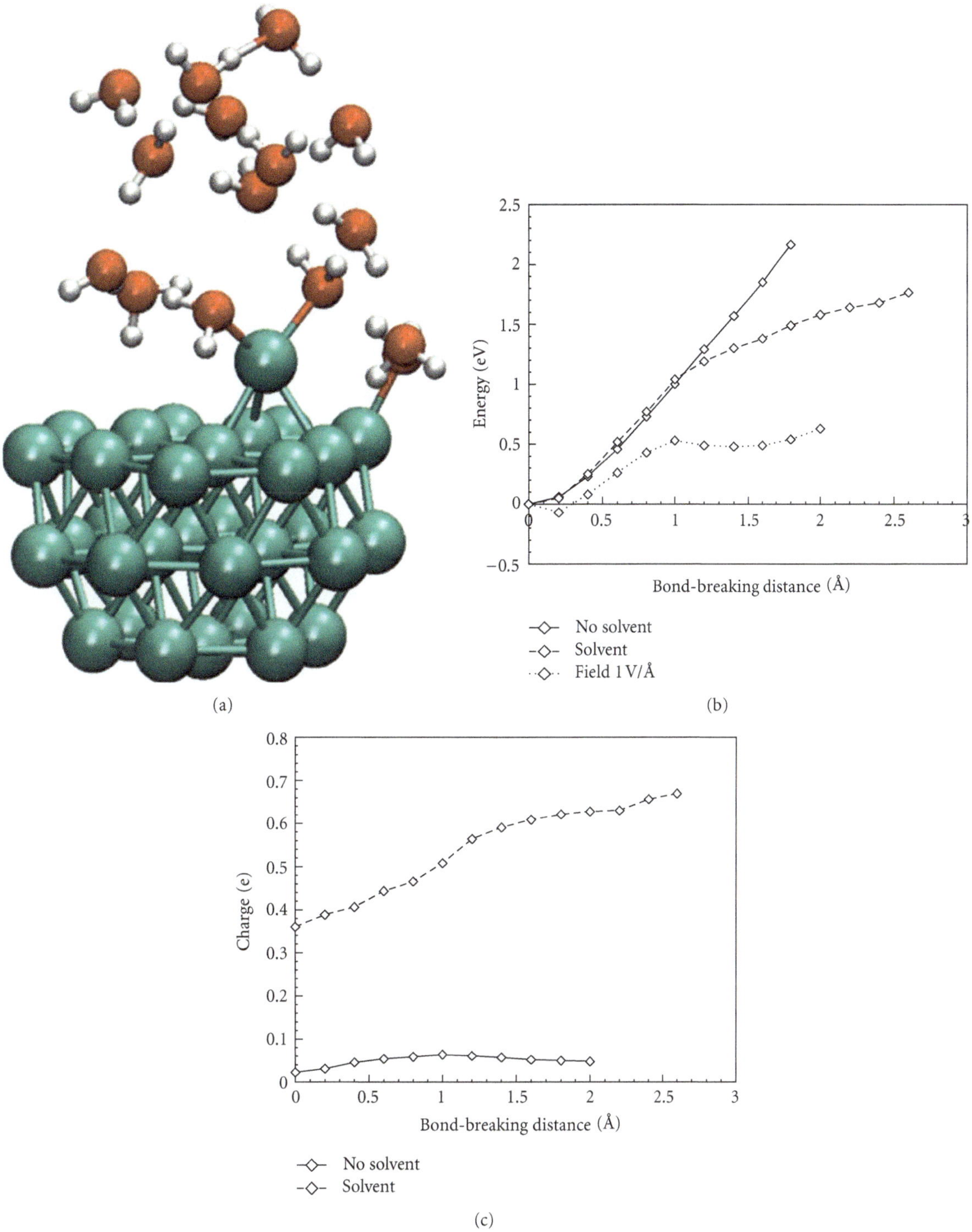

FIGURE 6: (a) Initial solvent-metal particle configuration, (b) electronic structure energies for dissolution/bond-breaking of a copper atom in proximity to a (111) copper surface, and (c) charges determined by the method of Bader for the copper adatom being "dragged" away from the (111) copper surface. Reproduced with permission from [111].

using our density functional theory slab calculations, is presented in Figure 7.

Figure 7 was calculated using a particular set of boundary conditions that apply to a saturated NH_4Cl solution, mimicking the case when a hygroscopic NH_4Cl salt film precipitates on the surface of a mild steel component. Similarly to the case of water of transition metal surfaces, there is a range of H, OH, and O stabilities. Chloride in the solution phase, introduced by the equilibration of NH_4Cl salt film with condensed water, can also be adsorbed on to the surface between the electrochemical potentials of −450 and −420 mV SHE. In this window, which corresponds to

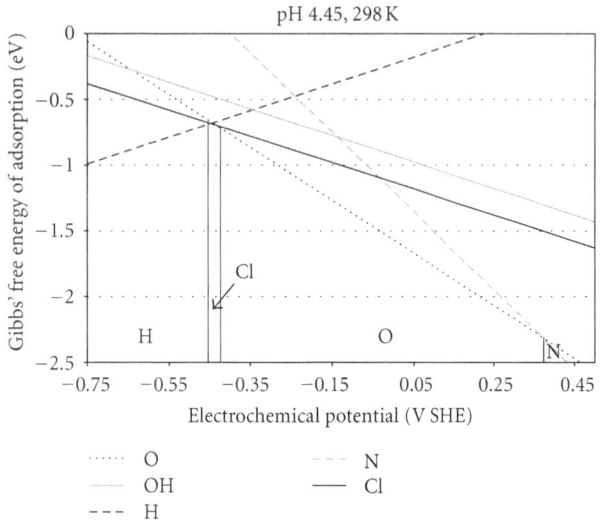

Figure 7: Free energy diagrams for surface coverage of the Fe(110) surface when exposed to a saturated NH_4Cl solution at pH 4.45 and 25°C (298 K). Reproduced with permission from [113].

the important prepassive state, chloride may be active in accelerating the corrosion of the mild steel [1]. Chloride may also inhibit passivation by displacing oxygen from the surface [116, 118]. The high concentrations of ammonia present also lead to the prediction of a nitrogenated surface beyond 370 mV SHE (although at this potential a bare iron surface is unlikely, unless cracking of the oxide film occurs, or the film is destabilized by some other chemical reaction). Similar diagrams can be constructed for other systems, in which the pH may be buffered to a more alkaline value or at different temperatures. The surface analysis generally supports the overall mechanisms proposed by Pletnev et al., Shimbarevich and Tseitlin, and Alvarez and Galvele [1, 116–118], although a more detailed model coupling surface thermodynamics, kinetics, and mass transport processes would be required to provide a more comprehensive understanding of the problem.

6. Perspectives

In this paper we have highlighted three specific cases where atomistic modeling has provided additional insight into the pathways via which important corrosion processes, including prepassivation and water chemistry, metal dissolution, and competitive adsorption, may occur. In particular, computational modeling of atomistic processes allows the scientist to directly assess the applicability of various corrosion mechanisms to a particular materials/environment combination and to characterize the nature of the interface at nanometer-subnanometer resolution.

At the same time, there is still some way to go for the computational science community to be able to represent the kinds of length and time scales associated with the macro- and mesoscopic corrosion processes. Some of these challenges include:

(1) capturing adequate dynamic time scales for the rearrangement of solvent particles around atoms undergoing dissolution,

(2) capturing not only details of the metal/environment interface but also the metal/oxide and oxide/environment interfaces,

(3) Incorporating the very important effects of alloying components in both the metallic and oxide states in both solid solution and as particle inclusions,

(4) Including the role of defect states, such as step edges, terraces, roughened surfaces, grain boundaries, and dislocations as potentially the most active sites at which corrosion events are going to occur.

The development of such advanced toolkits, using either state of the art potentials, such as the reactive force field [119], or multiscale modeling projects [28, 29, 37], is an active focus in our current portfolio of research projects.

That is not to say that the present toolkit does not allow the determination of meaningful insights into corrosion science. Using the kinds of density functional theory calculations, we have described above it is possible to do the following.

(1) Predict thermodynamic and kinetic parameters for fundamental processes on metal and metal-oxide surfaces, such as water dissociation and adsorption of molecular and atomic species (such as NH_3 and Cl).

(2) Understand the role local binding and morphology have on the thermodynamic cohesive energies of metal atoms exposed to the environment.

(3) Explore dissolution trajectories and the influence of environmental species in raising or lowering the activation barriers to such processes. Consider, for example, the metal atom dissolution described in Section 3, but replacing a water molecule with a different ligand, such as chloride or nitrate, for example.

(4) Adsorption and absorption of solution phase species, including hydrogen, for hydrogen embrittlement scenarios.

(5) Computation of fundamental atomistic parameters for utilization in higher-level kinetic Monte Carlo simulations or phase field models.

The above five examples are just a subset of the kinds of simulations currently being performed for corrosion-related problems in the materials science literature. We anticipate many new applications developing that go beyond the specifications outlined here, as more groups adopt atomistic modeling within their research toolkit, and new algorithms and synergies with continuum or mesoscale techniques are developed.

A final point should be made regarding experimental validation. First-principles modeling techniques are subject to variation depending upon the granularity and accuracy of the physics approximations used in their construction. For this reason, it is not uncommon to find variations, for example, in binding energies of molecules and materials calculated using different methods of 30–50 kJ/mol. Similarly, structural features, such as bond lengths and lattice parameters, can also be subject to variation depending upon the physics model selected. For these reasons, it is imperative that an equal, if not greater, effort continues to be applied in the characterization of interfacial phenomena using high-resolution, surface-specific probes (scanning tunneling microscopy and surface-enhanced Raman spectroscopy, e.g.). Improving resolution in both time and length scales is critical to provide the guidance needed for the improvement of the physics models that underlie the kinds of simulations described in this paper. As is the case in most research areas, advances in the understanding of corrosion processes can occur when a synergy is applied between the theoretical and experimental sciences.

Acknowledgments

The author is grateful for the illuminating and motivational discussions with Professor Robert Kelly (University of Virginia), Professor Matthew Neurock (University of Virginia), Professor Eliezer Gileadi (Tel-Aviv University), and Dr. Scott Lillard (Los Alamos National Laboratory). The author is also indebted to Michael Francis (U. Virginia, Los Alamos National Laboratory) for a careful reading of this paper. Los Alamos National Laboratory is operated by the Los Alamos National Security LLC for the National Nuclear Security Administration of the US Department of Energy under contract DE-AC52[28]-06NA2539.

References

[1] M. A. Pletnev, S. G. Morozov, and V. P. Alekseev, "Peculiar effect of chloride ions on the anodic dissolution of iron in solutions of various acidity," *Protection of Metals*, vol. 36, no. 3, pp. 202–208, 2000.

[2] P. Marcus, "Corrosion mechanics in theory and practice," in *Corrosion Technology*, P. A. Schweitzer, Ed., Marcel-Dekker, New York, NY, USA, 2nd edition, 2002.

[3] J. O. M. Bockris, A. K. N. Reddy, and M. Gamboa-Aldeco, *Modern Electrochemistry. Vol. 2A*, Kluwer Academic; Plenum Publishers, New York, NY, USA, 2000.

[4] C. D. Taylor, "Cohesive relations for surface atoms in the iron-technetium binary system," *Journal of Metallurgy*, vol. 2011, Article ID 954170, 8 pages, 2011.

[5] J. Postlethwaite, S. Nešič, G. Adamopoulos, and D. J. Bergstrom, "Predictive models for erosion-corrosion under disturbed flow conditions," *Corrosion Science*, vol. 35, no. 1–4, pp. 627–633, 1993.

[6] Y. Feng, K. S. Siow, W. K. Teo, K. L. Tan, and A. K. Hsieh, "Corrosion mechanisms and products of copper in aqueous solutions at various pH values," *Corrosion*, vol. 53, no. 5, pp. 389–398, 1997.

[7] S. P. Trasatti and E. Sivieri, "Corrosion behaviour of titanium in non-aqueous solvents," *Materials Chemistry and Physics*, vol. 92, no. 2-3, pp. 475–479, 2005.

[8] L. Stockert and H. Böhni, "Susceptibility to crevice corrosion and metastable pitting of stainless steels," *Materials Science Forum*, vol. 44-45, pp. 313–328, 1989.

[9] T. R. Beck and S. G. Chan, "Experimental observations and analysis of hydrodynamic effects on growth of small pits," *Corrosion*, vol. 37, no. 11, pp. 665–671, 1981.

[10] C. Y. Chao, L. F. Lin, and D. D. Macdonald, "A point defect model for anodic passive films: I. Film growth kinetics," *Journal of the Electrochemical Society*, vol. 128, no. 6, pp. 1187–1194, 1981.

[11] L. F. Lin, C. Y. Chao, and D. D. Macdonald, "A point defect model for anodic passive films: II. Chemical breakdown and pit initiation," *Journal of the Electrochemical Society*, vol. 128, no. 6, pp. 1194–1198, 1981.

[12] D. D. MacDonald, "The history of the point defect model for the passive state: a brief review of film growth aspects," *Electrochimica Acta*, vol. 56, no. 4, pp. 1761–1772, 2011.

[13] V. A. Bogdanovskaya, M. R. Tarasevich, L. A. Reznikova, and L. N. Kuznetsova, "Composition, surface segregation, and electrochemical properties of binary PtM/C (M = Co, Ni, Cr) catalysts," *Russian Journal of Electrochemistry*, vol. 46, no. 9, pp. 1011–1020, 2010.

[14] F. Friedersdorf and K. Sieradzki, "Film-induced brittle intergranular cracking of silver-gold alloys," *Corrosion (Houston)*, vol. 52, no. 5, pp. 331–336, 1996.

[15] S. Jain, N. D. Budiansky, J. L. Hudson, and J. R. Scully, "Surface spreading of intergranular corrosion on stainless steels," *Corrosion Science*, vol. 52, no. 3, pp. 873–885, 2010.

[16] E. M. Lehockey, A. M. Brennenstuhl, and I. Thompson, "On the relationship between grain boundary connectivity, coincident site lattice boundaries, and intergranular stress corrosion cracking," *Corrosion Science*, vol. 46, no. 10, pp. 2383–2404, 2004.

[17] D. N. Wasnik, V. Kain, I. Samajdar, B. Verlinden, and P. K. De, "Controlling grain boundary energy to make austenitic stainless steels resistant to intergranular stress corrosion cracking," *Journal of Materials Engineering and Performance*, vol. 12, no. 4, pp. 402–407, 2003.

[18] M. Baumgärtner and H. Kaesche, "Aluminum pitting in chloride solutions: morphology and pit growth kinetics," *Corrosion Science*, vol. 31, pp. 231–236, 1990.

[19] R. Ambat, A. J. Davenport, G. M. Scamans, and A. Afseth, "Effect of iron-containing intermetallic particles on the corrosion behaviour of aluminium," *Corrosion Science*, vol. 48, no. 11, pp. 3455–3471, 2006.

[20] F. Thébault, B. Vuillemin, R. Oltra, C. Allely, and K. Ogle, "Modeling bimetallic corrosion under thin electrolyte films," *Corrosion Science*, vol. 53, no. 1, pp. 201–207, 2011.

[21] M. A. Arafin and J. A. Szpunar, "A novel microstructure-Grain boundary character based integrated modeling approach of intergranular stress corrosion crack propagation in polycrystalline materials," *Computational Materials Science*, vol. 47, no. 4, pp. 890–900, 2010.

[22] V. Randle, "Grain boundary engineering: an overview after 25 years," *Materials Science and Technology*, vol. 26, no. 3, pp. 253–261, 2010.

[23] E. A. West and G. S. Was, "IGSCC of grain boundary engineered 316L and 690 in supercritical water," *Journal of Nuclear Materials*, vol. 392, no. 2, pp. 264–271, 2009.

[24] J. K. Nørskov, F. Abild-Pedersen, F. Studt, and T. Bligaard, "Density functional theory in surface chemistry and catalysis," *Proceedings of the National Academy of Sciences of the United States of America*, vol. 108, no. 3, pp. 937–943, 2011.

[25] S. A. Wasileski, C. D. Taylor, and M. Neurock, "Modeling electrocatalytic reaction systems from first principles," in *Device and Materials Modeling in PEM Fuel Cells*, S. J. Paddison, Ed., Springer, Berlin, Germany, 2009.

[26] R. A. van Santen and M. Neurock, *Molecular Heterogeneous Catalysis: A Conceptual and Computational Approach*, Wiley-VCH, Weinheim, Germany, 2006.

[27] E. Gileadi, "Problems in interfacial electrochemistry that have been swept under the carpet," *Journal of Solid State Electrochemistry*, vol. 15, pp. 1359–1371, 2011.

[28] T. J. Campbell, G. Aral, S. Ogata, R. K. Kalia, A. Nakano, and P. Vashishta, "Oxidation of aluminum nanoclusters," *Physical Review B*, vol. 71, no. 20, Article ID 205413, 14 pages, 2005.

[29] S. Serebrinsky, E. A. Carter, and M. Ortiz, "A quantum-mechanically informed continuum model of hydrogen embrittlement," *Journal of the Mechanics and Physics of Solids*, vol. 52, no. 10, pp. 2403–2430, 2004.

[30] C. Taylor, R. G. Kelly, and M. Neurock, "First-principles calculations of the electrochemical reactions of water at an immersed Ni (111) H_2O interface," *Journal of the Electrochemical Society*, vol. 153, no. 12, pp. E207–E214, 2006.

[31] C. Taylor, R. G. Kelly, and M. Neurock, "Theoretical analysis of the nature of hydrogen at the electrochemical interface between water and a Ni(111) single-crystal electrode," *Journal of the Electrochemical Society*, vol. 154, no. 3, pp. F55–F64, 2007.

[32] C. D. Taylor, R. G. Kelly, and M. Neurock, "First-principles prediction of equilibrium potentials for water activation by a series of metals," *Journal of the Electrochemical Society*, vol. 154, no. 12, pp. F217–F221, 2007.

[33] G. Jomard, T. Petit, A. Pasturel, L. Magaud, G. Kresse, and J. Hafner, "First-principles calculations to describe zirconia pseudopolymorphs," *Physical Review B*, vol. 59, no. 6, pp. 4044–4052, 1999.

[34] S. K. R. S. Sankaranarayanan and S. Ramanathan, "On the low-temperature oxidation and ultrathin oxide growth on zirconium in the presence of atomic oxygen: a modeling study," *Journal of Physical Chemistry C*, vol. 112, no. 46, pp. 17877–17882, 2008.

[35] J. Greeley, "First-principles investigations of electrocatalysis and corrosion," in *Proceedings of the 214th ECS Meeting Proton Exchange Membrane Fuel Cells (PEMFC '08)*, vol. 16, pp. 209–213, Honolulu, Hawaii, USA, October 2008.

[36] C. D. Taylor et al., "Calculation of surface properties of selected waste-form alloy systems," Report for DOE: Fundamental Waste Form Science, Los Alamos National Laboratory, Los Alamos, NM, USA, 2011.

[37] R. Spatschek, E. Brener, and A. Karma, "Phase field modeling of crack propagation," *Philosophical Magazine*, vol. 91, no. 1, pp. 75–95, 2011.

[38] R. M. Martin, *Electronic Structure: Basic Theory and Practical Methods*, Cambridge University Press, Cambridge, UK, 2004.

[39] E. Kaxiras, *Atomic and Electronic Structure of Solids*, Cambridge University Press, Cambridge, UK, 2003.

[40] I. N. Levine, *Quantum Chemistry*, Prentice Hall, Upper Saddle River, NJ, USA, 5th edition, 2000.

[41] P. B. Balbuena, K. P. Johnston, and P. J. Rossky, "Molecular dynamics simulation of electrolyte solutions in ambient and supercritical water. 1. Ion solvation," *Journal of Physical Chemistry*, vol. 100, no. 7, pp. 2706–2715, 1996.

[42] D. Domínguez-Ariza, C. Hartnig, C. Sousa, and F. Illas, "Combining molecular dynamics and ab initio quantum-chemistry to describe electron transfer reactions in electrochemical environments," *Journal of Chemical Physics*, vol. 121, no. 2, pp. 1066–1073, 2004.

[43] P. F. Weck, E. Kim, F. Poineau, and K. R. Czerwinski, "Structural evolution and properties of subnanometer Tc_n ($n = 2$–15) clusters," *Physical Chemistry Chemical Physics*, vol. 11, no. 43, pp. 10003–10008, 2009.

[44] A. B. Anderson and N. C. Debnath, "Mechanism of iron dissolution and passivation in an aqueous environment: active and transition ranges," *Journal of the American Chemical Society*, vol. 105, no. 1, pp. 18–22, 1983.

[45] A. B. Anderson and N. K. Ray, "Structures and reactions of H_3O^+, H_2O, and OH on an Fe electrode. Potential dependence," *Journal of Physical Chemistry*, vol. 86, no. 4, pp. 488–494, 1982.

[46] E. E. Ebenso, T. Arslan, F. Kandemirli, N. Caner, and I. Love, "Quantum chemical studies of some rhodanine azosulpha drugs as corrosion inhibitors for mild steel in acidic medium," *International Journal of Quantum Chemistry*, vol. 110, no. 5, pp. 1003–1018, 2010.

[47] E. E. Ebenso, T. Arslan, F. Kandemirli et al., "Theoretical studies of some sulphonamides as corrosion inhibitors for mild steel in acidic medium," *International Journal of Quantum Chemistry*, vol. 110, no. 14, pp. 2614–2636, 2010.

[48] M. C. Payne, M. P. Teter, D. C. Allan, T. A. Arias, and J. D. Joannopoulos, "Iterative minimization techniques for ab initio total-energy calculations: molecular dynamics and conjugate gradients," *Reviews of Modern Physics*, vol. 64, no. 4, pp. 1045–1097, 1992.

[49] I. Milas, B. Hinnemann, and E. A. Carter, "Diffusion of Al, O, Pt, Hf, and y atoms on α-Al_2O_3(0001): implications for the role of alloying elements in thermal barrier coatings," *Journal of Materials Chemistry*, vol. 21, no. 5, pp. 1447–1456, 2011.

[50] O. Runevall and N. Sandberg, "Self-diffusion in MgO—a density functional study," *Journal of Physics*, vol. 23, no. 34, pp. 345–402, 2011.

[51] D. A. Andersson, B. P. Uberuaga, P. V. Nerikar et al., "U and Xe transport in $UO_{2\pm x}$: density functional theory calculations," *Journal of Chemical Physics*, vol. 84, no. 5, Article ID 054105, 12 pages, 2011.

[52] C. D. Taylor, T. Lookman, and R. S. Lillard, "Ab initio calculations of the uranium-hydrogen system: thermodynamics, hydrogen saturation of α-U and phase-transformation to UH_3," *Acta Materialia*, vol. 58, no. 3, pp. 1045–1055, 2010.

[53] G. Lu and E. Kaxiras, "Hydrogen embrittlement of aluminum: the crucial role of vacancies," *Physical Review Letters*, vol. 94, no. 15, Article ID 155501, 4 pages, 2005.

[54] G. Lu, D. Orlikowski, I. Park, O. Politano, and E. Kaxiras, "Energetics of hydrogen impurities in aluminum and their effect on mechanical properties," *Physical Review B*, vol. 65, no. 6, Article ID 064102, pp. 641021–641028, 2002.

[55] G. Lu, Q. Zhang, N. Kioussis, and E. Kaxiras, "Hydrogen-enhanced local plasticity in aluminum: an ab initio study," *Physical Review Letters*, vol. 87, no. 9, Article ID 095501, 4 pages, 2001.

[56] J. Greeley and J. K. Nørskov, "Electrochemical dissolution of surface alloys in acids: thermodynamic trends from first-principles calculations," *Electrochimica Acta*, vol. 52, no. 19, pp. 5829–5836, 2007.

[57] A. Bouzoubaa, D. Costa, B. Diawara, N. Audiffren, and P. Marcus, "Insight of DFT and atomistic thermodynamics on the adsorption and insertion of halides onto the hydroxylated NiO(111) surface," *Corrosion Science*, vol. 52, no. 8, pp. 2643–2652, 2010.

[58] M. I. Baskes and R. A. Johnson, "Modified embedded atom potentials for HCP metals," *Modelling and Simulation in Materials Science and Engineering*, vol. 2, no. 1, article 011, pp. 147–163, 1994.
[59] S. M. Foiles, M. I. Baskes, and M. S. Daw, "Embedded-atom-method functions for the fcc metals Cu, Ag, Au, Ni, Pd, Pt, and their alloys," *Physical Review B*, vol. 33, no. 12, pp. 7983–7991, 1986.
[60] B. J. Lee and M. I. Baskes, "Second nearest-neighbor modified embedded-atom-method potential," *Physical Review B*, vol. 62, no. 13, pp. 8564–8567, 2000.
[61] J. W. Halley, A. Mazzolo, Y. Zhou, and D. Price, "First-principles simulations of the electrode—electrolyte interface," *Journal of Electroanalytical Chemistry*, vol. 450, no. 2, pp. 273–280, 1998.
[62] J. W. Halley, B. B. Smith, S. Walbran et al., "Theory and experiment on the cuprous-cupric electron transfer rate at a copper electrode," *Journal of Chemical Physics*, vol. 110, no. 13, pp. 6538–6552, 1999.
[63] D. L. Price and J. W. Halley, "Molecular dynamics, density functional theory of the metal-electrolyte interface," *The Journal of Chemical Physics*, vol. 102, no. 16, pp. 6603–6612, 1995.
[64] B. B. Smith and J. W. Halley, "Simulation study of the ferrous ferric electron transfer at a metal-aqueous electrolyte interface," *The Journal of Chemical Physics*, vol. 101, no. 12, pp. 10915–10924, 1994.
[65] E. Spohr, "Computer simulation of the water/platinum interface. Dynamical results," *Chemical Physics*, vol. 141, no. 1, pp. 87–94, 1990.
[66] E. Spohr, "Molecular dynamics simulations of water and ion dynamics in the electrochemical double layer," *Solid State Ionics*, vol. 150, no. 1-2, pp. 1–12, 2002.
[67] E. Spohr, "Some recent trends in computer simulations of aqueous double layers," *Electrochimica Acta*, vol. 49, no. 1, pp. 23–27, 2003.
[68] E. Spohr and K. Heinzinger, "Molecular dynamics simulation of a water/metal interface," *Chemical Physics Letters*, vol. 123, no. 3, pp. 218–221, 1986.
[69] E. Spohr, G. Tóth, and K. Heinzinger, "Structure and dynamics of water and hydrated ions near platinum and mercury surfaces as studied by MD simulations," *Electrochimica Acta*, vol. 41, no. 14, pp. 2131–2144, 1996.
[70] X. M. Bai, A. F. Voter, R. G. Hoagland, M. Nastasi, and B. P. Uberuaga, "Efficient annealing of radiation damage near grain boundaries via interstitial emission," *Science*, vol. 327, no. 5973, pp. 1631–1634, 2010.
[71] T. P. Schulze, "Efficient kinetic Monte Carlo simulation," *Journal of Computational Physics*, vol. 227, no. 4, pp. 2455–2462, 2008.
[72] A. F. Voter, "Introduction to the kinetic monte carlo method," in *Radiation Effects in Solids*, K. E. Sickafus and E. A. Kotomin, Eds., Springer, NATO Publishing Unit, Dordrecht, The Netherlands, 2005.
[73] B. Temel, H. Meskine, K. Reuter, M. Scheffler, and H. Metiu, "Does phenomenological kinetics provide an adequate description of heterogeneous catalytic reactions?" *Journal of Chemical Physics*, vol. 126, no. 20, Article ID 204711, 2007.
[74] B. Diawara, M. Legrand, J. J. Legendre, and P. Marcus, "Use of quantum chemistry results in 3D modeling of corrosion of iron-chromium alloys," *Journal of the Electrochemical Society*, vol. 151, no. 3, pp. B172–B178, 2004.
[75] D. M. Artymowicz, J. Erlebacher, and R. C. Newman, "Relationship between the parting limit for de-alloying and a particular geometric high-density site percolation threshold," *Philosophical Magazine*, vol. 89, no. 21, pp. 1663–1693, 2009.
[76] J. Erlebacher, "An atomistic description of dealloying porosity evolution, the critical potential, and rate-limiting behavior," *Journal of the Electrochemical Society*, vol. 151, no. 10, pp. C614–C626, 2004.
[77] S. A. Policastro, J. C. Carnahan, G. Zangari et al., "Surface diffusion and dissolution kinetics in the electrolyte-metal interface," *Journal of the Electrochemical Society*, vol. 157, no. 10, pp. C328–C337, 2010.
[78] W. M. Young and E. W. Elcock, "Monte Carlo studies of vacancy migration in binary ordered alloys: I," *Proceedings of the Physical Society*, vol. 89, no. 3, article 329, pp. 735–746, 1966.
[79] R. S. Lillard, G. F. Wang, and M. I. Baskes, "The role of metallic bonding in the cristallographic pitting of magnesium," *Journal of the Electrochemical Society*, vol. 153, no. 9, pp. B358–B364, 2006.
[80] C. D. Taylor and M. Neurock, "Theoretical insights into the structure and reactivity of the aqueous/metal interface," *Current Opinion in Solid State and Materials Science*, vol. 9, no. 1-2, pp. 49–65, 2005.
[81] M. A. Henderson, "The interaction of water with solid surfaces: fundamental aspects revisited," *Surface Science Reports*, vol. 46, no. 1–8, pp. 1–308, 2002.
[82] S. Izvekov, A. Mazzolo, K. VanOpdorp, and G. A. Voth, "Ab initio molecular dynamics simulation of the Cu(110)-water interface," *Journal of Chemical Physics*, vol. 114, no. 7, pp. 3248–3257, 2001.
[83] M. F. Toney, J. N. Howard, J. Richer et al., "Voltage-dependent ordering of water molecules at an electrode-electrolyte interface," *Nature*, vol. 368, no. 6470, pp. 444–446, 1994.
[84] A. Michaelides, A. Alavi, and D. A. King, "Insight into H_2O-ice adsorption and dissociation on metal surfaces from first-principles simulations," *Physical Review B*, vol. 69, no. 11, Article ID 113404, 4 pages, 2004.
[85] A. Michaelides, V. A. Ranea, P. L. De Andres, and D. A. King, "General model for water monomer adsorption on close-packed transition and noble metal surfaces," *Physical Review Letters*, vol. 90, no. 21, Article ID 216102, 4 pages, 2003.
[86] A. Michaelides, A. Alavi, and D. A. King, "Different surface chemistries of water on Ru{0001}: from monomer adsorption to partially dissociated bilayers," *Journal of the American Chemical Society*, vol. 125, no. 9, pp. 2746–2755, 2003.
[87] S. M. Bruemmer and L. E. Thomas, "High-resolution analytical electron microscopy characterization of corrosion and cracking at buried interfaces," *Surface and Interface Analysis*, vol. 31, no. 7, pp. 571–581, 2001.
[88] C. D. Taylor and M. Rossi, "A mechanistic approach to iodine induced stress corrosion cracking of zircaloy cladding: introductory assessment and preliminary results," Report prepared for the Consortium for the Advanced Simulation of LWRs—Materials Performance and Optimization (C.R. Stanek), Los Alamos National Laboratory, Los Alamos, NM, USA, 2011.
[89] G. Kresse and J. Hafner, "Ab initio molecular dynamics for liquid metals," *Physical Review B*, vol. 47, no. 1, pp. 558–561, 1993.
[90] G. Kresse and J. Hafner, "Ab initio molecular-dynamics simulation of the liquid-metalamorphous- semiconductor transition in germanium," *Physical Review B*, vol. 49, no. 20, pp. 14251–14269, 1994.

[91] G. Kresse and J. Furthmüller, "Efficiency of ab-initio total energy calculations for metals and semiconductors using a plane-wave basis set," *Computational Materials Science*, vol. 6, no. 1, pp. 15–50, 1996.

[92] G. Kresse and J. Furthmüller, "Efficient iterative schemes for ab initio total-energy calculations using a plane-wave basis set," *Physical Review B*, vol. 54, no. 16, pp. 11169–11186, 1996.

[93] G. Kresse and D. Joubert, "From ultrasoft pseudopotentials to the projector augmented-wave method," *Physical Review B*, vol. 59, no. 3, pp. 1758–1775, 1999.

[94] S. Walbran, A. Mazzolo, J. W. Halley, and D. L. Price, "Model for the electrostatic response of the copper-water interface," *Journal of Chemical Physics*, vol. 109, no. 18, pp. 8076–8080, 1998.

[95] J. Weissenrieder, A. Mikkelsen, J. N. Andersen, P. J. Feibelman, and G. Held, "Experimental evidence for a partially dissociated water bilayer on Ru{0001}," *Physical Review Letters*, vol. 93, no. 19, pp. 1–196102, 2004.

[96] C. D. Taylor, S. A. Wasileski, J. S. Filhol, and M. Neurock, "First principles reaction modeling of the electrochemical interface: consideration and calculation of a tunable surface potential from atomic and electronic structure," *Physical Review B*, vol. 73, Article ID 165402, 16 pages, 2006.

[97] A. Seyeux, V. Maurice, L. H. Klein, and P. Marcus, "In situ scanning tunnelling microscopic study of the initial stages of growth and of the structure of the passive film on Ni(111) in 1 mM NaOH(aq)," *Journal of Solid State Electrochemistry*, vol. 9, no. 5, pp. 337–346, 2005.

[98] C. D. Taylor, R. G. Kelly, and M. Neurock, "A first-principles analysis of the chemisorption of hydroxide on copper under electrochemical conditions: a probe of the electronic interactions that control chemisorption at the electrochemical interface," *Journal of Electroanalytical Chemistry*, vol. 607, no. 1-2, pp. 167–174, 2007.

[99] C. D. Taylor, *First principles modeling of the structure and reactivity of water at the metal/water interface*, Dissertation, University of Virginia, Charlottesville, Va, USA, 2006.

[100] G. Niaura, "Surface-enhanced Raman spectroscopic observation of two kinds of adsorbed OH^- ions at copper electrode," *Electrochimica Acta*, vol. 45, no. 21, pp. 3507–3519, 2000.

[101] J. Rossmeisl, J. K. Nørskov, C. D. Taylor, M. J. Janik, and M. Neurock, "Calculated phase diagrams for the electrochemical oxidation and reduction of water over Pt(111)," *Journal of Physical Chemistry B*, vol. 110, no. 43, pp. 21833–21839, 2006.

[102] E. Gileadi, "Can an electrode reaction occur without electron transfer across the metal/solution interface?" *Chemical Physics Letters*, vol. 393, no. 4–6, pp. 421–424, 2004.

[103] E. Gileadi, "Charge and mass transfer across the metal/solution interface," *Israel Journal of Chemistry*, vol. 48, no. 3-4, pp. 121–131, 2008.

[104] A. Michaelides, V. A. Ranea, P. L. De Andres, and D. A. King, "General model for water monomer adsorption on close-packed transition and noble metal surfaces," *Physical Review Letters*, vol. 90, no. 21, Article ID 216102, 4 pages, 2003.

[105] C. D. Taylor, M. Neurock, and J. R. Scully, "First-principles investigation of the fundamental corrosion properties of a model Cu_{38} nanoparticle and the (111), (113) surfaces," *Journal of the Electrochemical Society*, vol. 155, no. 8, pp. C407–C414, 2008.

[106] R. S. Lillard, G. F. Wang, and M. I. Baskes, "The role of metallic bonding in the cristallographic pitting of magnesium," *Journal of the Electrochemical Society*, vol. 153, no. 9, pp. B358–B364, 2006.

[107] W. Schmickler, K. Pötting, and M. Mariscal, "A new simulation model for electrochemical metal deposition," *Chemical Physics*, vol. 320, no. 2-3, pp. 149–154, 2006.

[108] M. I. Baskes, S. G. Srinivasan, S. M. Valone, and R. G. Hoagland, "Multistate modified embedded atom method," *Physical Review B*, vol. 75, no. 9, Article ID 094113, 2007.

[109] P. Biedermann, E. Torres, and A. Blumenau, "Oxygen reduction at thiol/Au(111) SAMs, atomistic modelling and experiment," in *Proceedings of the 212th Meeting of the Electrochemical Society on Modeling and Simulation of Dissolution and Corrosion Processes*, The Electrochemical Society, Washington, DC, USA, 2007.

[110] S. M. Valone and S. R. Atlas, "Electron correlation, reference states and empirical potentials," *Philosophical Magazine*, vol. 86, no. 17-18, pp. 2683–2711, 2006.

[111] C. D. Taylor, "The transition from metal-metal bonding to metal-solvent interactions during a dissolution event as assessed from electronic structure," *Chemical Physics Letters*, vol. 469, no. 1–3, pp. 99–103, 2009.

[112] American Petroleum Institute, *Damage Mechanisms Affecting Fixed Equipment in the Refining Industry*, American Petroleum Institute, Washington, DC, USA, 2003.

[113] C. D. Taylor, "Predictions of surface electrochemistry of saturated and alkaline NH_4Cl solutions interacting with Fe(110) from ab initio calculations," *Corrosion*. In press.

[114] R. A. White, *Materials Selection for Petroleum Refineries and Gathering Facilities*, NACE International, Houston, Tex, USA, 1998.

[115] O. Forsén, J. Aromaa, and M. Tavi, "Corrosion resistance of different materials in dilute ammonium chloride-bearing environments," *Corrosion Science*, vol. 35, no. 1–4, pp. 297–301, 1993.

[116] V. A. Shimbarevich and K. L. Tseitlin, "Influence of ammonia concentration and temperature on corrosion of carbon steel in ammonium chloride solutions," *Protection of Metals*, vol. 15, no. 5, pp. 455–458, 1979.

[117] V. A. Shimbarevich and K. L. Tseitlin, "nfluence of ammonia on corrosion of stainless steels and titanium in ammonium chloride solutions at 200°C," *Protection of Metals*, vol. 17, no. 2, pp. 144–148, 1981.

[118] M. G. Alvarez and J. R. Galvele, "The mechanism of pitting of high purity iron in NaCl solutions," *Corrosion Science*, vol. 24, no. 1, pp. 27–48, 1984.

[119] A. C. T. Van Duin, S. Dasgupta, F. Lorant, and W. A. Goddard, "ReaxFF: a reactive force field for hydrocarbons," *Journal of Physical Chemistry A*, vol. 105, no. 41, pp. 9396–9409, 2001.

9

Inhibition of Mild Steel Corrosion by 3-Benzoylmethyl Benzimidazolium Hexafluoroantimonate in Acidic Solution

Ayssar Nahlé, Ideisan I. Abu-Abdoun, and Ibrahim Abdel-Rahman

Department of Chemistry, College of Sciences, University of Sharjah, P.O. Box 27272, Sharjah, UAE

Correspondence should be addressed to Ayssar Nahlé, anahle@sharjah.ac.ae

Academic Editor: Flavio Deflorian

The corrosion inhibition of mild steel in 1 M HCl solution by a synthesized compound (3-benzoylmethyl benzimidazolium hexafluoroantimonate) was investigated electrochemically and by weight loss experiments. The concentration of this inhibitor ranged from 5×10^{-7} M to 5×10^{-4} M. The effect of temperature (from 303 to 343 K) and concentrations (from 5×10^{-7} M to 5×10^{-4} M) were investigated. The percentage inhibition increased with the increase of the concentration of the inhibitor and reached about 98% at the concentration of 5×10^{-4} M at 303 K. The percentage inhibition decreased with the increase of temperature. The thermodynamic parameters for the adsorption of this inhibitor on the metal surface were calculated. This compound was found to be a very good corrosion inhibitor due to the presence of nitrogen in benzimidazole and phenyl ring.

1. Introduction

The corrosion of metals is a serious problem in many industries, installations, and civil services such as water and sewage supplies. One of the most useful and practical methods that used to control and protect metals against corrosion is the use of inhibitors, especially in acidic media. Not only the costs due to corrosion increase rapidly, but also the efficiency of the plants and the quality of the products are also reduced.

Most inhibitors are organic compounds containing polar groups strongly adsorbed on the metal surface [1, 2]. These inhibitors, which include the organic N, P, S, and OH groups, are known to be similar to catalytic poisons, as they decrease the reaction rate at the metal/solution interface without, in general, being involved in the reaction considered. It is generally accepted that most organic inhibitors act via adsorption at the metal/solution interface. The mechanism by which an inhibitor decreases the corrosion current is achieved by interfering with some of the steps for the electrochemical process.

The corrosion inhibition of mild steel in aggressive acidic solutions has been widely investigated. In industries, hydrochloric acid solutions are often used in order to remove scale and salts from steel surfaces, and cleaning tanks and pipelines. This treatment may be prerequisite for coating by electroplating, galvanizing, or painting techniques. The acid must be treated to prevent an extensive dissolution of the underlying metal. This treatment involves the addition of some organic inhibitors to the acid solution that adsorb at the metal/solution interface by displacing water molecules on the surface and forming a compact barrier film.

Nitrogen-containing compounds as metal corrosion inhibitors have been studied thoroughly by many authors. These compounds included benzimidazole and imidazole derivatives [3–13], bipyrazole [14], stilbazole [15], quaternary ammonium salts [16–23], polyaminobenzoquinone polymers [24], substituted aniline-N-salicylidenes [25], amides [26], heterocyclic compounds [27, 28], and cationic surfactants [29, 30]. Other authors worked on phosphorous-containing and sulfur-containing inhibitors [15, 31–39]. Other studies involved the effect of addition of some ions on the inhibition efficiency of some organic compounds. These ions included chromium [40], iodide [41–43], and chloride [10, 44]. Structural effect of organic compounds as corrosion inhibitors was also studied [8, 42, 43, 45–47]. In all these studies, the nitrogen, phosphorous, and sulfur atom(s) in the compounds as well as the unsaturated double bonds like phenyl rings showed to be able to adsorb very well on the metal surface and form protective layer, which in turn increased the corrosion inhibition with the increase in the concentration of the inhibitor, in some cases reaching 99% inhibition [15, 18, 36, 37].

No studies have been reported on 3-benzoylmethyl benzimidazolium hexafluoroantimonate (BMBI) (Scheme 1), in terms of studying both the electrochemical and the temperature effect on the corrosion inhibition of mild steel in 1.0 M HCl solution. Mild steel was chosen in our studies since high temperature aggressive acids are widely used in industries in connection to mild and low alloy steels.

The aim of this work is

(i) to synthesize a high percentage yield of pure 3-benzoylmethyl benzimidazolium hexafluoroantimonate (BMBI);

(ii) to study, using potentiodynamic and weight loss measurements, the effect of temperature on the corrosion inhibition of mild steel in 1.0 M HCl solution by BMBI and to calculate the thermodynamic parameters.

The output of this study is intended to be the building block or the nucleus for a new family or group of imidazole onium derivatives in all studies of corrosion inhibitors.

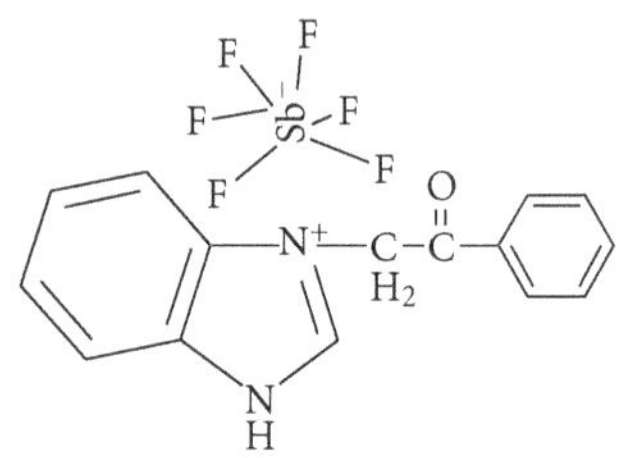

Scheme 1: 3-Benzoylmethyl benzimidazolium hexafluoroantimonate.

2. Experimental Details

2.1. Synthesis of 3-Benzoylmethyl Benzimidazolium Hexafluoroantimonate (BMBI). The 3-benzoylmethyl benzimidazolium bromide salt (3) was prepared by refluxing an equimolar mixture of the 1-vinylmidazole and phenacyl bromide (2) in acetone as follows.

2.50 grams (0.021 mol) from benzimidazole (1) and 4.20 grams from phenacyl bromide (0.023 mol) (2), in 100 mL acetone, were refluxed for two hours. After cooling to room temperature, a crystalline solid of salt (3) was collected by filtration then washed with excess diethyl ether and dried.

The bromide salt (3) was converted to the hexafluoroantimonate salt (4) by treatment with potassium hexafluoroantimonate ($KSbF_6$) in methanol/water mixture 70/30% by volume. After stirring for 20 min, the reaction product was filtered, washed several times with water then with diethyl ether, and dried under vacuum at room temperature. Dissolution in chloroform and addition of excess diethyl ether gave the pure hexafluoroantimonate salt in 80 yield percent as shown in Scheme 2.

3. Electrochemistry

3.1. Electrode Preparation (Electrochemistry). A 5 mm diameter piece cut from a mild steel rod (IS 226 containing 0.18% C, 0.6% Mn, and 0.35% Si) supplied by "Reliable Steel Traders," Sharjah, UAE, formed the working electrode and was mounted, using Araldite epoxy resin, in a glass tube that fits in the electrochemical cell. Prior to each experiment, the working mild steel electrode was abraded using a series of carborundum papers starting with 600 grades and ending with 1200 grades. The electrode surface was then polished with 0.3 μm alumina on cloth, washed with deionized distilled water, and rinsed with pure ethanol before being transferred to the electrochemical cell that contained deaerated fresh electrolyte.

3.2. Instrumentation (Electrochemistry). The electrochemical cell (Figure 1) consisted of a mild steel working electrode (WE), a saturated calomel electrode (SCE) as a reference electrode (RE), and platinum gauze counter electrode (CE). Prior to each experiment, the electrolyte was deaerated by nitrogen bubbling. The cell was designed in a way that the nitrogen was allowed to escape into the solution, precluding its collection at the electrode surface.

In order to protect the working electrode from any substance that may be produced at the counter electrode during the electrochemical reactions, the counterelectrode compartment was separated from the working-electrode compartment with a glass frit. The following electrochemical instrument was used a PC controlled AUTOLAB PGSTAT128N Modular Potentiostat (electrochemical workstation) (supplied from Metrohm) capable of driving currents up to ±800 mA with an output potential across the cell of up to ±10 V.

3.3. Measuring Procedure (Electrochemistry). Electrochemical corrosion measurements (Tafel plots) were carried out on the mild steel electrode, prepared as described before, in 1.0 M HCl and in 1.0 M HCl containing various concentrations of the prepared BMBI inhibitor. The concentration of the inhibitor ranged from 5×10^{-7} M to 5×10^{-4} M. Due to the restricted solubility of BMBI inhibitor, higher concentrations could not be prepared.

The electrode potential was scanned between over potentials of −100 mV to +100 mV versus SCE at a sweep rate of $1\ mV \cdot s^{-1}$.

3.4. Results and Discussions (Electrochemistry). Figure 2 shows the anodic and cathodic polarization curves (Tafel plot) of the mild steel electrode in deaerated 1.0 M HCl solution with and without the addition of various concentrations of 3-benzoylmethyl benzimidazolium hexafluoroantimonate (BMBI). The presence of the inhibitor affected both the anodic and cathodic branches of the curve as it can be clearly seen in Figure 2. The corrosion current was found to decrease with the increase of the concentration of BMBI inhibitor as shown in Table 1. In the absence of inhibitor (in

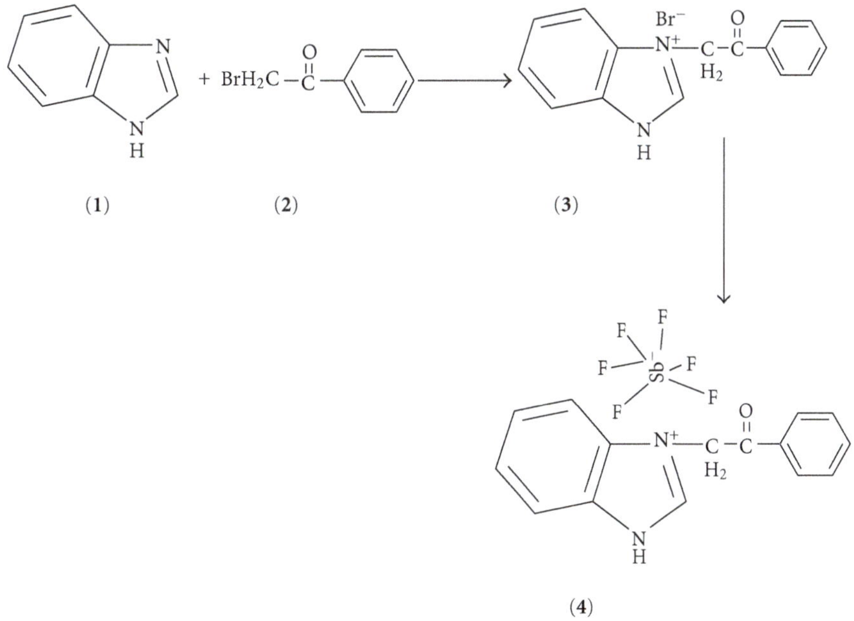

Scheme 2: Synthesis of 3-benzoylmethyl benzimidazolium hexafluoroantimonate.

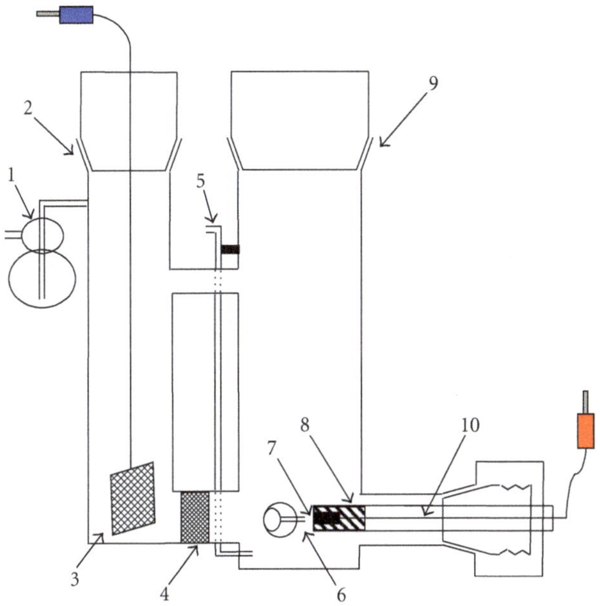

(1) Gas bubbler
(2) B 12 glass socket
(3) Platinum gauze (counter electrode)
(4) Glass frit
(5) Inlet for nitrogen gas
(6) luggin capillary (reference electrode)
(7) Iron rod (working electrode)
(8) Epoxy resin
(9) B 24 glass socket
(10) Copper wire

Figure 1: Electrochemical cell.

1.0 M HCl), the corrosion current was found to be 0.74 mA, which decreased sharply to 0.026 mA at 5×10^{-4} M of BMBI inhibitor in 1.0 M HCl (Table 1).

The percentage inhibition values of BMBI at various concentrations in 1.0 M HCl were calculated according to the following equation (1) and the results are shown in Table 1:

$$\text{Percentage Inhibition} = \frac{(I_{\text{Corr.}})_{\text{Uninh.}} - (I_{\text{Corr.}})_{\text{Inh.}}}{(I_{\text{Corr.}})_{\text{Uninh.}}} \times 100, \quad (1)$$

where $(I_{\text{Corr.}})_{\text{Uninh.}}$ is corrosion current in the uninhibited solution and $(I_{\text{Corr.}})_{\text{Inh.}}$ is corrosion current in inhibited solution.

Figure 3 shows the plot of the percentage inhibition versus the concentration of BMBI. In this figure, the percentage inhibition increases gradually from about 54% (with 5×10^{-7} M inhibitor) to reach about 96% with 5×10^{-4} M inhibitor.

4. Effect of Temperature

4.1. Specimen Preparation (Effect of Temperature). Rectangular specimens (1.0 cm × 2.3 cm × 0.3 cm) cut from large sheet of 3 mm thick mild steel (IS 226 containing 0.18% C, 0.6% Mn, and 0.35% Si) supplied by "Reliable Steel Traders," Sharjah, UAE, were used for weight loss measurements. A 2 mm diameter hole was drilled close to the upper edge of the specimen and served to be hooked with a glass rod for immersion purposes. Prior to each experiment, the specimens were polished with 600 grade emery paper, rinsed with distilled water, degreased with acetone, dried, and finally weighed precisely on an accurate analytical balance.

Table 1: Tafel corrosion currents and percent inhibitions of 3-benzoylmethyl benzimidazolium hexafluoroantimonate (BMBI) at various concentrations in 1.0 M HCl at room temperature.

	1.0 M HCl	1.0 M HCl + 5×10^{-7} M BMBI	1.0 M HCl + 5×10^{-6} M BMBI	1.0 M HCl + 5×10^{-5} M BMBI	1.0 M HCl + 5×10^{-4} M BMBI
E_{rest}/V versus SCE	−0.459	−0.472	−0.472	−0.461	−0.583
$I_{corrosion}$/mA	0.74	0.34	0.27	0.16	0.026
Percent Inhibition	—	54.05	63.51	78.38	96.49

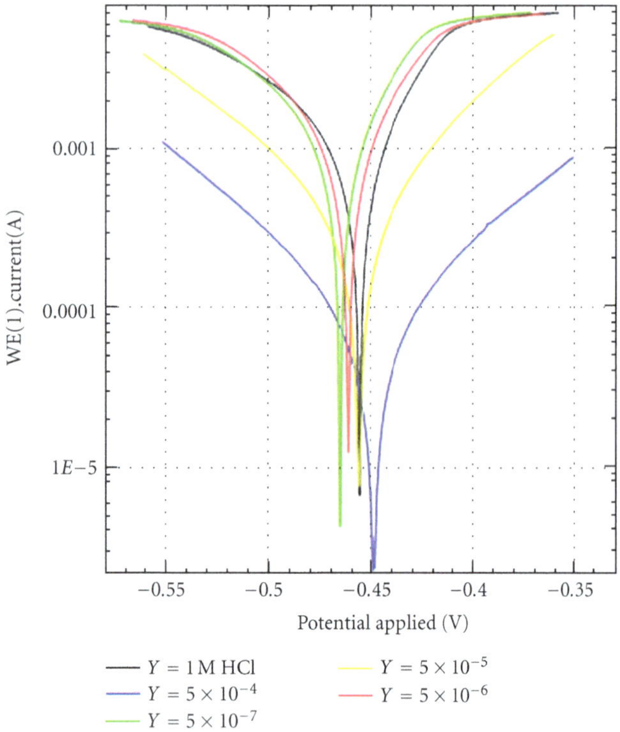

Figure 2: Anodic and cathodic polarization curves of mild steel in an uninhibited 1.0 M HCl solution and in 1.0 M HCl containing various concentrations of 3-benzoylmethyl benzimidazolium hexafluoroantimonate (BMBI) at ambient temperature. (1) 1.0 M HCl; (2) 1.0 M HCl + 5.0 × 10^{-7} M inhibitor; (3) 1.0 M HCl + 5.0 × 10^{-6} M inhibitor; (4) 1.0 M HCl + 5.0 × 10^{-5} M inhibitor; (5) 1.0 M HCl + 5.0 × 10^{-4} M inhibitor.

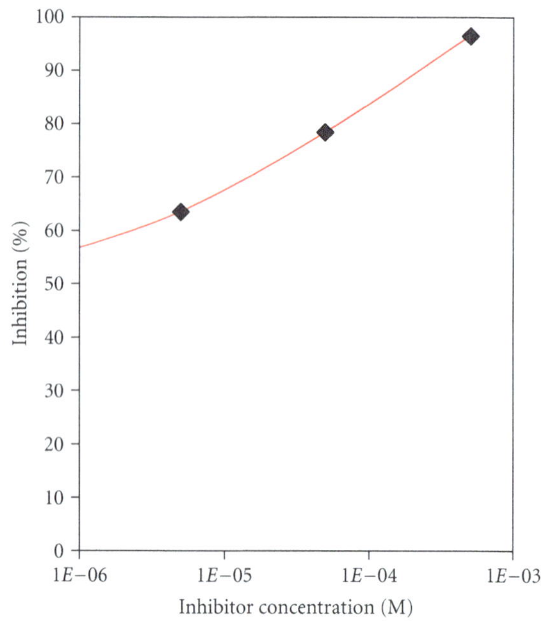

Figure 3: Percentage inhibition versus the concentration of 3-benzoylmethyl benzimidazolium hexafluoroantimonate (BMBI) at ambient temperature.

4.2. Instrumentation (Effect of Temperature). The experimental setup consisted of a 250 mL round bottom flask fitted with a reflux condenser and a long glass rod on which the specimen was hooked and in turn immersed in a thermally controlled water bath.

4.3. Measuring Procedure (Effect of Temperature). The flask was filled with 100 mL of 1.0 M HCl solution either with or without BMBI of various concentrations, then placed in water bath. As soon as the required working temperature was reached, the precisely weighed mild steel specimen was immersed in the solution and left there for exactly six hours, after which the sample was removed, rinsed with distilled deionized water, degreased with acetone, dried, and finally weighed precisely on an accurate analytical balance. This procedure was repeated with all the samples with a variety of inhibitor concentrations ranging from 5 × 10^{-7} M up to 5 × 10^{-4} M and at temperatures ranging from 303 K to 343 K.

4.4. Results and Discussions (Effect of Temperature). Weight loss corrosion tests were carried out on the mild steel in 1 M HCl in the absence and presence of (BMBI) over a period of 6 hours. Table 2 represents the corrosion rates [$mg \cdot cm^{-2} \cdot h^{-1}$], and the percentage efficiencies [%] for the studied inhibitor with concentrations varying from 5 × 10^{-7} M to 5×10^{-4} M at 303, 313, 323, 333, and 343 K, respectively. The percentage efficiency was calculated according to the following expression (2):

$$\%\ \text{Inhibition} = \frac{W_{\text{Uninh.}} - W_{\text{Inh.}}}{W_{\text{Uninh.}}} \times 100, \quad (2)$$

where $W_{\text{Uninh.}}$ is corrosion rate without inhibitor; and $W_{\text{Inh.}}$ is corrosion rate with inhibitor.

Figures 4 and 5 show the plots of the corrosion rate of (BMBI) as a function of concentration at 303, 313, 323, 333, and 343 K extracted from Table 2. At 303 K (Table 2, Figure 4), the corrosion rate dropped sharply from the value 0.961 $mg \cdot cm^{-2} \cdot h^{-1}$ (1 M HCl in the absence of the

TABLE 2: Effect of concentration of 3-Benzoylmethyl benzimidazolium hexafluoroantimonate (BMBI) on the corrosion rate ($mg \cdot cm^{-2} \cdot h^{-1}$) and percentage efficiency of mild steel in 1.0 M HCl at various temperatures.

Concentration of Inhibitor	Temperature/K									
	303		313		323		333		343	
	Corr. Rate	% Efficiency	Corr. Rate	% Efficiency	Corr. Rate	% Efficiency	Corr. Rate	% Efficiency	Corr. Rate	% Efficiency
1 M HCl	0.961	—	1.394	—	4.671	—	12.225	—	26.280	—
1 M HCl + 5×10^{-7} M	0.363	62.3	0.686	50.8	2.673	42.8	8.539	30.2	21.874	16.8
1 M HCl + 5×10^{-6} M	0.229	76.2	0.508	63.6	2.177	53.4	6.619	45.9	17.227	34.4
1 M HCl + 5×10^{-5} M	0102	89.4	0.182	86.9	1.179	74.8	4.440	63.7	11.835	55.0
1 M HCl + 5×10^{-4} M	0.018	98.1	0.052	96.0	0.256	94.5	2.321	81.0	7.503	71.5

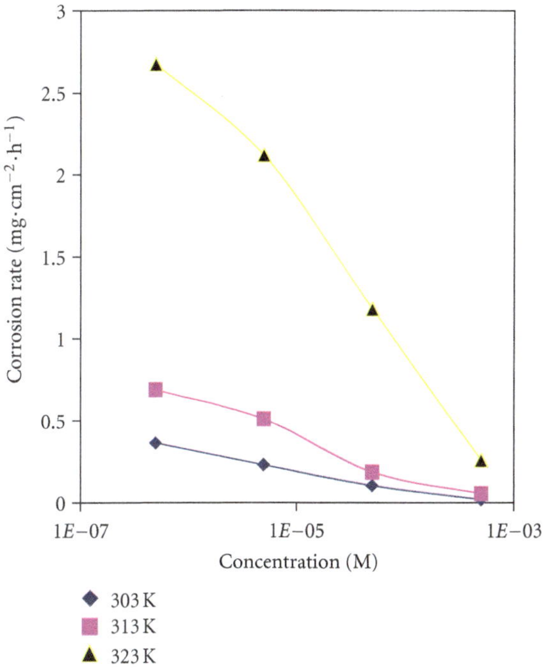

FIGURE 4: Effect of concentration of 3-Benzoylmethyl benzimidazolium hexafluoroantimonate (BMBI) on the corrosion rate ($mg \cdot cm^{-2} \cdot h^{-1}$) of mild steel in 1.0 M HCl at various temperatures.

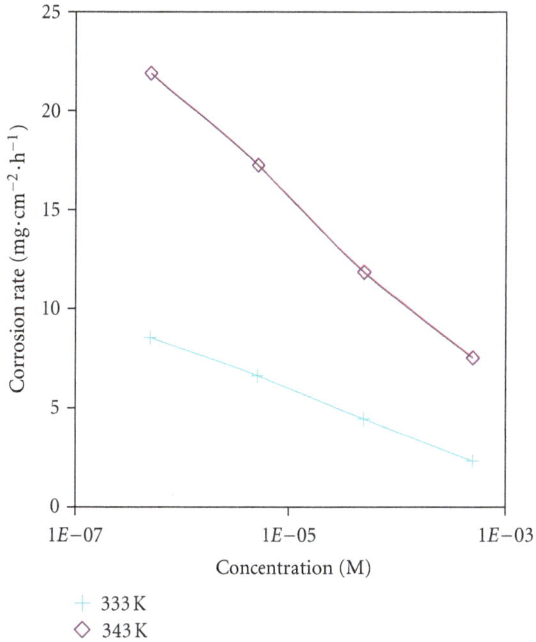

FIGURE 5: Effect of concentration of 3-Benzoylmethyl benzimidazolium hexafluoroantimonate (BMBI) on the corrosion rate ($mg \cdot cm^{-2} \cdot h^{-1}$) of mild steel in 1.0 M HCl at various temperatures.

inhibitor) to the value 0.363 $mg \cdot cm^{-2} \cdot h^{-1}$ (62.3% inhibition) when 5×10^{-7} M of (BMBI) was present in the 1 M HCl. This corrosion rate continued to decrease gradually and linearly to reach 0.018 $mg \cdot cm^{-2} \cdot h^{-1}$ (98.1% inhibition) at 5×10^{-4} M. At 313 K (Table 2, Figure 4); the curve showed to have similar shape as that obtained at 303 K. The corrosion rate decreased gradually and linearly from 0.686 $mg \cdot cm^{-2} \cdot h^{-1}$ (50.8% inhibition) when 5×10^{-7} M of (BMBI) to reach about 0.052 $mg \cdot cm^{-2} \cdot h^{-1}$ (96.0%) at 5×10^{-4} M.

At 323 K (Table 2, Figure 4), the corrosion rate decreased also linearly but more steeply from 2.673 $mg \cdot cm^{-2} \cdot h^{-1}$ (42.8% inhibition) when 5×10^{-7} M of (BMBI) to reach about 0.256 $mg \cdot cm^{-2} \cdot h^{-1}$ (94.5%) at 5×10^{-4} M.

In Figure 5, the same behavior is shown as in Figure 4. The corrosion rate at 333 and 343 K are shown as a function of the concentration of (BMBI). It can be clearly seen that the presence of the (BMBI) inhibitor at these high temperatures has acted as a corrosion inhibitor reaching a percent inhibition of 81.0% and 71.5% when 5×10^{-4} M inhibitor was employed at 333 K and 343 K, respectively.

Figure 6 shows the plots of the % inhibition versus the concentration of the inhibitor at temperatures of 303, 313, 323, 333, and 343 K, respectively. This figure showed that the % inhibition increased as the temperature decreased at the same inhibitor concentration. The presence of the inhibitor has greatly increased the % inhibition at all temperatures.

Table 3: The data obtained from the weight loss measurements for Arrhenius equation: (I/T) against ln (corrosion rate).

$(1/T) \times 10^3\ K^{-1}$	Ln corrosion rate ($mg \cdot cm^{-2} \cdot h^{-1}$)				
	1 M HCl	1 M HCl + 5×10^{-7} M	1 M HCl + 5×10^{-6} M	1 M HCl + 5×10^{-5} M	1 M HCl + 5×10^{-4} M
3.30	−0.03978	−1.01611	−1.47403	−2.28278	−4.01738
3.19	0.332177	−0.37688	−0.67727	−1.70375	−2.96423
3.10	1.541373	0.9832	0.77795	0.16467	−1.36258
3.00	2.503483	2.14464	1.88994	1.49065	0.842
2.92	3.268808	3.0853	2.84648	2.47106	2.0153

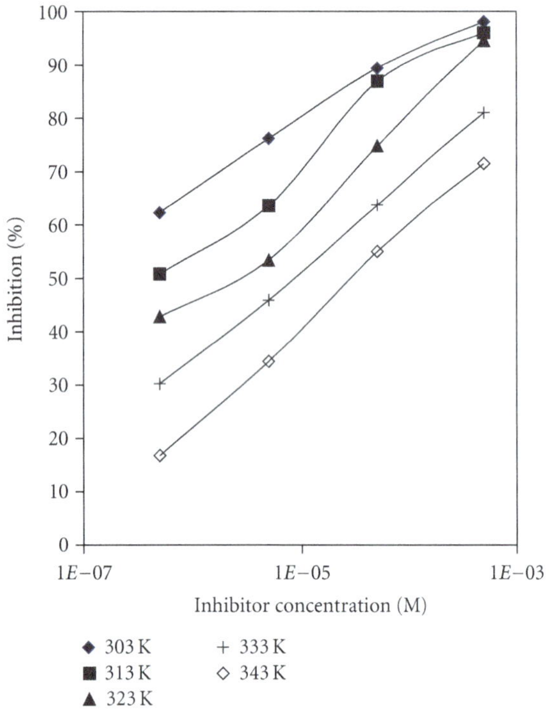

Figure 6: Effect of concentration of 3-Benzoylmethyl benzimidazolium hexafluoroantimonate (BMBI) triphenyl phosphonium bromide on the percent inhibition of mild steel in 1.0 M HCl at various temperatures.

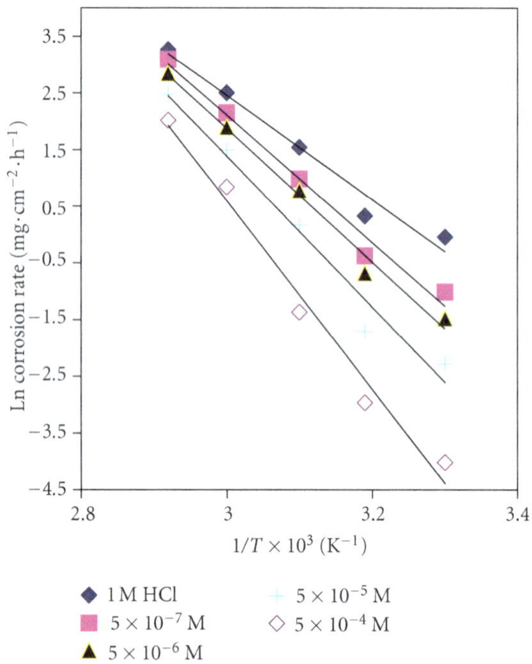

Figure 7: Effect of temperature on the corrosion rate of mild steel in 1.0 M HCl solution with and without the presence of various concentrations of 3-benzoylmethyl benzimidazolium hexafluoroantimonate (BMBI).

Table 4: The activation energy (E_a) for the corrosion of mild steel in 1.0 M HCl with and without 3-benzoylmethyl benzimidazolium hexafluoroantimonate inhibitor at various concentrations.

System	Activation energy, E_a ($kcal \cdot mol^{-1}$)			
	5×10^{-4} M	5×10^{-5} M	5×10^{-6} M	5×10^{-7} M
1 M HCl	18.27	18.27	18.27	18.27
1 M HCl + BMBI	33.06	26.43	23.35	22.31

The data obtained from the weight loss measurements (Table 3) were plotted in accordance to Arrhenius equation (3):

$$\ln \text{rate} = -\frac{E_a}{RT} + \text{const.}, \quad (3)$$

where E_a is activation energy [$kcal \cdot mol^{-1}$], R is gas constant [$kcal \cdot mol^{-1}$], T is absolute temperature [K], and const. is constant.

Figure 7 represents the Arrhenius plot of the corrosion of mild steel in 1 M HCl solution (Ln corrosion rate as a function of $1/T$) with and without the presence of (BMBI) at concentrations ranging from 5×10^{-4} M to 5×10^{-7} M. From this figure, the slope ($-E_a/R$) of each individual line was determined and used to calculate the activation energy according to (3) and taking $R = 1.987 \times 10^{-3}\ kcal \cdot mol^{-1}$ (Table 4). The increase of concentration of (BMBI) (from 5×10^{-7} M to 5×10^{-4} M) increased the activation energies for the corrosion of mild steel in 1 M HCl (initially $18.27\ kcal \cdot mol^{-1}$) (Table 4). The increase in the activation energies for the corrosion is attributed to a decrease in the adsorption of the inhibitor on the metal surface as the temperature increased; subsequently, an increase in the corrosion rate will result due to the greater exposed area of the metal surface to the acid.

The results in Table 4 show that the activation energy (E_a) for the corrosion of mild steel in the presence of the inhibitor is higher compared to the activation energy in the absence of

Table 5: Effect of concentration of 3-benzoylmethyl benzimidazolium hexafluoroantimonate on surface coverage for mild steel in 1.0 M HCl at various temperatures.

	Temperature/K				
Concentration of inhibitor	303	313	323	333	343
	Surface coverage θ	Surface coverage θ	Surface coverage θ	Surface coverage θ	Surface coverage θ
1 M HCl + 5×10^{-7} M	0.623	0.508	0.428	0.302	0.168
1 M HCl + 5×10^{-6} M	0.762	0.636	0.534	0.459	0.344
1 M HCl + 5×10^{-5} M	0.894	0.869	0.748	0.637	0.550
1 M HCl + 5×10^{-4} M	0.981	0.960	0.945	0.810	0.715

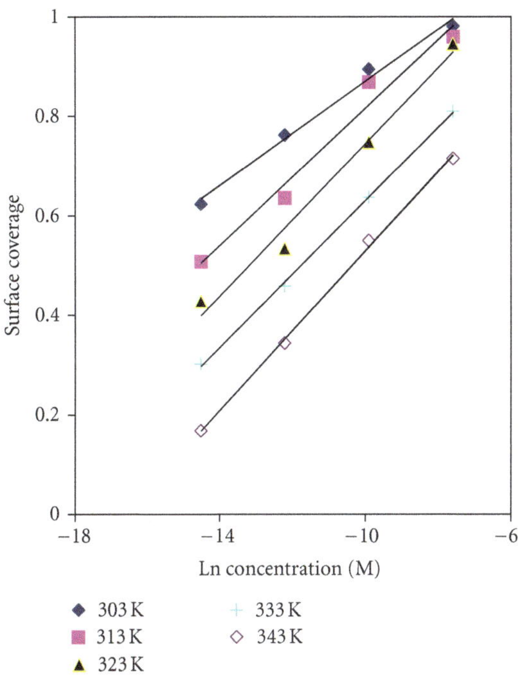

Figure 8: Effect of concentration of 3-benzoylmethyl benzimidazolium hexafluoroantimonate (BMBI) on the surface coverage of mild steel in 1.0M HCl at various temperatures.

inhibitor at all concentrations ranging from 5×10^{-4} M to 5×10^{-7} M (from about 33 to about 22 kcal·mol^{-1}, resp.). This can be attributed to the fact that higher values of E_a in the presence of inhibitor compared to its absence are generally consistent with a physisorption, while unchanged or lower values of E_a in inhibited solution suggest charge sharing or transfer from the organic inhibitor to the metal surface to form coordinate covalent bonds [42, 43].

Table 5 shows the surface coverage, θ, of various concentrations of (BMBI) (from 5×10^{-7} M to 5×10^{-4} M) on mild steel surface as a function of temperature. These values were extracted from the corresponding % efficiency values reported earlier in Table 2. The plot of surface coverage against the natural logarithm of the concentration, ln C, for mild steel in the presence of the various inhibitor concentrations is shown in Figure 8. After examining the data and adjusting them to different theoretical adsorption isotherms, it was concluded that all inhibitors were adsorbed on the mild steel surface according to Temkin isotherm:

$$-2a\theta = \ln KC, \tag{4}$$

where a is molecular interaction constant, θ is degree of coverage, K is equilibrium constant for the adsorption reaction, and C is concentration of the inhibitor.

The equilibrium constant for the adsorption reaction, K, is related to the standard free energy of adsorption via the following equation given by Damaskin et al. [1]:

$$K = \frac{1}{55.5}\exp\left(-\frac{\Delta G}{RT}\right), \tag{5}$$

where K is equilibrium constant for the adsorption reaction, 55.5 is concentration of water [mol·L^{-1}], ΔG is standard free energy [kcal·mol^{-1}], R is gas constant [kcal·mol^{-1}], and T is absolute temperature [K].

According to (4), the straight lines shown in Figure 8 will have the following slopes and intercepts:

$$\text{Slope} = -\frac{1}{2a}, \tag{6}$$

$$\text{Intercept} = -\frac{1}{2a}\ln K. \tag{7}$$

Combination of (6) and (7) leads to the following relationship:

$$\text{Intercept} = \text{Slope} \times (\ln K), \tag{8}$$

from which the equilibrium constant for the adsorption reaction, K, can be calculated:

$$K = e^{(\text{Intercept/slope})}. \tag{9}$$

The standard free energy of adsorption of the inhibitor, ΔG^0, can be calculated from the results in Figure 8 used to calculate the equilibrium constant, K, and (5) at various temperatures (303 to 343 K).

The enthalpy of adsorption, ΔH^0, for the inhibitor can be calculated from the following equation:

$$\Delta H^0 = E_a - RT. \tag{10}$$

The entropy, ΔS^0, can be calculated at various temperatures for the inhibitor using the following equation:

$$\Delta G^0 = \Delta H^0 - T\Delta S^0. \tag{11}$$

TABLE 6: The free energy of adsorption (ΔG_{ads}) for mild steel in 1.0 M HCl in the presence of 3-benzoylmethyl benzimidazolium hexafluoroantimonate (BMBI) at various temperatures (303 K–343 K).

ΔG, kcal·mol^{-1}				
303 K	313 K	323 K	333 K	343 K
−18.55	−16.07	−15.33	−15.06	−14.07

TABLE 7: The enthalpy of adsorption (ΔH) for mild steel in 1.0 M HCl in the presence of 2×10^{-5} M 3-benzoylmethyl benzimidazolium hexafluoroantimonate (BMBI) inhibitor at various temperatures (303 K–343 K).

ΔH, kcal·mol^{-1}				
303 K	313 K	323 K	333 K	343 K
32.46	32.44	32.42	32.40	32.38

TABLE 8: The change in entropy (ΔS) for mild steel in 1.0 M HCl in the presence of 3-benzoylmethyl benzimidazolium hexafluoroantimonate (BMBI) inhibitor at various temperatures (303 K–343 K).

ΔS, kcal·K^{-1}·mol^{-1}				
303 K	313 K	323 K	333 K	343 K
0.168	0.155	0.148	0.143	0.135

Tables 6, 7, and 8 show the thermodynamic data obtained in the presence of the inhibitor at 5×10^{-4} M. These thermodynamic quantities represent the algebraic sum of the values for adsorption and desorption. The negative value of ΔG^0 indicates the spontaneous adsorption of inhibitor on the surface of the mild steel. The standard free energy, ΔG^0, increases from about -19 kcal·mol^{-1}·K^{-1} at 303 K to about -14 kcal·mol^{-1}·K^{-1} at 343 K. The adsorption process is believed to be exothermic and associated with a decrease in entropy (ΔS) of solute, while the opposite is true for the solvent. The gain in entropy which accompanies the substitutional adsorption process is attributed to the increase in the solvent entropy. This agrees with the general suggestion that the values of ΔG^0 increase with the increase of inhibition efficiency as the adsorption of organic compound is accompanied by desorption of water molecules off the surface [33, 45].

The high inhibition efficiency may be attributed to the preferred flat orientation of this compound on the metal surface. The interaction occurs between the delocalized π-electrons of the benzimidazole and phenyl rings, the carbonyl group (C=O), and the lone pair of electrons on N and O atoms with the positively charged metal surface.

5. Conclusion

3-Benzoylmethyl benzimidazolium hexafluoroantimonate was found to be a highly efficient inhibitor for mild steel in 1.0 M HCl solution, reaching about 98% at 5.0×10^{-4} M and room temperature, a concentration considered to be very moderate.

3-Benzoylmethyl benzimidazolium hexafluoroantimonate is a potential corrosion inhibitor since it contains not only nitrogen atom, but also phenyl and benzimidazole ring together with carbonyl group (C=O). It was apparent from the molecular structure that this compound would be adsorbed onto the metal surface through the lone pair of electron of nitrogen and oxygen and pi electrons of the aromatic rings (phenyl and benzimidazole) and the carbonyl group.

The percentage of inhibition in the presence of this inhibitor decreased as the temperature increased which indicates that physical adsorption was the predominant inhibition mechanism, because the quantity of adsorbed inhibitor decreases with increasing temperature.

Acknowledgment

The authors would like to thank the "College of Graduate Studies and Research" at the University of Sharjah for financially supporting this research.

References

[1] B. B. Damaskin, O. A. Pietrij, and W. W. Batrokov, "Adsorpcja organiczeskich sojedinienij na electrodach," *Moskva*, 1968.

[2] G. Okamoto, M. Nagayama, J. Kato, and T. Baba, "Effect of organic inhibitors on the polarization characteristics of mild steel in acid solution," *Corrosion Science*, vol. 2, no. 1, pp. 21–27, 1962.

[3] O. Benali, L. Larabi, M. Traisnel, L. Gengembre, and Y. Harek, "Electrochemical, theoretical and XPS studies of 2-mercapto-1-methylimidazole adsorption on carbon steel in 1 M $HClO_4$," *Applied Surface Science*, vol. 253, no. 14, pp. 6130–6139, 2007.

[4] E. S. H. E. Ashry, A. E. Nemr, S. A. Essawy, and S. Ragab, "Corrosion inhibitors part V: QSAR of benzimidazole and 2-substituted derivatives as corrosion inhibitors by using the quantum chemical parameters," *Progress in Organic Coatings*, vol. 61, no. 1, pp. 11–20, 2008.

[5] K. F. Khaled, "The inhibition of benzimidazole derivatives on corrosion of iron in 1 M HCl solutions," *Electrochimica Acta*, vol. 48, no. 17, pp. 2493–2503, 2003.

[6] A. Popova, "Temperature effect on mild steel corrosion in acid media in presence of azoles," *Corrosion Science*, vol. 49, no. 5, pp. 2144–2158, 2007.

[7] A. Popova, M. Christov, S. Raicheva, and E. Sokolova, "Adsorption and inhibitive properties of benzimidazole derivatives in acid mild steel corrosion," *Corrosion Science*, vol. 46, no. 6, pp. 1333–1350, 2004.

[8] A. Popova, M. Christov, and A. Zwetanova, "Effect of the molecular structure on the inhibitor properties of azoles on mild steel corrosion in 1 M hydrochloric acid," *Corrosion Science*, vol. 49, no. 5, pp. 2131–2143, 2007.

[9] V. S. Sastri, J. R. Perumareddi, M. Lashgari, and M. Elboujdaini, "Application of ligand field theory in corrosion inhibition," *Corrosion*, vol. 64, no. 4, pp. 283–288, 2008.

[10] M. Scendo and M. Hepel, "Inhibiting properties of benzimidazole films for Cu(II)/Cu(I) reduction in chloride media studied by RDE and EQCN techniques," *Journal of Electroanalytical Chemistry*, vol. 613, no. 1, pp. 35–50, 2008.

[11] S. Vishwanatham and A. Kumar, "Corrosion inhibition of mild steel in binary acid mixture," *Corrosion Reviews*, vol. 23, no. 2-3, pp. 181–194, 2005.

[12] D. Q. Zhang, L. X. Gao, and G. D. Zhou, "Inhibition of copper corrosion in aerated hydrochloric acid solution by heterocyclic compounds containing a mercapto group," *Corrosion Science*, vol. 46, no. 12, pp. 3031–3040, 2004.

[13] D. Q. Zhang, L. X. Gao, G. D. Zhou, and K. Y. Lee, "Undecyl substitution in imidazole and its action on corrosion inhibition of copper in aerated acidic chloride media," *Journal of Applied Electrochemistry*, vol. 38, no. 1, pp. 71–76, 2008.

[14] K. Tebbji, A. Aouniti, A. Attayibat et al., "Inhibition efficiency of two bipyrazole derivatives on steel corrosion in hydrochloric acid media," *Indian Journal of Chemical Technology*, vol. 18, pp. 244–253, 2011.

[15] A. Nahlé, I. Abu-Abdoun, and I. Abdel-Rahman, "Electrochemical studies of the effect of trans-4-hydroxy-4'-stilbazole on the corrosion inhibition of mild steel in HCl solution," *Anti-Corrosion Methods and Materials*, vol. 54, no. 4, pp. 244–248, 2007.

[16] S. M. Beloglazov, Z. I. Dzhafarov, V. N. Polyakov, and N. N. Demushia, "Quaternary ammonium salts as corrosion inhibitors of steel in the presence of sulfate-reducing bacteria," *Protection of Metals USSR*, vol. 27, no. 6, pp. 810–813, 1991.

[17] A. V. Fokin, M. V. Pospelov, and A. N. Levichev, "Structural and protective capacities of organic corrosion inhibitors. 2. Alkylethylene diamines and quaternary ammonium salts," *Protection of Metals USSR*, vol. 19, no. 2, pp. 242–244, 1983.

[18] A. H. Nahlé, "Electrochemical studies of corrosion inhibition of a series of quaternary ammonium salts on iron in HCl solution," *Corrosion Prevention and Control*, vol. 44, no. 4, pp. 99–105, 1997.

[19] A. H. Nahlé, "Inhibition of iron in HCl using benzyl trimethyl- and triethyl-ammonium chlorides," *Corrosion Prevention and Control*, vol. 45, no. 4, pp. 124–130, 1998.

[20] A. Nahle, "Effect of triethanolamine on the electrochemical dissolution of solder in NaOH solution," *Bulletin of Electrochemistry*, vol. 18, no. 3, pp. 105–110, 2002.

[21] A. Nahlé and F. C. Walsh, "Electrochemical studies of two corrosion inhibitors for iron in HCl: cetyltrimethyl ammonium bromide and tetraphenyl phosphonium chloride," *Corrosion Prevention and Control*, vol. 42, no. 2, pp. 30–34, 1995.

[22] B. V. Savithri and S. M. Mayanna, "Tetrabutyl ammonium iodide, cetyl pyridinium bromide and cetyl trimethyl ammonium bromide as corrosion inhibitors for mild steel in sulphuric acid," *Indian Journal of Chemical Technology*, vol. 3, no. 5, pp. 256–258, 1996.

[23] T. Vasudevan, S. Muralidharan, S. Alwarappan, and S. V. K. Iyer, "The influence of N-hexadecyl benzyl dimethyl ammonium chloride on the corrosion of mild steel in acids," *Corrosion Science*, vol. 37, no. 8, pp. 1235–1244, 1995.

[24] S. Muralidharan, K. L. N. Phani, S. Pitchumani, S. Ravichandran, and S. V. K. Iyer, "Polyamino-benzoquinone polymers: a new class of corrosion inhibitors for mild steel," *Journal of the Electrochemical Society*, vol. 142, no. 5, pp. 1478–1483, 1995.

[25] J. D. Talati, M. N. Desai, and N. K. Shah, "Meta-Substituted aniline-N-salicylidenes as corrosion inhibitors of zinc in sulphuric acid," *Materials Chemistry and Physics*, vol. 93, no. 1, pp. 54–64, 2005.

[26] T. Tüken, B. Yazici, and M. Erbil, "The effect of nicotinamide on iron corrosion in chloride solutions," *Turkish Journal of Chemistry*, vol. 26, no. 5, pp. 735–742, 2002.

[27] A. A. A. Fattah, E. M. Mabrouk, R. M. A. Elgalil, and M. M. Ghoneim, "N-heterocyclic compounds as corrosion inhibitors for Zn in HCl acid solutions," *Bulletin de la Societé Chimique de France*, vol. 1, pp. 48–53, 1991.

[28] S. L. Granese, B. M. Rosales, C. Oviedo, and J. O. Zerbino, "The inhibition action of heterocyclic nitrogen organic compounds on Fe and steel in HCl media," *Corrosion Science*, vol. 33, no. 9, pp. 1439–1453, 1992.

[29] H. A. Al-Lohedan, E. Khamis, and Z. A. Issa, "Studies on the influence of temperature on the adsorption of some cationic surfactants on to steel," *Adsorption Science and Technology*, vol. 13, no. 3, pp. 137–152, 1996.

[30] L. G. Qiu, A. J. Xie, and Y. H. Shen, "A novel triazole-based cationic gemini surfactant: synthesis and effect on corrosion inhibition of carbon steel in hydrochloric acid," *Materials Chemistry and Physics*, vol. 91, no. 2-3, pp. 269–273, 2005.

[31] B. G. Ateya, B. E. El-Anadouli, and F. M. El-Nizamy, "The effect of thiourea on the corrosion kinetics of mild steel in H_2SO_4 ," *Corrosion Science*, vol. 24, no. 6, pp. 497–507, 1984.

[32] B. G. Ateya, B. E. El-Anadouli, and F. M. El-Nizamy, "The adsorption of thiourea on mild steel," *Corrosion Science*, vol. 24, no. 6, pp. 509–515, 1984.

[33] A. S. Fouda, M. N. Moussa, F. I. Taha, and A. I. Elneanaa, "The role of some thiosemicarbazide derivatives in the corrosion inhibition of aluminium in hydrochloric acid," *Corrosion Science*, vol. 26, no. 9, pp. 719–726, 1986.

[34] A. Nahle, "Effect of temperature on the corrosion inhibition of carbon steel in HCl solutions," *Bulletin of Electrochemistry*, vol. 17, no. 5, pp. 221–226, 2001.

[35] A. Nahlé, I. Abdel-Rahman, and M. Alfarouk, "Effect of temperature on the inhibition of corrosion of carbon steels by semicarbazides and thiosemicarbazides," *Bulletin of Electrochemistry*, vol. 21, no. 8, pp. 353–361, 2005.

[36] A. Nahlé, I. Abu-Abdoun, and I. Abdel-Rahman, "Corrosion Inhibition by (Anthraquinone-2-ylmethyl) triphenyl phosphonium bromide," *Bulletin of Electrochemistry*, vol. 23, pp. 201–209, 2007.

[37] A. Nahlé, I. Abu-Abdoun, and I. Abdel-Rahman, "Inhibition of carbon steel corrosion by 4-vinylbenzyl triphenyl phosphonium chloride in HCl solution," *Anti-Corrosion Methods and Materials*, vol. 55, no. 4, pp. 217–224, 2008.

[38] S. N. Raicheva, B. V. Aleksiev, and E. I. Sokolova, "The effect of the chemical structure of some nitrogen- and sulphur-containing organic compounds on their corrosion inhibiting action," *Corrosion Science*, vol. 34, no. 2, pp. 343–350, 1993.

[39] S. H. Sanad, A. A. Ismail, and A. A. El-Meligi, "The effect of temperature on the corrosion and corrosion inhibition of steel alloys in hydrochloric acid solutions," *Bulletin of Electrochemistry*, vol. 11, no. 10, pp. 462–469, 1995.

[40] F. Zucchi, G. Trabanelli, and G. Brunoro, "The influence of the chromium content on the inhibitive efficiency of some organic compounds," *Corrosion Science*, vol. 33, no. 7, pp. 1135–1139, 1992.

[41] Y. L. Huang, C. N. Cao, M. Lu, and H. C. Lin, "Inhibition effects of I^- and I_2 on stress corrosion cracking of stainless steel in acidic chloride solutions," *Corrosion*, vol. 49, no. 8, pp. 644–649, 1993.

[42] A. Popova, M. Christov, and T. Deligeorgiev, "Influence of the molecular structure on the inhibitor properties of benzimidazole derivatives on mild steel corrosion in 1 M hydrochloric acid," *Corrosion*, vol. 59, no. 9, pp. 756–764, 2003.

[43] A. Popova, E. Sokolova, S. Raicheva, and M. Christov, "AC and DC study of the temperature effect on mild steel corrosion in acid media in the presence of benzimidazole derivatives," *Corrosion Science*, vol. 45, no. 1, pp. 33–58, 2003.

[44] M. Yamaguchi and H. Nishihara, "The inhibitive effect of organic cations on passive film breakdown of iron in a chloride-containing borate buffer solution," *Corrosion Science*, vol. 36, no. 7, pp. 1133–1141, 1994.

[45] A. S. Fouda, A. Abd El-Aal, and A. B. Kandil, "The effect of some phthalimide derivatives on the corrosion behaviour of copper in nitric acid," *Anti-Corrosion Methods and Materials*, vol. 52, no. 2, pp. 96–101, 2005.

[46] K. Kobayashi, K. Shimizu, and M. Iida, "Structural effects of organic compounds as corrosion inhibitors for hydrogen entry into iron in sulphuric acid," *Corrosion Science*, vol. 35, no. 5-8, pp. 1431–1435, 1993.

[47] L. D. Skryler, E. A. Streltsova, and T. L. Skryleva, "Hydrocarbon chain length and their effect on corrosion inhibition by alkylammonium chlorides," *Protection of Metals USSR*, vol. 27, no. 6, pp. 755–758, 1991.

10

Study of Zn-Ni Alloy Coatings Modified by Nano-SiO_2 Particles Incorporation

Olfa Hammami,[1,2] Leila Dhouibi,[1] Patrice Berçot,[2] El Mustafa Rezrazi,[2] and Ezzeddine Triki[1]

[1] *Unit of Research "COPROMET" ENIT, University of Tunis-El Manar, BP 371002 Belvedere, Tunisia*
[2] *Institut UTINAM, CNRS UMR 6213, Équipe Sonochimie et Réactivité des Surfaces, Université de Franche-Comté, 16 Route de Gray, 25030 Besançon Cedex, France*

Correspondence should be addressed to Olfa Hammami, holfa@rocketmail.com

Academic Editor: F. J. M. Pérez

The aim of this research work was to codeposit nano-SiO_2 particles into Zn-Ni alloy coatings in order to improve some surface properties. It had been investigated the effect of loading the plating bath with nanoparticles on composition, morphology, phase structure of deposits, and their subsequent influence on the corrosion process in corrosive solution of 3% NaCl and the thermal stability of deposits at 200°C. It was found that Zn-Ni alloy composites contain a higher percentage of Ni with incorporation in the deposit of 1.54% of silica. X-ray diffraction measurements revealed that the alloys consisted of two phases, pure zinc and γ phase. Electrochemical characterization of the composites had been carried out through potentiodynamic polarization and electrochemical impedance spectroscopy. The composites exhibited higher values of microhardness and better corrosion resistance in corrosive media.

1. Introduction

The need for coatings with improved resistance to highly aggressive environments is high as a result of a growing demand for extended safe service life of industrial objects. Composite electrodeposition is a valuable new surface intensification technology to obtain composite coatings allowing to codeposit inorganic and organic particles into the coatings in order to improve the surface properties. Several studies showed that electrochemically embedded particles impart special properties to composite coatings produced, which can meet the industrial request such as wear resistance, self-lubrication, and corrosion resistance [1–3]. Currently and for the foreseeable future, the most important types of nanoparticles are simple oxides, such as Al_2O_3, TiO_2, and SiO_2 used in established applications [4–6]. The electrodeposition of Zn-Ni-SiO_2 has generated amount of interest in the past years due to the attractive properties of Zn-Ni-SiO_2 coating, the introduction of this alloy into industrial practice has been extremely successful. In addition, it has been observed that varying the electroplating bath chemistry, conditions, and deposition parameters may give rise to a wide variation in properties of the resulting deposits.

Zn-Ni-SiO_2 coating has recently been used to replace cadmium plating which is widely used in the aerospace industries (i.e., landing gear, hydraulic actuators, high-performance fasteners, break, and fuel lines) because of its excellent corrosion resistance. The relation between our work and existing works on Zn-Ni-SiO_2 composite coating is to replace the coating with Zn-Ni-SiO_2 alloy as this coating shows great promise from the corrosion resistance point of view, representing a real challenge. Zn-Ni-SiO_2 coating system may also find a wide range of advanced applications in marine industry, so we found it interesting to study this type of composite coating in 3% NaCl.

The particles of SiO_2 are hydrophilic and hence incorporate hardly into the surface of cathode. The main purpose of this research work was to codeposit nano-SiO_2 particles into the Zn-Ni alloy coatings under direct current in order to improve the surface properties and the corrosion resistance in aggressive media.

Table 1: Solution composition and conditions for alloy electroplating [7].

Electrolyte ingredients	Concentration ($g \cdot L^{-1}$)	Plating parameters	$R = [Zn^{2+}]/[Ni^{2+}]$
$ZnSO_4 \cdot 7H_2O$	57.5		
$NiSO_4 \cdot 6H_2O$	52.5	pH = 2, 5	
H_3BO_3	9.3	Temperature (°C): 30 ± 0.5	1.1
Na_2SO_4	56.8		
H_2SO_4	0.53		

Then, the composite coatings had been characterized from compositional (EDX), morphological (SEM), structural (XRD), and depth profile (GDOES) points of view. In the meantime, electrochemical properties of the composite coatings had been studied by potentiodynamic polarization and electrochemical impedance spectroscopy.

Table 2: Electrophoretic mobility and Zeta potential measurements by "Rank Brother II".

pH of bath solution	Effective charge of SiO_2 particles	Electrophoretic mobility ($cm^2 \cdot V^{-1} \cdot s^{-1}$)	Zeta potential (mV)
6.7	—	1.45×10^{-8}	18.6

2. Experimental

2.1. Electroplating Process. The chemical compositions of the basic electrolyte of Zn-Ni alloys deposition is given in Table 1 [7]. All experiments were carried out in duplicates, and the ability to reproduce of these measurements was found satisfactory.

Galvanostatic measurements were performed on steel rod cathode which was placed in a PTFE mount for obtaining cross-sectional area of 0.2 cm^2 in contact with solution. The deposition was done on steel (STUB 100CR6) substrates. The chemical composition of STUB 100CR6 steel was 98.6% Fe, 0.95% C, 0.2%, Mn 0.15% Si, and 0.015% S (wt%). Initially, the steel samples were mechanically polished with successively finer grades of emery paper. They were then washed in deionized water. This procedure was repeated until a clear and smooth surface was obtained. To obtain fixed hydrodynamic conditions, the Zn-Ni coatings were deposited on a rotating disc electrode (RDE) with a constant rotation velocity of 1000 rpm. The deposition experiments were performed at a current density of 25 mA/cm^2, and the plating time was 15 min. The SiO_2 powder (AEROSIL 200) with a mean diameter of 12 nm was used as received without any pretreatment.

Before electrodeposition, the nanoparticles were dispersed in the bath by ultrasonic wave for 30 min. Particles with concentration of 5 g/L were maintained in an electrolytic bath in suspension by continuous magnetic stirring of 200 rpm for at least 24 hours before deposition.

2.2. Elesctrochemical Measurements. For electrochemical measurements (anodic polarization curves and impedance spectra), the coated samples were exposed to 3% NaCl solution (pH = 7 ± 0.1, $T = 25$°C). These studies were performed respectively with potentiostat PGP 201 and interface Solartron SI 1287/SI 1250 in a thermostatic three-electrode cell. The working electrode was a coated sample, the counter electrode was platinum with an area of 1 cm^2 and the reference electrode was saturated calomel SCE ($Hg/Hg_2Cl_2/KCl$ saturated). The deposition and potentiodynamic stripping processes were performed in a nonagitated electrolyte at room temperature.

2.3. Characterization Studies. An X-ray diffraction investigation of Zn-Ni electrodeposits was carried out using an X-ray diffractometer D8 (Advance Bruker) equipped with a copper anode generating Ni-filtered CuKα radiation (l = 1.5405 Å, 45 KV, 40 mA). The quantity of Zn and Ni in the coatings and their thicknesses was determined by means of Fischerscope X-Ray XDAL fluorescence. Surface morphology of the deposits was followed with electron microscope at 15 KV. Measurements of the Vickers microhardness (HV) of deposits were performed in the surface by using as HMV-M3 SHIMADZU Microhardness tester under 25 g to 100 g load for 10 s, and the corresponding final values were determined as the average of 5 measurements.

Glow discharge optical emission spectroscopy (GDOES); the distribution of species in the deposit, was determined by depth profiling using a Jobin Yvon GD-Profiler instrument equipped with a 4 mm diameter anode and operating after optimization at a pressure of 650 Pa and a power of 30 W in an argon atmosphere. This low power was retained to decrease the speed of abrasion of the deposits with low thickness and to obtain maximum information at the surface. Quantified compositional results were evaluated automatically utilizing the standard Jobin Yvon quantum Intelligent Quantification software. The instrument was calibrated with standard known composition. Depths were calculated using relative sputter rates, obtained from the sputter yields of each major element with corrections for composition and discharge conditions.

3. Results and Discussion

3.1. Behavior of SiO_2 Nanoparticles in the Plating Solution. In Table 2, we regrouped zeta potential and electrophoretic mobility measurements of particles suspended in the electroplated solution diluted 100 times. By observing the movement of particles under the influence of an electric field in

(a) (b)

FIGURE 1: SEM micrographs and EDS of the surfaces of Zn-Ni-coated steel at 25 mA/cm^2 for 15 min in plating bath: (a) Zn-Ni alloy coating, (b) Zn-Ni alloy composite (with 5 g/L SiO_2).

the airframe microelectrophoresis "Rank Brothers II," we can determine the effective charge of particles and calculate the electrophoretic mobility of the particle and its corresponding zeta potential.

We noted negatively charged particles of SiO_2. This type of oxide is hydrophilic, so it always interacts with the electrolyte, and therefore chemical and physical adsorption of electrolyte ions onto the particle occurs. This adsorption and the initial particle surface composition determine the particle surface charge, which induces a double layer of electrolyte ions around the particle. In electrolytes, double layers play a major role in the interactions between particles and also between particles and the electrode [8]. SiO_2 has a negative surface charge at pH 6.7, and therefore zinc or nickel ions should surround the particles to give them an overall positive surface charge. According to some authors, this may lead to codeposition at the negatively charged cathode [9].

The absolute value of the zeta potential is a very important factor for the degree of particles incorporation. In fact, zeta potential determined the static electricity force among nano-SiO_2 particles and influenced greatly the character either agglomeration or dispersion among the nanoparticles [10]. When the nanoparticles of SiO_2 became close to the cathode, the repulsion force among the double-charge layers of nanoparticles and increased the possibility of conglobation among nanoparticles decreased. In addition, the positive/negative of the zeta potential value on the nanoparticle surfaces had the crucial influence to the interaction among nanoparticles and cathode surface during electrodeposition. So the stability of the composite electrodeposition solution could be characterized by the zeta potential on the nanoparticle surfaces. It can be seen from Table 2 that, in pH 6.7, the value of zeta potential was lower than 25 mV, showing that, in the composite electrodeposition solution, the conglobation trend among nanoparticles increased and the dispersibility of the composite electrodeposition solution was weakened [10].

3.2. Compositional Analysis, Morphology, and Depth Profile of Deposits. Table 3 showed a comparison between Zn-Ni deposits composition electroplated at 25 mA/cm^2 for 15 min.

TABLE 3: Composition analysis by X-ray fluorescence of Zn-Ni deposits electroplated at 25 mA/cm^2 for 15 min onto steel substrate.

Coating systems	Ni (±0.2 wt.%)	Zn (±0.2 wt.%)	R' = Zn/Ni
Zn-Ni alloy coating	6.3	93.7	14.9
Zn-Ni/nano-SiO_2 alloy composite	12.3	87.7	7.1

The X-ray fluorescence analysis of the composite coatings revealed about a double percentage of Ni in the alloy composite in comparison with Zn-Ni alloy coating. The Ni content of Zn-Ni alloy deposits plays a major role in controlling the phases present and the grain size [11]. In the electrodeposition of Zn-Ni/SiO_2 composites, a phenomenon of induced codeposition was observed, since SiO_2 increased the percentage of Ni in the alloys, Ni cations simultaneously accelerated the codeposition of SiO_2 [12].

The ratio of the less noble metal Zn to the more noble metal Ni in the deposit (R') is larger than in the bath (R). This phenomenon is known as "anomalous codeposition," fully described by Brenner [13].

The surface morphology of Zn-Ni alloy coating showed branched acicular crystallite structure (Figure 1(a)), as proved in our anterior results [7]. The surface of the composite coating (Figure 1(b)) seems to keep the same appearance with piled and smooth grains. EDS analysis of the composite coating surface revealed a silica content of 1.54%.

Figure 2 showed the GDOES depth profile of the wt.% amount of the selected elements: zinc, nickel, silicon and iron for the Zn-Ni composite coating. The distributions of the different species are revealed clearly with sharply defined interface between the deposit and the substrate. The thickness of the composite measured by GDOES correlate very well with the one obtained by X-ray fluorescence (approximately 7 μm). Zinc and nickel species are incorporated relatively uniformly along the cross-section as depicted in Figure 2. Just in the outer region of the film, we confirm the presence of silicon which decreased until a depth of approximately 0.5 μm.

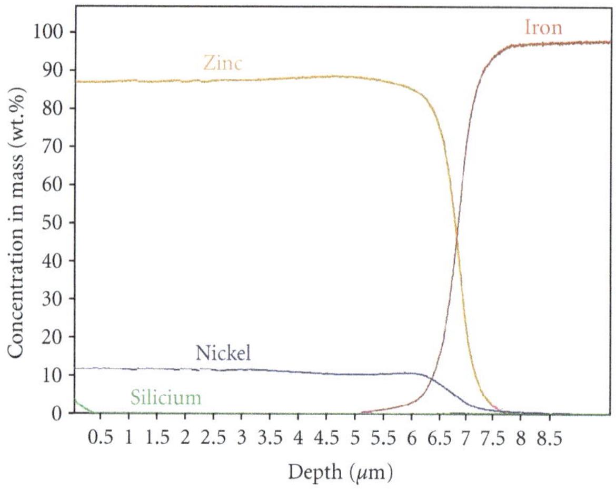

FIGURE 2: GDOES depth quantified profile of the Zn-Ni alloy composite (with 5 g/L SiO_2) electrodeposited with (25 mA/cm^2, 15 min) onto steel substrate.

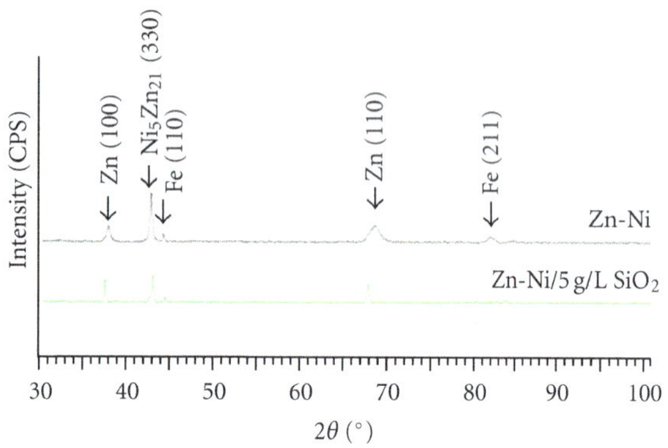

FIGURE 3: X-ray diffractograms for Zn-Ni alloy coating and Zn-Ni alloy composite (with 5 g/L SiO_2) electroplated at 25 mA/cm^2 for 15 min.

3.3. Phase Structure and Microhardness. Figure 3 showed the XRD patterns for the plated Zn-Ni alloy coatings. It is clear that no significant change of phases content in spectra resulted from loading the plating bath with nano-SiO_2 particles. Only the diffraction lines of zinc and γ phase of Zn-Ni alloys were observed. However, the peak of Fe (211) in the Zn-Ni/5 g/L SiO_2 alloy composite diffractogram disappeared giving rise probably to a deposit covering the surface of steel better than Zn-Ni alloy coating. We noted that Zn-Ni alloy coating and composite coating had approximately the same thicknesses.

Crystallite sizes of the coatings were calculated from the X-ray peak broadening of the (330) diffraction peak using Scherrer's formula [14]:

$$D = K\lambda/(\beta cos\theta), \quad (1)$$

where K is the Scherrer factor ≈ 1, D is the crystallite size, λ is the incident radiation wavelength, β is the integral breadth of the structurally broadened profile, and θ is the angular position.

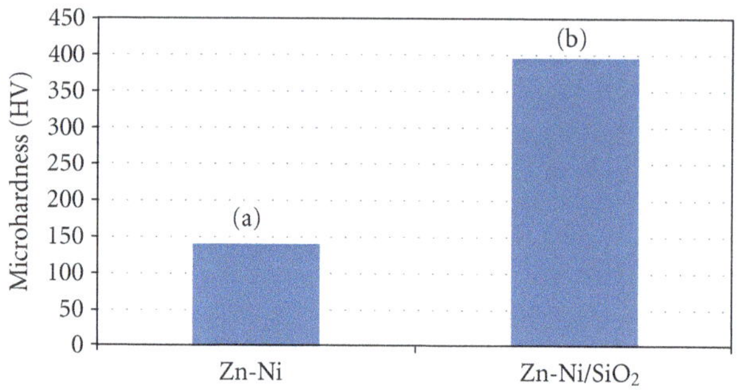

FIGURE 4: Evolution of microhardness of Zn-Ni alloy coatings (a) and Zn-Ni alloy composites (b) electroplated onto steel substrate at 25 mA/cm^2 for 15 min.

Zn-Ni alloy deposit exhibited a crystallite size of 50 nm, whereas the crystallite size of Zn-Ni composite coating decreased to 26 nm. We could explain this evolution by the fact that the nanoparticles of silica might perturb the crystal growth by increasing the number of nucleation sites and consequently a reduction in the grain size occurs as reported by Pavlatou et al. [1]; namely, the growth of the electrodeposited layer is a competition between the nucleation and crystal growth. Nanoparticles provide more nucleation sites and hence retard the crystal growth; subsequently the corresponding Zn-Ni matrix in the composite coating has a smaller crystal size.

Vickers microhardness (HV) values of Zn-Ni alloy deposits are shown in Figure 4.

The evolution of coatings hardness showed an increase from about 140 HV for pure Zn-Ni alloy to about 396 HV for samples prepared with SiO_2 particles. We can conclude that the microhardness was affected by the incorporation of this type of nanoparticles into deposits. In turn, crystallite size is an important variable affecting the hardness value [15]. In fact, smaller crystallite size implies a greater number of grain boundaries that impede dislocation motion and then creates harder materials.

3.4. Thermal Stability. The thermal stability of Zn-Ni alloy electrodeposits is important to their general evaluation for future applications, particularly in applications where the coatings are expected to perform at elevated temperature such as on some automotive parts. In this study, in order to evaluate the stability of the deposits, we followed the microhardness evolution of deposits treated at 200°C continuously for 30 min, 2 hours, and 24 hours (Figure 5).

After annealing at 200°C, deposits revealed reduced hardness values after 30 min of thermal treatment, especially the hardness of composites alloy. The reduced hardness could be attributed to surface oxidation phenomena or a rapid grain growth and a decrease of internal stresses as reported by Apachitei et al. [16]. Further heating for 24 hours at 200°C led to stable Vickers microhardness of composite deposits. These values were maintained higher than those of Zn-Ni alloy coatings.

TABLE 4: E_{corr} and peaks positions of stripping response in 3% NaCl solution for Zn-Ni alloy deposits elaborated under DC (25 mA/cm², 15 min).

Samples		Peak 1		Peak 2		Peak 3	
	$E_{corr} \pm 5$ mV/SCE	$E_p \pm 4$ mV/SCE	$i_p \pm 2$ mA/cm²	$E_p \pm 4$ mV/SCE	$i_p \pm 2$ mA/cm²	$E_p \pm 4$ mV/SCE	$i_p \pm 2$ mA/cm²
Zn-Ni alloy coating	−1020	−800	15	−615	8	−445	5
Zn-Ni alloy composite	−941	−815	5	−544	13	−404	10.5

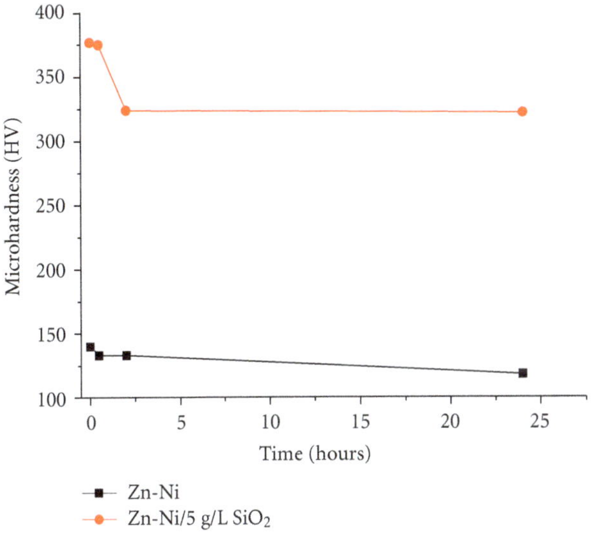

FIGURE 5: Variation of microhardness of Zn-Ni alloy deposit and coating composite electroplated onto steel substrate at 25 mA/cm² for 15 min and heat treated at 200°C for 30 min, 2 h, and 24 h.

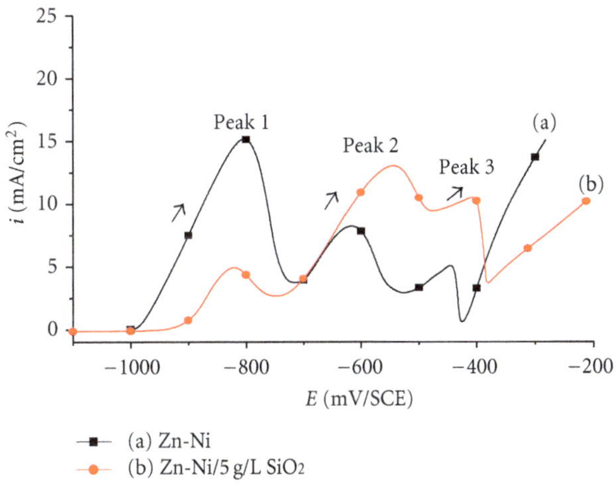

FIGURE 6: Anodic polarization curves of steel substrates coated with Zn-Ni alloy coating (a) and Zn-Ni alloy composite (b) elaborated under DC (25 mA/cm², 15 min); immersion in 3% NaCl.

3.5. Corrosion Behavior of Zn-Ni Alloy Deposits in 3% NaCl Solution

3.5.1. Potentiodynamic Polarization Studies. Figure 6 showed the anodic polarization curves elaborated at $v = 25$ mV/min of steel substrates coated with Zn-Ni alloy coating immerged in 3% NaCl. Distinct peaks were observed which could be attributed to the dissolution of Zn-Ni alloy components [17, 18] in the corroding media.

Based on previous results [7], we could make a qualitative estimate of Zn-Ni alloy structure. The two first peaks corresponded to the dissolution (dealloying) of Zn from η and γ phases which were situated at, respectively, −800 and −615 mV/SCE, whereas the third peak corresponded to the dissolution of Ni from their phases (Table 4). We noted that the intensity of the first peak relative to Zn-Ni alloy composite decreased. This evolution suggests a reduce of the electrochemical kinetics of Zn-Ni composite. In fact, an increase in life of the protective coating would simply depend on the life of the more electronegative phase, which is a Zn-rich η phase [19]. Dissolution of this phase would expose the nickel-rich phase to the environment.

As mentioned in Table 4, there was a clear shift for second and third peaks toward the noble direction (Table 4), giving rise to an increase in corrosion resistance of Zn-Ni composite coating. Moreover, the corrosion potential shifted to positive values (Table 4) indicating a change in the content ratio in the alloy and higher corrosion resistance compared with the Zn-Ni alloy coating.

We followed the loss of metal from Zn-Ni alloy deposit by stripping with X-ray fluorescence. In comparison with Zn-Ni alloy coatings, we noted that the dezincification rate of Zn-Ni alloy composites decreased from 24% to 2% until the last remaining phase (α-phase).

3.5.2. Electrochemical Impedance Spectroscopy. EIS was used to evaluate the barrier properties of the coatings after 24 hours of immersion in 3% NaCl and to determine their polarization resistance. Figure 7(a) presented a comparison of Nyquist responses obtained for Zn-Ni alloy coatings and Zn-Ni alloy composites elaborated under the same quantities of electricity. Three semicircles are generally observed in the complex-plane representation of the EIS data; the enlargement of the curves at high frequency (Figure 7(b)) showed the third semicircle.

The Nyquist-like diagram of the samples is characteristic of electrode process under kinetic and charge transfer control. The polarization resistance values could be approximately determined by fitting the Nyquist response to an equivalent circuit consisting of three RC components: the first

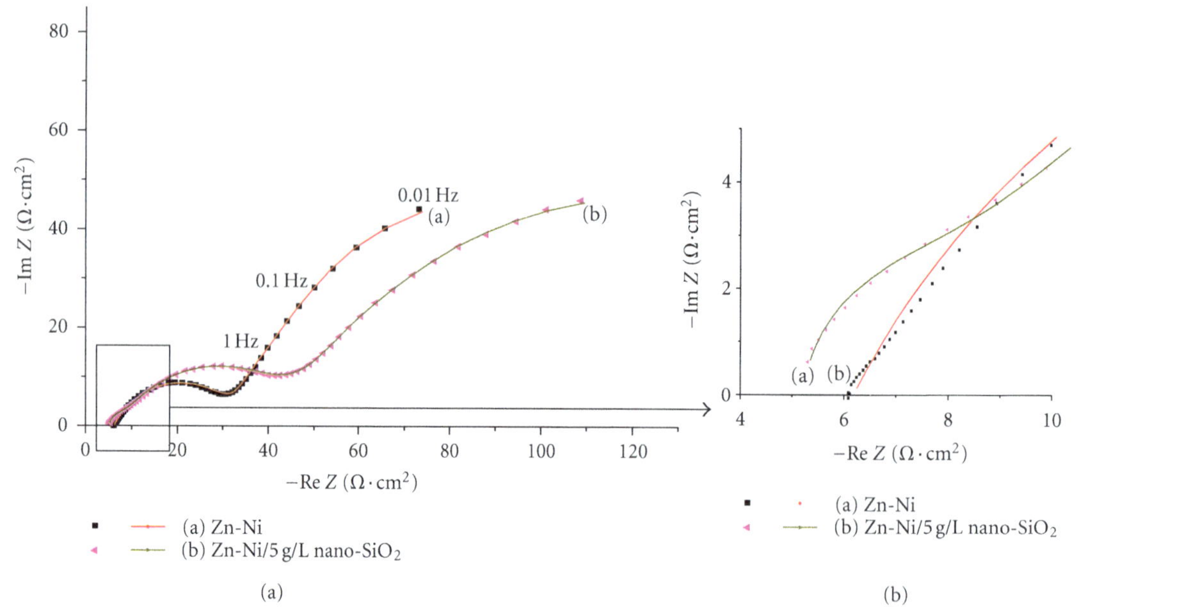

Figure 7: Nyquist plots obtained for Zn-Ni alloy coating (a) and Zn-Ni alloy composite (b), after 24 h immersion in 3% NaCl solution. Symbols are the experimental data, and lines were modeled using fitting model.

Table 5: Data obtained by electrochemical impedance spectroscopy of coatings immersed in 3% NaCl solution for 24 h.

Coating systems	R_{pc} (ohm · cm²)	C_c (F · cm⁻²)	R_p (ohm · cm²)	Composition analyses (±0.2 wt.%)				
				Zn	Ni	Cl	Fe	Si
Zn-Ni alloy coating	0.4	10^{-5}	320	71	6	10.8	1.30	—
Zn-Ni alloy composite	6.5	3×10^{-6}	560	85.5	9.73	4	0.47	0.8

time constant (R_{pc}, C_c) was attributed to the protective coating properties, (R_t, C_{dl}) an intermediate relaxation attributed to charge transfer which produced in the film defaults with R_t was the charge transfer resistance, and C_{dl} was the double-layer capacitance at the interface. This time constant was probably due to the heterogeneity of the film which can result from the existence of high- and low-density zones inside the film. The third time constant (R_F, C_F) in the low-frequency part (1–10^{-2} Hz) could be attributed to the deposited corrosion products [20]. In all cases, the proposed model was in total agreement with the experimental data. The R_p values of Zn-Ni alloy coatings, determined by fitting with the model and summarized in Table 5, was estimated from the addition of (R_{pc} + R_t + R_F). From Table 5, we noted that R_{pc} values of this system, which reached 6.5 $\Omega \cdot cm^2$, were seen to improve, while their capacitance C_c decreases. In the meantime, we noted an increase of R_p value of Zn-Ni alloy composite denoting the decrease of the corrosion intensity.

Composition analyses of deposits immersed in 3% NaCl for 24 h (Table 5) showed that composites dissolve slower than Zn-Ni alloy coatings and keep a higher percentage of Ni in the deposit.

MEB micrographs indicated that Zn-Ni alloy coating immersed in 3% NaCl for 24 hours suffered from defects with grains of salt deposited in clusters on the surface (Figure 8(a)). However, micrographs of Zn-Ni alloy composite showed a reduction in the defects sizes (Figure 8(b)) and the deposit seemed to keep the same appearance represented in Figure 1(b). EDS analysis showed that the nanoparticles of SiO_2 were still incorporated in the composite, while their percentage decreased to 0.8%. In fact, the incorporation of nanoparticles in the deposit may be helpful to segregate the corrosion products appeared in a thin layer in Figure 8(b). This is confirmed by composition analyses showed in Table 5, which proved a decrease of the percentage of Fe and Cl in the composite in comparison with Zn-Ni alloy coating. Several hypotheses have been developed assuming that the decrease of corrosion intensity may be related to the structural and/or electronic properties of the passive oxide particles [21].

4. Conclusion

This work represented the electrodeposition and the corrosion behavior of Zn-Ni alloy deposits elaborated in absence or in presence of 5 g/L of nano-SiO_2 in the plating bath. This study was made to evaluate the influence of nanoparticles addition on some properties such as hardness, morphologic,

(a)

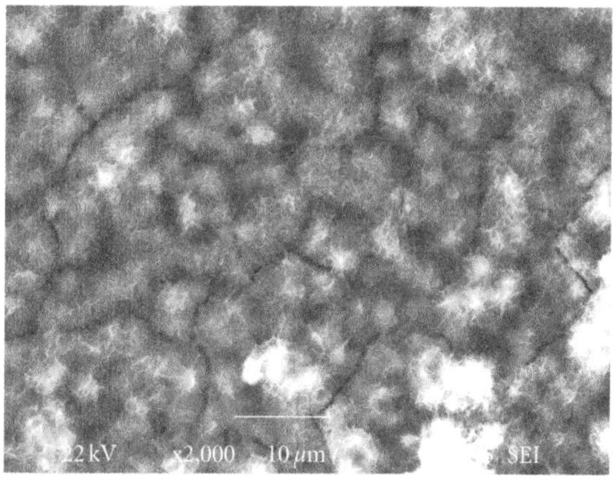

(b)

FIGURE 8: Comparison of SEM micrographs of the surfaces of deposits elaborated at 25 mA/cm^2 for 15 min onto steel substrate and immerged for 24 hours in 3% NaCl solution: (a) Zn-Ni alloy coating, (b) Zn-Ni alloy composite.

structure characteristics, and corrosion resistance. In comparison with Zn-Ni alloy coatings, the results revealed that Zn-Ni alloy composites:

(i) revealed a higher percentage of Ni and an incorporation of 1.54% of silica in the deposit,

(ii) formed a mixture of two phases, zinc and γ phases, with smaller crystallite size,

(iii) exhibited higher values of microhardness and were thermally stable up to 24 hours at 200°C,

(iv) revealed better corrosion resistance in corrosive media of 3% NaCl. This behavior indicates that Ni content, morphology, and the content of nanoparticles in the composite are responsible for the observed corrosion behavior.

Acknowledgments

The authors would like to acknowledge the financial support provided by "Action Intégrée Franco-Tunisienne du Ministère des Affaires Etrangères et Européennes Français et du Ministère de l'Enseignement Supérieur, de la Recherche Scientifique et de la Technologie Tunisien."

References

[1] E. A. Pavlatou, M. Stroumbouli, P. Gyftou, and N. Spyrellis, "Hardening effect induced by incorporation of SiC particles in nickel electrodeposits," *Journal of Applied Electrochemistry*, vol. 36, no. 4, pp. 385–394, 2006.

[2] P. Gyftou, M. Stroumbouli, E. A. Pavlatou, P. Asimidis, and N. Spyrellis, "Tribological study of Ni matrix composite coatings containing nano and micro SiC particles," *Electrochimica Acta*, vol. 50, no. 23, pp. 4544–4550, 2005.

[3] N. K. Shrestha, D. B. Hamal, and T. Saji, "Composite plating of Ni–P–Al_2O_3," *Surface and Coatings Technology*, vol. 183, no. 2-3, pp. 247–253, 2004.

[4] S. Kim, J. J. Gislason, R. W. Morton, X. Q. Pan, H. P. Sun, and R. M. Laine, "Liquid-feed flame spray pyrolysis of nanopowders in the alumina-titania system," *Chemistry of Materials*, vol. 16, no. 12, pp. 2336–2343, 2004.

[5] P. A. Janeway, "Nanotechnology—it's more than size," *American Ceramic Society Bulletin*, vol. 82, no. 4, pp. 31–38, 2003.

[6] B. Losiewicz, A. Stępień, D. Gierlotka, and A. Budniok, "Composite layers in Ni-P system containing TiO_2 and PTFE," *Thin Solid Films*, vol. 349, no. 1, pp. 43–50, 1999.

[7] O. Hammami, L. Dhouibi, and E. Triki, "Influence of Zn-Ni alloy electrodeposition techniques on the coating corrosion behaviour in chloride solution," *Surface and Coatings Technology*, vol. 203, no. 19, pp. 2863–2870, 2009.

[8] A. Hovestad and L. J. J. Janssen, "Electrochemical codeposition of inert particles in a metallic matrix," *Journal of Applied Electrochemistry*, vol. 25, no. 6, pp. 519–527, 1995.

[9] M. Azizi, W. Schneider, and W. Plieth, "Electrolytic co-deposition of silicate and mica particles with zinc," *Journal of Solid State Electrochemistry*, vol. 9, no. 6, pp. 429–437, 2005.

[10] P. C. Hiemenz, *Principles of Colloid and Surface Chemistry*, vol. 2, Marcel Dekker, 1986.

[11] A. M. Alfantazi, A. M. El-Sherik, and U. Erb, "The role of nickel in the morphology evolution of pulse plated ZnNi alloy coatings," *Scripta Metallurgica et Materiala*, vol. 30, no. 10, pp. 1245–1250, 1994.

[12] T. R. Khan, A. Erbe, M. Auinger, F. Marlow, and M. Rohwerder, "Electrodeposition of zinc-silica composite coatings: challenges in incorporating functionalized silica particles into a zinc matrix," *Science and Technology of Advanced Materials*, vol. 12, no. 5, Article ID 055005, 2011.

[13] A. Brenner, *Electrodeposition of alloys, Principles and Practice*, vol. 2, Academic Press, 1963.

[14] H. P. Klug and L. E. Alexander, *X-Ray Diffraction Procedures*, vol. 7, Chapman and Hall, 1959.

[15] A. Portinha, V. Teixeira, J. O. Carneiro, S. N. Dub, R. Shmegera, and C. J. Tavares, "Hard ZrO_2/Al_2O_3 nanolaminated PVD coatings evaluated by nanoindentation," *Surface and Coatings Technology*, vol. 200, no. 1–4, pp. 765–768, 2005.

[16] I. Apachitei, F. D. Tichelaar, J. Duszczyk, and L. Katgerman, "The effect of heat treatment on the structure and abrasive wear resistance of autocatalytic NiP and NiP-SiC coatings," *Surface and Coatings Technology*, vol. 149, no. 2-3, pp. 263–278, 2002.

[17] S. K. Zečević, J. B. Zotović, S. Lj. Gojković, and V. Radmilović, "Electrochemically deposited thin films of amorphous Fe-P alloy Part I. Chemical composition and phase structure characterization," *Journal of Electroanalytical Chemistry*, vol. 448, no. 2, pp. 245–252, 1998.

[18] S. Swathirajan, "Electrodeposition of zinc + nickel alloy phases and electrochemical stripping studies of the anomalous

codeposition of zinc," *Journal of Electroanalytical Chemistry*, vol. 221, no. 1-2, pp. 211–228, 1987.

[19] M. M. Abou-Krisha, H. M. Rageh, and E. A. Matter, "Electrochemical studies on the electrodeposited Zn-Ni-Co ternary alloy in different media," *Surface and Coatings Technology*, vol. 202, no. 15, pp. 3739–3746, 2008.

[20] D. Zhu and W. J. van Ooij, "Corrosion protection of AA 2024-T3 by bis-[3-(triethoxysilyl) propyl]tetrasulfide in sodium chloride solution.Part 2: mechanism for corrosion protection," *Corrosion Science*, vol. 45, no. 10, pp. 2177–2197, 2003.

[21] M. A. M. H. Al-Rabea, F. F. M. Saeed, A. W. K. Rajih, and F. M. Hussoun, "Using (Zn-Ni-SiO_2) electrocomposite coating to prevent carbon steel against acidic environments," Conference Paper, Corrosion in Babylon University Atlanta, 6 pages, March 2009.

11

Inhibition of Mild Steel Corrosion in Hydrochloric Acid Solution by Ciprofloxacin Drug

Inemesit A. Akpan and Nnanake-Abasi O. Offiong

Corrosion and Materials Science Unit, Department of Chemistry, University of Uyo, PMB 1017, Akwa Ibom State, Uyo 520001, Nigeria

Correspondence should be addressed to Inemesit A. Akpan; iaakpanchem2007@yahoo.com

Academic Editor: Jerzy A. Szpunar

The inhibition of mild steel corrosion in hydrochloric acid solution by ciprofloxacin drug as an eco-friendly and commercially available inhibitor was studied at room temperature by weight loss technique. It was found that the test drug has a promising inhibitory action against corrosion of mild steel in the medium investigated. The inhibition efficiency was found to increase with a corresponding increase in the concentration of the inhibitor. It was also found that the adsorption as well as the inhibition process followed a first-order kinetics and obeyed Langmuir's adsorption isotherm.

1. Introduction

Mild steel is extensively used in industries and as a result corrodes when exposed to various industrial environments and conditions. The application of inhibitors has been said to be amongst the most practicable ways for protection of metals against corrosion, especially in acidic media [1–3].

The inhibitive reactivity of an inhibitor is fundamentally affected by the molecular structure of the inhibiting molecules [4, 5]. Most prominent corrosion inhibitors are organic compounds containing nitrogen, sulphur, oxygen, and phosphorus in their functional groups [6–9]. The mechanism of these compounds has been proposed to be the adsorption, by means of lone pairs of electron, of the organic functional groups on the metal surfaces [10]. Furthermore, a good number of drugs are known to posses most of these qualities, and current research has been geared to identify cheap drugs, which are environmentally safe as corrosion inhibitors [11, 12].

The objective of this present work is to study the inhibition of mild steel corrosion in HCl solution by ciprofloxacin drug by weight loss method.

2. Experimental

2.1. Materials. A commercially available grade of mild steel (purity = 98% Fe) identified and obtained locally was employed in this study. The sheets of metal were mechanically press cut into coupons of 3 cm × 3 cm × 0.1 cm dimensions. A small hole of about 5 mm diameter near the upper edge of the coupons was made to help hold them with grass hooks and suspend them into the corrosive medium. The coupons were used without further polishing. However, they were degreased in acetone, washed in doubly distilled water, and stored in moisture-free desiccators before use.

The inhibitor used in this study was an antibiotic drug with common name: ciprofloxacin, while its systematic name is 1-cyclopropyl-6-fluoro-4-oxo-7-(piperazin-1-yl)-quinoline-3-carboxylic acid. The molecular formula of the drug is $C_{17}H_{18}FN_3O_3$ with molecular mass of 331.346 g/moL. Ciprofloxacin has the chemical structure shown in Figure 1.

The tablets of ciprofloxacin were obtained from a local pharmacy and were used without further purification or modification. From the mass of the drug samples and it's

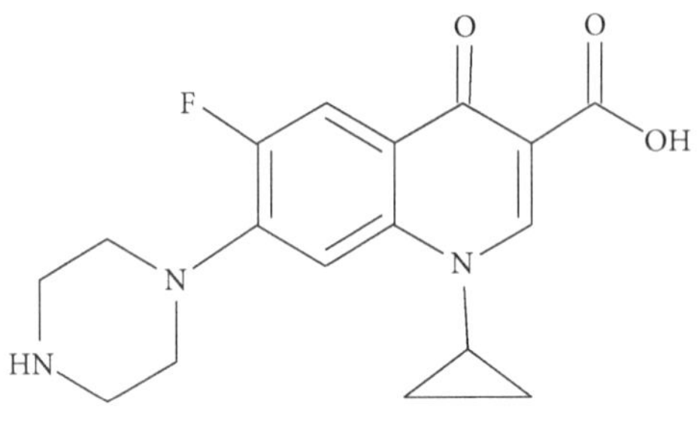

FIGURE 1: Chemical structure of ciprofloxacin drug.

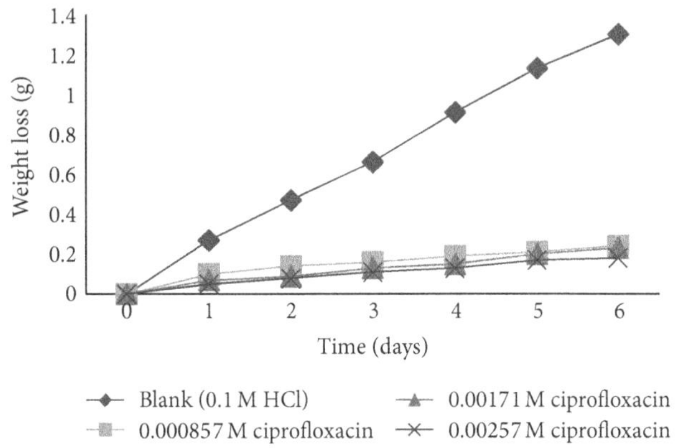

FIGURE 2: Variation of weight loss with time of exposure for corrosion mild steel in 0.1 M HCl solution in absence and presence of different concentrations of inhibitors.

molecular weight relation, appropriate concentrations of the drug were prepared by dilution.

The corrosive medium used was 0.1 M solution of hydrochloric acid. It was prepared by appropriate dilution of analytical grade of the acid reagent with doubly distilled deionised water without further purification.

2.2. Weight Loss Measurements. In the weight loss experiment, five plastic containers of 250 mL capacity were labelled A to E, each containing 0.1 M of HCl solution. The first beaker was reserved as blank while each of the four remaining beakers contained the drug at different concentrations all placed at room temperature (about 30°C). The metal coupons were immersed in the experimental solutions with the help of glass hooks and monitored daily (after 24 hours). The weights of the specimens were noted before immersion. After every immersion time of 24 hours, the specimens were removed, polished with emery papers, washed in double distilled water, degreased with acetone, dried in warm air, and reweighed. From the initial and final weights of the specimens, the loss of weights was calculated, and the corrosion rate (in mpy^{-1}—millimetre penetration per year) was computed from the following equation [13, 14]:

$$\text{Corrosion rate, CR} = \frac{534W}{DAt}, \quad (1)$$

where W is the weight loss (g), D is the density of the specimen (7.85 g/cm^3), A is the surface area of specimen (cm^3) and t is the immersion time (days).

The efficiency of the inhibitor was computed using the following equation [15–17]:

$$\text{Inhibition efficiency, \%IE} = \frac{W_o - W_1}{W_o} \times 100, \quad (2)$$

where W_0 is the weight loss without inhibitor and W_1 is the weight loss with inhibitor.

3. Results and Discussions

3.1. Weight Loss Measurements. The inhibition of mild steel corrosion in 0.1 M solution of hydrochloric acid by a commercially available and environmentally safe antibiotic drug, ciprofloxacin, was studied at room temperature by weight loss technique. Results obtained from weight loss measurements are as shown in Table 1. The result indicates that introduction of ciprofloxacin into the corrosive medium caused a significant reduction in the corrosion of mild steel. The corrosion inhibition efficiency increased with a corresponding increase in the concentration of the inhibitor. This may have resulted due to sufficient adsorption and wider coverage by the inhibitor molecules.

The calculated values for the corrosion rates (CR) are shown in Table 2. It has been observed that the corrosion rates of the mild steel in the corrodent medium were reduced on addition of different concentrations of the inhibitor.

It has also been observed from Figure 2 that the weight loss increase as the time of exposure increased.

3.2. Kinetics of Corrosion Inhibition. Chemical kinetic treatment of the data was necessary in order to obtain information about the order of the reaction. If the concentration of the corroding metallic material is estimated in terms weight loss per volume (g/L) of the corrodent, and later converted to molar concentrations via mass of metal-molar mass of iron relation, then, the kinetics of the system may be proposed. Following the work of K. K. Sharma and L. K. Sharma [18], we assume that if a mol/L is the initial concentration of the mild steel (Fe) and after time t, x mol/L of MS had decomposed into corrosion products. Therefore, the remaining concentration of MS at time $t = (a - x)$ mol/L. If a plot of $\log(a - x)$ or log [MS] against t gives a straight line graph, then the reaction can be said to be a first order reaction. It was based on this that we calculated for the reacted concentration of MS from weight loss measurements and obtained a graph shown in Figure 3. The shape of the graph in Figure 3 shows that the system under consideration followed a first order kinetics.

3.3. Adsorption Isotherm. Since the action of corrosion inhibitors are in most cases believed to be by adsorption on the metal surface by the inhibitor molecules using their adsorption centres, it is a good practice to find out the possible adsorption mode by testing the experimental data with several adsorption isotherms. The degree of surface coverage (θ)

Table 1: Weight loss values and calculated inhibition efficiency for mild steel corrosion in 0.1 M HCl in the presence and absence of different concentrations of inhibitors.

Inhibitor concentration	1 day		2 days		3 days		4 days		5 days		6 days	
	Weight loss	IE%	Weight loss	IE%	Weight loss	IE%	Weight loss	IE%	Weight loss	IE%	Weight loss	IE%
Blank	0.27	—	0.47	—	0.66	—	0.91	—	1.13	—	1.30	—
0.857×10^{-3} M	0.10	63	0.14	70	0.16	76	0.19	79	0.21	81	0.24	81
1.71×10^{-3} M	0.07	74	0.09	81	0.13	80	0.15	84	0.20	82	0.23	82
2.570×10^{-3} M	0.05	81	0.08	83	0.11	83	0.13	86	0.17	85	0.18	86

Table 2: Calculated values of corrosion rate of mild steel in 0.1 M HCl solution in absence and presence of different concentrations of inhibitors.

Time (days)	Corrosion rate, CR (mp/y)			
	Blank	0.857×10^{-3} M	1.71×10^{-3} M	2.57×10^{-3} M
1	0.0399	0.0148	0.0103	0.0074
2	0.0347	0.0103	0.0066	0.0059
3	0.0325	0.0079	0.0064	0.0054
4	0.0340	0.0070	0.0062	0.0048
5	0.0333	0.0062	0.0059	0.0050
6	0.0320	0.0059	0.0057	0.0044

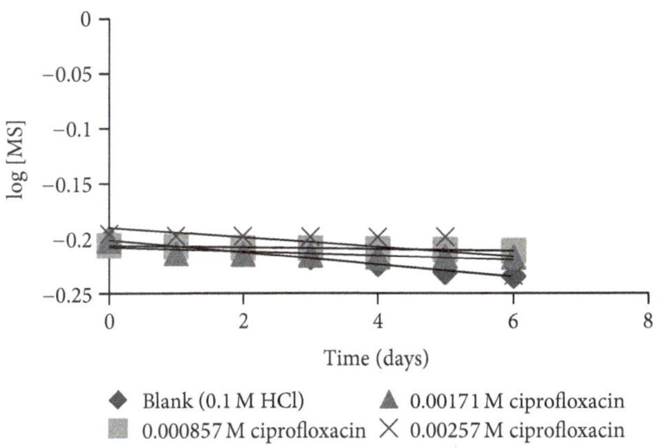

Figure 3: Linear plots of log of concentration of mild steel against time.

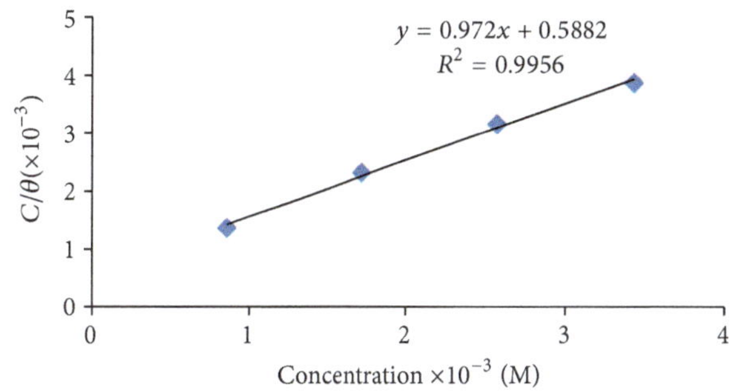

Figure 4: Langmuir's adsorption isotherm for mild steel in 0.1 M HCl containing various concentrations of ciprofloxacin as inhibitor.

at different concentrations of the inhibitor is one of the factors considered in this test and was computed from weight loss measurements using (3) [19, 20]:

$$\theta = \frac{W_o - W}{W_o}, \quad (3)$$

where W_o is the weight loss without inhibitor and W is the weight loss with inhibitor:

$$\frac{C}{\theta} = \frac{1}{k} + C, \quad (4)$$

where C is the concentration of the corrosion inhibitor, θ is the degree of surface coverage, and k is the adsorption equilibrium constant.

On consideration of the Langmuir adsorption isotherm, which is well described by (4) [21–25], it has been found that the experimental data gave a straight line graph on a plot of C/θ versus C and fitted the adsorption isotherm shown in Figure 4.

The Langmuir isotherm assumes a monolayer adsorption of the inhibitor molecules on the metal surface [26].

3.4. Mechanism of Inhibition. The inhibitive action of organic compounds depends of their structure and functional groups, nature of the metal, and aggressive medium [27, 28]. The inhibitory action of ciprofloxacin drug may be due to any or combination of the following mechanisms:

(1) physical adsorption of the molecules on the metal surface;

(2) electrostatic interactions between protonated nitrogen atoms and already adsorbed oxygen atoms;

(3) coordination due to donor-acceptor interactions between the unshared electron pairs of oxygen, fluorine, and possibly nitrogen;

(4) the π-electrons from the aromatic rings may also interact with the vacant d-orbitals of the atoms of the metal at the interface;

(5) since the molecular weight of the compound is large, inhibition be may also be effected by hindering of attacks by aggressive species due to wider surface covered by large molecules of the inhibitor.

4. Conclusion

In the study of the inhibitory effect of an eco-friendly and commercially available antibiotic drug, ciprofloxacin, on the corrosion of mild steel in 0.1 M HCl via weight loss technique at room temperature (30°C), the following conclusions may be drawn:

(1) the corrosion rate of mild steel reduced on addition of the test drug;

(2) the inhibition efficiency increased with an increase in the concentration of the inhibitor;

(3) the inhibition process followed a first-order kinetics;

(4) the inhibition process obeyed the Langmuir adsorption isotherm.

References

[1] P. S. Desai and S. M. Kapopara, "Inhibiting effect of anisidines on corrosion of aluminium in hydrochloric acid," *Indian Journal of Chemical Technology*, vol. 16, no. 6, pp. 486–491, 2009.

[2] A. S. Fouda, G. Y. Elewady, and M. N. El-Haddad, "Corrosion inhibition of carbon steel in acidic solution using some azodyes," *Canadian Journal on Scientific and Industrial Research*, vol. 2, no. 1, pp. 1–18, 2011.

[3] M. Ramananda Singh, K. Bhrara, and G. Singh, "The inhibitory effect of diethanolamine on corrosion of mild steel in 0.5M sulphuric acid medium," *Portugaliae Electrochimica Acta*, vol. 26, pp. 479–492, 2008.

[4] H. Ashassi-Sorkhabi and S. A. Nabavi-Amri, "Corrosion inhibition of carbon steel in petroleum/water mixtures by N-containing compounds," *Acta Chimica Slovenica*, vol. 47, no. 4, pp. 507–517, 2000.

[5] I. El Ouali, B. Hammouti, A. Aouniti et al., "Thermodynamic characterisation of steel corrosion in HCl in the presence of 2-phenylthieno (3, 2-b) quinoxaline," *Journal of Materials and Environmental Science*, vol. 1, no. 1, pp. 1–8, 2010.

[6] H. P. Sachin, M. H. M. Khan, and N. S. Bhujangaiah, "Surface modification of mild steel by orthophenylenediamine and its corrosion study," *International Journal of Electrochemical Science*, vol. 4, no. 1, pp. 134–143, 2009.

[7] S. A. Umoren, U. M. Eduok, and E. E. Oguzie, "Corrosion inhibition of mild steel in 1M H_2SO_4 by polyvinyl pyrrolidone and synergistic iodide additives," *Portugaliae Electrochimica Acta*, vol. 26, pp. 533–546, 2008.

[8] A. Kumar, "Corrosion inhibition of mild steel in hydrochloric acid by Sodium Lauryl Sulfate (SLS)," *E-Journal of Chemistry*, vol. 5, no. 2, pp. 275–280, 2008.

[9] A. K. Maayta, M. M. Fares, and A. F. Al-Shawabkeh, "Influence of linear alkylbenzene sulphonate on corrosion of iron in presence of magnetic field: kinetic and thermodynamic parameters," *International Journal of Corrosion*, vol. 2010, Article ID 156194, 9 pages, 2010.

[10] P. M. Niamien, A. Trokourey, and D. Sissouma, "Copper corrosion inhibition in 1M HNO_3 by 2-thiobenzylbenzimidazole: adsorption and chemical modeling of inhibition efficiency," *International Journal of Research in Chemistry and Environment*, vol. 2, no. 4, pp. 204–214, 2012.

[11] S. U. Ofoegbu and P. U. Ofoegbu, "Corrosion inhibition of mild steel in 0.1M hydrochloric acid media by chloroquine diphosphate," *ARPN Journal of Engineering and Applied Sciences*, vol. 7, no. 3, pp. 272–276, 2012.

[12] S. Hari Kumar and S. Karthikeyan, "Inhibition of mild steel corrosion in hydrochloric acid solution by cloxacillin drug," *Journal of Materials and Environmental Science*, vol. 3, no. 5, pp. 925–934, 2012.

[13] J. I. Bhat and V. Alva, "Corrosion inhibition of aluminium by 2-chloronicotinic acid in HCl medium," *Indian Journal of Chemical Technology*, vol. 16, no. 3, pp. 228–233, 2009.

[14] M. Abdallah, H. E. Megahed, M. A. Radwan, and E. Abdfattah, "Polyethylene glycol compounds as corrosion inhibitors for aluminium in 0.5M hydrochloric acid solution," *Journal of American Science*, vol. 8, no. 11, pp. 49–55, 2012.

[15] I. A. Akpan and N. O. Offiong, "Effect of ethanolamine and ethylamine on the entropy content of the corrosion of mild steel in tetraoxosulphate (VI) acid solution," *Chemistry and Materials Research*, vol. 2, no. 7, pp. 40–47, 2012.

[16] I. A. Akpan, "Inhibitory action of bile salt on the deterioration of asbestos in acid rain," *Bulletin of Pure and Applied Sciences*, vol. 31, no. 2, pp. 49–58, 2012.

[17] B. S. Shylesha, T. V. Venkatesha, and B. M. Praveen, "Corrosion inhibition studies of mild steel by new inhibitor in different corrosive medium," *Research Journal of Chemical Sciences*, vol. 1, no. 7, pp. 46–50, 2011.

[18] K. K. Sharma and L. K. Sharma, *A Textbook of Physical Chemistry*, Vikas Publishing House, New Delhi, India, 1999.

[19] H. Cang, Z. Fei, J. Shao, W. Shi, and Q. Xu, "Corrosion inhibition of mild steel by Aloes extracts in HCl solution medium," *International Journal of Electrochemical Science*, vol. 8, pp. 720–734, 2013.

[20] B. M. Mistry, N. S. Patel, and S. Jauhari, "Heterocyclic organic derivatives as corrosion inhibitos for mild steel in 1N HCl," *Archives of Applied Science Research*, vol. 3, no. 5, pp. 300–308, 2011.

[21] M. Mobin, M. Parveen, and M. Alam Khan, "Inhibition of mild steel corrosion in HCl solution using amino acid L-tryptophan," *Recent Research in Science and Technology*, vol. 3, no. 12, pp. 40–45, 2011.

[22] M. A. Quraishi and R. Sardar, "Effect of some nitrogen and sulphur based synthetic inhibitors on corrosion inhibition of mild steel in acid solutions," *Indian Journal of Chemical Technology*, vol. 11, no. 1, pp. 103–107, 2004.

[23] L. A. Nnanna, V. U. Obasi, O. C. Nwadiuko, K. I. Mejeh, N. D. Ekekwe, and S. C. Udensi, "Inhibition by *Newbouldia leavis* leaf extract of the corrosion of aluminium in HCl and H_2SO_4 solutions," *Archives of Applied Science Research*, vol. 4, no. 1, pp. 207–217, 2012.

[24] A. A. Khadom, A. S. Yaro, A. S. Altaie, and A. A. H. Kadum, "Electrochemical, activations and adsorption studies for the corrosion inhibition of low carbon steel in acidic media," *Portugaliae Electrochimica Acta*, vol. 27, no. 6, pp. 699–712, 2009.

[25] O. R. M. Khalifa, A. K. Kassab, H. A. Mohamed, and S. Y. Ahmed, "Corrosion inhibition of copper and copper alloy in 3M nitric acid solution using organic inhibitors," *Journal of American Science*, vol. 6, no. 8, pp. 487–498, 2010.

[26] B. Joseph, S. John, A. Joseph, and B. Narayana, "Imidazolidine-2-thione as corrosion inhibitor for mild steel in hydrochloric acid," *Indian Journal of Chemical Technology*, vol. 17, no. 5, pp. 366–374, 2010.

[27] S. Chitra, K. Parameswari, C. Sivakami, and A. Selvaraj, "Sulpha Schiff Bases as corrosion inhibitors for mild steel in 1M sulphuric acid," *Chemical Engineering Research Bulletin*, vol. 14, pp. 1-6, 2010.

[28] R. V. Saliyan and A. V. Adhikari, "Corrosion inhibition of mild steel in acid media by quinolinyl thiopropano hydrazone," *Indian Journal of Chemical Technology*, vol. 16, no. 2, pp. 162–174, 2009.

12

Six-Year Evaluation of Thermal-Sprayed Coating of Zn/Al in Tropical Marine Environments

Orlando Salas, Oladis Troconis de Rincón, Daniela Rojas, Adriana Tosaya, Nathalie Romero, Miguel Sánchez, and William Campos

Centro de Estudios de Corrosión, Facultad de Ingeniería, Universidad del Zulia, Maracaibo, Zulia, Venezuela

Correspondence should be addressed to Oladis Troconis de Rincón, oladis1@yahoo.com

Academic Editor: Blanca M. Rosales

The main objective of this research was to evaluate the performance of thermal-sprayed coating of Zn/Al (double layer) after six years of exposure, with and without the use of sealant (wash primer) in tropical marine environments of very high aggressiveness: La Voz Station (located at the Peninsula de Paraguaná/Falcón State) and Lake Maracaibo Crossing Station (located at Zulia State), in Venezuela. To that effect, carbon steel coupons (100 mm × 150 mm × 2 mm) were sprayed by flame process. The coupons were characterized by means of initial weight, thickness, metallographic, adherence, and roughness, being evaluated monthly by visual inspection during six years. After removal, the coupons were evaluated by microscopic analysis to determine the morphology of attack, microstructure, penetration of contaminants, composition, and morphology of corrosion products. The results showed that after six years, the double-layer system represents an excellent choice for corrosion protection of steel by combining the galvanic protection of zinc with the erosion resistance of aluminum. However, due to the erosion-corrosion effect, a sealant such as wash primer can be used in order to extend its service life.

1. Introduction

Worldwide, many organizations have made numerous efforts in terms of studies and research looking for new coating systems to improve the durability of structures exposed to marine and coastal marine environments, providing protection to the steel during its service life and producing a minimal impact on the environment. In marine and coastal marine environments, where a number of factors are present such as high-chloride concentrations, strong winds, changing conditions of relative humidity and temperature (especially in tropical climates like Venezuela), organic coatings do not perform properly. In this regard, thermal-sprayed coatings appear as an alternative to protect steel structures in those environments. In the United States, Europe, and Japan [1, 2], these coatings are gaining importance for their promising low environmental impact, corrosion resistance, cost benefits, durability, and other advantages over other conventional paint systems.

In 1974, the American Welding Society [3] performed a study that was considered a pioneer in thermal sprayed coating where it reported that these coatings with Zn and Al, with and without sealer, exposed in marine and industrial environments, provided protection against corrosion for 19 years. Moreover, in 1987, the Laque Center for Corrosion Technology, INC. [4] reported that the degree of protection of the coatings, after 34 years of continuous exposure in a moderately aggressive marine atmosphere, varies considerably with alloy content, type of material applied, and method of application. They also highlighted that the best thermal sprayed system is given by the coating of Zn/Al, double layer, and Zn/Al alloy (with a higher content of aluminum).

In tropical marine environments, for example, Venezuela, little is known about the application and performance of this coating technology. Venezuela's environment is completely different from the above-mentioned places given the high incidence of winds, high temperatures, and relative humidity during all the year. Only the PATINA Network [5] is known to have studied the effect of these environments on coating systems. In it, the behavior under corrosion of a wide variety of conventional and new generation coatings in the atmospheres of the Ibero-American region

Table 1: Characterization of coating system [10].

Coating	Thickness (μm)	Adhesion (MPa)	Porosity (%)		Anchor profile (μm)
Zn/Al by flame process	215.9–469.9	2.90	23 (Zn)	8 (Al)	37.85 ± 4.01

was evaluated. The performance of metal-based sacrificial coatings (aluminum, zinc and their alloys, among them thermal sprayed aluminum and Zn/15Al with vinyl sealer and wash primer) was also evaluated. The Zn/15Al with sealer gave the best performance in the atmosphere at La Voz. Recent research [6–13] in Venezuela (at Centro de Estudios de Corrosión, Faculty of Engineering, Universidad del Zulia) on thermal sprayed coatings with different materials (Zn, Al, Zn/85Al, and Zn/Al Double Layer) applied by two processes (flame and electric arc) with and without the use of different sealants (wash primer 1, wash primer 2, phenolic, and polyurethane), indicates that thermal sprayed coatings applied with only zinc or aluminum do not provide adequate protection in the exposure environments evaluated. However, the Zn/Al double-layer system has performed excellently in comparison with other thermal-sprayed systems. This coating system represents one of the best options for countering these aggressive environments. This paper presents the results after six years of evaluation.

2. Experimental Procedure

2.1. Preparing the Coupons [6–10]

2.1.1. Preparing the Surface. The company applying the thermal spray coating (flame) used its own steel coupons (100 mm × 150 mm × 2 mm) and its own criteria for preparing the surface; it did not supply the latter information. Because of this, indirect measurements of the anchor profile left on the substrate were made by using optical microscopy at 200x magnification, in transversal sections in some of the coupons, sweeping and measuring the distance between the peaks and valleys left on the substrate, with an average value of the measurements being reported.

2.1.2. Coatings and Application Methods. Two systems were evaluated: (1) a dual system (Zn/Al) formed by a first coat of thermal sprayed Zn as a primer and a second coat of thermal sprayed aluminum as a top coat, applied by the flame process without sealer; (2) the same coating but with the application of two additional layers of wash primer sealer.

In each of the test stations, 5 coupons per coating were installed. Each year, one coupon per coating was removed for evaluation in the laboratory to complete the total exposure period of six years, presented in this paper.

2.2. Characterization of Coatings [6–10]. Before being set up in the test stations, the coupons were swabbed with acetone to remove any oil and grease, then, cleaned with distilled water and dried immediately with an air pistol. After cleaning, they were characterized with respect to thickness, microscopic analysis, porosity, and adhesion.

2.2.1. Thickness. This measurement was carried out in accordance with Standard ASTM D 1186 [14], using a nondestructive instrument ("constant pressure probe DFT", "gauge (type II)") that measures the dry-film thickness of nonmagnetic coatings on ferrous metallic substrates. Due to the variability of thicknesses (nonuniform thicknesses) in this type of coating even on the same coupon, it was decided to use a template with the same dimensions (100 mm × 150 mm) with ten evenly distributed orifices and located at 2 cm away from the edges so that the measurements would be always taken on the same spot.

2.2.2. Metallographic Analysis and Porosity. An optical microscope with magnifications of 200x and 400x was used to analyze microstructure and porosity as well as to verify the thickness of these coatings. Samples were prepared by means of 2.5 cm × 2.5 cm cuts and then encapsulated in Bakelite with a cupping machine. It was then surfaced with emery paper and burnished with alumina to a specular aluminum finish. Specifically, Standard ASTM E 2109-01 [15] was used to determine the percentage of porosity in the thermal sprayed coatings with an image analysis program.

2.2.3. Adhesion. Adhesion tests were run by the "Pulloff test" method of tensile strength according to ISO Standard 4624 [16]/ASTM4541 [17], using an adhesion tester.

2.2.4. Incisions. In order to evaluate the galvanic effect of the coatings, a horizontal incision of approximately 6 cm was made into each coupon to reach the metal base in the lower part; a 2 cm space was left between the incision and the lateral and lower borders for observation.

Table 1 shows a summary of the initial characterization of the coupons.

2.3. Evaluating the Coatings [6–10]

2.3.1. Visual Inspection. The coating systems were examined and photographed monthly, collecting data on registration forms. The presence or absence of corrosion products detected during these inspections indicated whether they were coating corrosion products or substrate corrosion products, both on the general surface of the coupon as well as on the edges and incisions.

2.3.2. Analysis under the Scanning Electron Microscope. After the coupons were taken out from the test stations, cuts were made in the places of interest, using a diamond-point cutter, for subsequent encapsulation in conducting Bakelite to enable the execution of the elementary chemical analyses. The chemical analyses and the morphology of the transversal area of the coatings were carried out using the Energy Dispersion Spectroscopy (EDS) technique on a PHILLIPS Model XL 30 Scanning Electron Microscope.

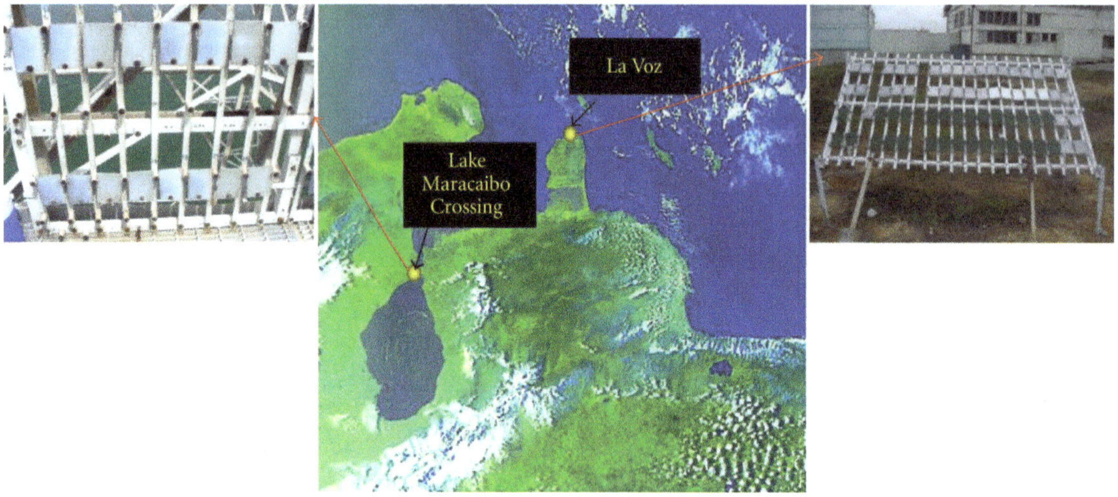

Figure 1: Geographic location of the Lake Maracaibo Crossing Station and La Voz Station.

2.3.3. Analysis by X-Ray Diffraction (XRD). This analysis was carried out to identify compounds and oxides formed after exposure, using a BRUKER AXS equipment of eight radiation sources of copper. Each sputtering lasted 45 minutes.

2.4. Test Stations. Two very aggressive tropical marine environments with high wind impingement were selected.

2.4.1. Lake Maracaibo Crossing Station (LMC). This station is located at the crossing with Lake Maracaibo, Venezuela, at 117 m above lake level (Figure 1). Besides being a marine atmosphere, it is characterized by strong winds, as well as changing conditions of high humidity and high temperatures. The station is also surrounded by two industrial complexes, both of which release sulfur dioxide into the atmosphere from fossil combustion processes.

2.4.2. La Voz Station. It is Located on the Peninsula of Paraguaná, Falcón State (Figure 1), in the direction of the prevailing winds, 21 nautical km away from the refinery at Aruba, and 300 m from the Caribbean. This atmosphere is classified as a special atmosphere in accordance with the MICAT Project (Latin American Map of Atmospheric Corrosion), because of its high aggressivity and strong prevailing winds, reaching corrosion rates of 921 μm/year for steel and 27 μm/year for zinc [18].

2.4.3. Chemical Pollutant Levels and Meteorological Variables. Chemical pollutant levels were measured according to ISO 9225 [19]. For the measurement of meteorological variables (relative humidity, temperature, time of wetness, wind speed, and rainfall), the data belonging to the station nearby was used [10]. Corrosivity of atmospheres was determined using ISO 9223 [20].

2.4.4. Location of Coupons on the Test Bank. The test bank at La Voz Station is placed at an angle of 45° to avoid the accumulation of water on the coupons and they are located facing the prevailing winds. At the Lake Maracaibo Crossing Station, the test bank is fastened to the railing of a tower at a height of 117 m above sea level, with the coupons in vertical position and facing east. The incidence of the prevailing winds at this station is particularly on the right edge of the test coupons.

3. Results and Discussion

3.1. Characterization of Environments. The results of the characterization of the environments under study are shown in Tables 2 and 3; notice their high aggressiveness (high–very high) where it is evident a more corrosive environment in La Voz Station (C_5/C_4) in relation to the Lake Maracaibo Crossing Station (C_4).

3.2. Behavior of the Thermal-Sprayed Systems in La Voz Station

3.2.1. Double-Layer System without Sealer. During the first twelve months of evaluation, the double-layer system Zn/Al without sealing (Figure 2(a)) had an excellent performance, presenting corrosion only at the coating in the bottom border. It must be pointed out that the area of the coupons hit by the prevailing winds, where the erosive effect is manifested, is different for each station. In the case of LMC station, winds impinge on the right edge whereas impingement is frontal in La Voz station.

However, for the second period of evaluation (24 months) [8], it was reported that these are susceptible to present border flaws, as after 14 months of evaluation, one of the valued samples evidenced detachment of the coating in the lower-left corner (Figure 3). This type of flaw in the border of the probe was possibly caused by an inadequate application of the coating due to the geometry of the probe (90° degrees angle/Figure 3(c1)), as being a critical and difficult area to cover; however, this can be corrected by guaranteeing a good thickness in that area or diminishing the angle of the

TABLE 2: Meteorological data during four and six years of exposure in La Voz Station and Lake Maracaibo Crossing Station (LMC), respectively.

Station	Time (years)	Temperature (°C)	Relative humidity (%)	Precipitation (mm)	Wind speed (m/s)
La Voz	1	28.3	82.4	834.6	5.6
	2	28.6	80.4	758.4	6.0
	3	28.7	82.5	538.4	7.3
	4	28.4	82.4	482.6	7.0
LMC	1	28.1	81.2	794.6	6.4
	2	27.8	83.4	1036.3	6.6
	3	28.0	80.3	544.3	6.5
	4	27.0	83.6	656.3	6.6
	5	26.7	80.4	698.4	6.6
	6	26.7	82.6	197.6	6.4

Source: For the LMC Station, the first 3 years correspond to meteorological data supplied by the "Air Base Rafael Urdaneta" (BARU), Zulia State, located at 30 km from the station. The rest of data for that station for the following three years was provided by a pilot station installed in situ at 110 m of height in the tower of the mentioned station.

TABLE 3: Average levels of air chemical agents during four and six years of exposure in La Voz Station and LMC Stations, respectively.

Station	Time (years)	Sulphur compounds mg/(m^2·d)	Chloride mg/(m^2·d)	Time of Wetness fraction	Corrosion category according to ISO 9223
La Voz	1	27.2	284.23	0.79	P_1S_2 T_5/C_5
	2	16.3	178.13	0.70	P_1S_2 T_5/C_5
	3	9.0	320.76	0.58	P_0S_3 T_5/C_5
	4	12.2	178.63	0.59	P_1S_2 T_5/C_5
LMC	1	16.4	125.3	0.48	P_1S_2 T_4/C_4
	2	17.8	89.49	0.46	P_1S_2 T_4/C_4
	3	13.1	100.44	0.52	P_1S_2 T_4/C_4
	4	45.0	80.72	0.71	P_2S_2 T_5/C_5
	5	—	—	0.62	—
	6	—	—	0.63	—

edge to reduce the drag coefficient, as the ones applied for the process of electric arch (Figure 3(c2)).

After 36 months of an excellent performance, the system started to diminish its protection capacity as it was affected by the erosive effect (Figures 2(c) and 2(d)), specially the border effect (lower-left corner); a worn area was noticed as well as some coating oxidation products; there were also deposits of salts and dust and even some substrate corrosion spots. In relation to the incision, an excellent galvanic protection was observed covered by the corrosion products from the coating, mixed with dust.

3.2.2. Double Layer System Sealed with Wash Primer. After 4 years of exposure, this system (Figure 4), contrary to the one without sealer, presented an excellent appearance, with some corrosion spots of the aluminum spread over the exposed surface and dust deposits in the low border. At the incision, the coating provided an excellent galvanic protection. It was totally covered by the corrosion products of the coating and dust, without signs of damages to the steel base.

According to the observations done, it can be said that the use of the sealer wash primer improves significantly the performance of this system (aluminum as finishing), as it offers an additional barrier to the erosive effect of the environment as well as it enhances its esthetic appearance. The reason for this improvement is the chemical composition of the sealant. This chemical analysis was carried out in another research [9] at the Centro de Estudios de Corrosión, Faculty of Engineering, Universidad del Zulia, on samples of thermal sprayed aluminum sealed with the same wash primer (wash primer 1) used in this evaluation and other sealer (wash primer 2), of another manufacturer, obtaining different performances.

Table 4 shows the chemical analysis for EDS that was made on the surface of a thermal sprayed sample with aluminum, sealed with wash primer 1 and its comparison with wash primer 2, on which some localized corrosion in the pores of the coatings was observed.

It was determined [9] that the improvement of this system is due to the chemical composition of this sealer (wash primer 1) as it presents, titanium (4,84%), a high chromium percentage (11,54%) and aluminum phosphates that are formed on the surface which increase the resistance of the passive aluminum film to the erosive effect. It should be noted that the performance of this system was better than each of the separate coatings (Al and Zn) [6–11].

(a) (b) (c) (d)

FIGURE 2: Thermal-sprayed coating of Zn/Al by the flame process (470 μm). (a) 12 months of exposure. (b) 24 months of exposure. (c) 36 months of exposure. (d) 48 months of exposure.

(a) (b) (c1) (c2)

FIGURE 3: (a) Thermal-sprayed coating of Zn/Al, Thickness = 445 μm. (b) Details of the edge flaw in the lower border. (c) Transversal cutting: details of the edge of the sample applied by the flame process (90° angle (c1) and acute (c2)), sample applied by electrical arch.

3.3. Behavior of the Thermal-Sprayed Systems in the Lake Maracaibo Crossing Station

3.3.1. Double-Layer System without Sealing. The environment in LMC Station is less aggressive (C_4) than in La Voz Station (C_5), which means that flaws will appear after a longer time of exposure. The thermal sprayed system with Zn/Al double layer without sealing, up to 36 months of evaluation, presented a good performance, without signs of corrosion at the surface substrate of the probe and with a good galvanic effect; it showed only a slight accumulation of environmental dust (Figure 5).

At 71 months of exposure, this coating began to surrender to the erosive attack, so some zinc corrosion products were noticed on the right edge (Figure 6), near the area of the border. This area, as it has been established in previous researches [6–9], is critical because coupons coated by the flame process presented an edge with a 90° angle which implies that the coating did not acquire the appropriate thickness as it was difficult to apply the product in that area; this opens the possibility for the aluminum fail to provide the necessary protection against the erosive effect and, therefore, to expose zinc, a material highly affected by the erosive action of the environment. The last inspection completed in this

(a) (b) (c) (d)

FIGURE 4: Thermal-sprayed coating of Zn/Al by flame process, sealed with wash primer (445 μm). (a) 12 months of exposure. (b) 24 months of exposure. (c) 36 months of exposure. (d) 48 months of exposure.

(a) (b) (c) (d)

FIGURE 5: Thermal-sprayed coating of Zn/Al by flame process (318 μm) (a) 12 months of exposure. (b) 24 months of exposure. (c) 36 months of exposure. (d) 41 months of exposure.

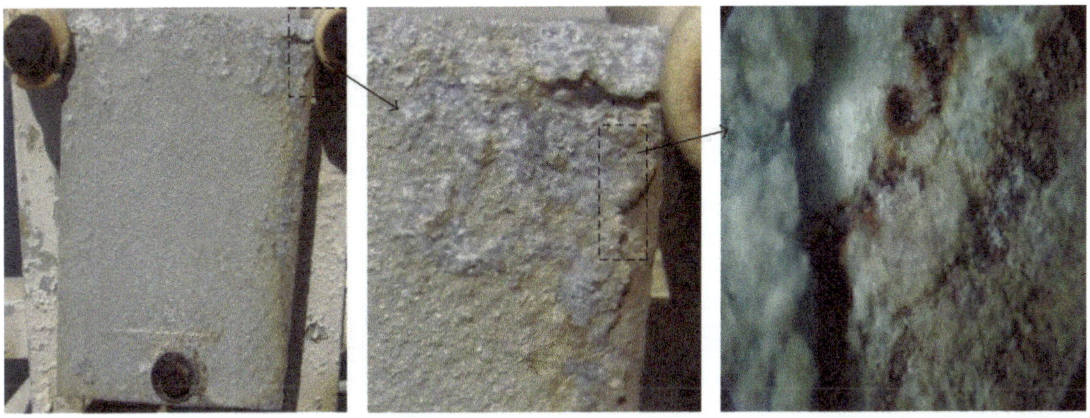

(a) (b) (c)

FIGURE 6: Thermal-sprayed coating of Zn/Al by flame process (318 μm). (a) 71 months of exposure. (b) Delamination detail of the coating in the right edge. Notice the formation of corrosion products of Zn (white color) and the substrate corrosion (brown color). (c) A magnifying stereoscopic view 45x.

Table 4: Elemental analysis: wash primer sealer manufacturer 1 and 2.

Chemical analysis (Weight %)	
Wash primer 1	Wash primer 2
Al = 23.31	Na = 6.49
P = 13.52	Al = 30.36
Cl = 3.24	Si = 15.63
Ca = 1.45	P = 18.59
Ti = 4.84	Cl = 12.38
Cr = 11.54	K = 2.46
Fe = 9.66	Ca = 1.66
Zn = 32.42	Cr = 1.69
—	Fe = 6.21
—	Zn = 4.53

environment was almost after six years of exposure; some substrate corrosion was observed as the coating at the right border was slightly starting to delaminate.

3.3.2. Double-Layer System Sealed with Wash Primer. After 71 months of exposure, the probe was practically intact, with an excellent appearance and galvanic effect (Figure 7). Of all the systems evaluated in the project, this one has had the best performance, only a slight accumulation of environmental dust, a product of the last months of drought.

At La Voz Station, and just as previously explained, wash primer 1 drastically improved the resistance of the coating, as it helped it to face these very aggressive environments from the corrosion-erosion point of view.

Through the evaluation of this project, from the 5 studied samples of this system in this environment, only one presented an edge flaw at the lower-left corner after 13 months of exposition (Figure 8), the crack advanced to a longitude of 3.5 cm.

In spite of this, and similarly to the results of the previous station, the performance of the system was better than the one of the other systems of Zn and Al separately (Figure 9) [6–11].

3.3.3. Analysis by Scanning Electron Microscope (SEM). During the different phases of the project, some cuttings were made in areas which presented good or problem conditions (e.g., with flaws at the edges) to analyze through SEM the integrity of the coating, its microstructure changes, the penetration of polluting agents, the morphology of corrosion products, and their corrosion mechanisms. One of the analyses is presented below, specifically, the one done in a sample after 36 months of exposure (Figure 10) at the LMC Station; the area analyzed is localized at the right edge; it was chosen so that the integrity of the coating in the proximities of the edge of incidence could be verified.

The elemental analysis in zone 1 shows an attack to the zinc layer (corrosion products/light gray color). In this area, two things can be observed: first, a high concentration of pollutants (Table 5); and, second, a group of imperfections or interconnected pores that form a kind of channels, through

Table 5: Chemical analysis for thermal-sprayed Zn/Al, double layer, without sealer.

Chemical analysis (weight %)			
Area 1	Area 2	Area 3	Area 4
Al = 10.42	Al = 3.09	Mn = 0.46	Al = 98.85
S = 6.79	S = 3.18	Fe = 90.67	Fe = 0.58
Cl = 8.74	Cl = 3.46	Zn = 8.72	Zn = 0.57
Fe = 0.50	Zn = 90.52	Ga = 0.15	—
Zn = 73.23	Ga = 0.26		—
Ga = 0.33			—

Table 6: Chemical analysis of the right edge.

Chemical analysis (weight %)
Al = 13.03
Si = 6.37
Cl = 0.76
K = 0.44
Ca = 1.08
Fe = 1.42
Zn = 76.89

which pollutants easily ingress toward the internal layers of the coating. On the other hand, in zone 2, there are high concentrations of chlorides and sulphur compounds. The high zinc content of the area suggests the presence of corrosion products of zinc. As observed in Figure 10, these zinc corrosion products are over the aluminum layer, as they are deposited on their discontinuities. Zone 3 is located in the interface area of steel/zinc (the furthest zone), as seen in Table 5, without the presence of pollutants for that time of exposition. However, in Figure 10, some microcracks and discontinuities of the zinc layer can be detected in the proximities of the zone; through them, aggressive agents could possibly ingress when reaching the level of the interface aluminum/zinc. Finally, in zone 4, the aluminum remains intact (Table 5), as expected; this element has remained in excellent conditions up to now.

The microscopic analysis revealed important aspects such as (a) differences in the thickness of the layers, which gave as a result a considerable aluminum thickness (200 μm approximately), clearly higher than zinc (100 μm); (b) interconnected porosity: there were appreciated some areas with interconnected pores through which aggressive agents entered through the zinc covered zone; (c) changes in zinc microstructure and penetration of polluting agents.

In relation to the analysis done on the right edge of the coupon (Figure 11), contrary to the observed in the border, there are two not-well-defined layers of zinc and aluminum. The chemical analysis (Table 6) shows that there is a low percentage of aluminum in the area, but it is rich in zinc. This inequality, and just as it was observed in the visual inspection, is responsible for the border flaws in these areas. Also, as it has been previously explained, it would be different if there was a uniform layer with a sufficient quantity of

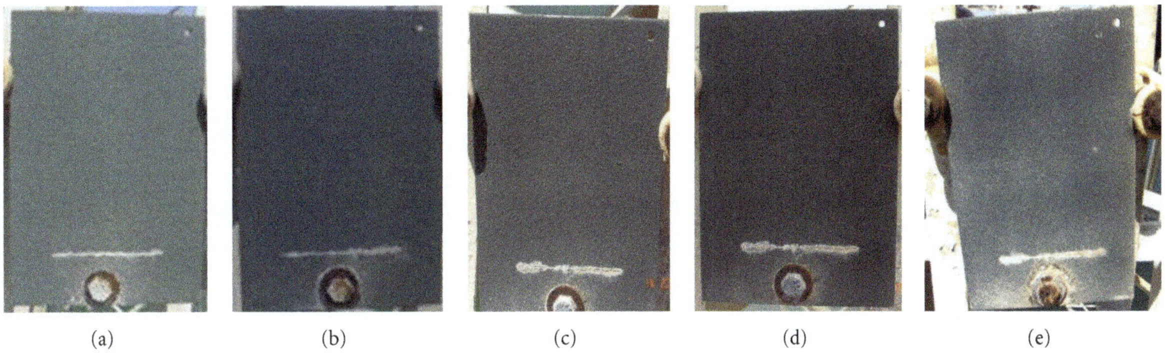

FIGURE 7: Thermal-sprayed coating of Zn/Al, sealed with wash primer 1 (345 μm). (a) 12 months of exposure. (b) 24 months of exposure. (c) 36 months of exposure. (d) 41 months of exposure. (e) 71 months of exposure.

FIGURE 8: Thermal-sprayed coating of Zn/Al/wash primer 1 (325 μm). Notice the flaw at the edge on the left corner.

aluminum. Furthermore, the chemical analysis of the area reveals the presence of silicon, chlorine, and calcium (Table 6), suggesting deposits of environmental dust.

3.4. Analysis by X-Rays Diffraction (XRD). The XRD assays were carried out for the unsealed and sealed systems with wash primer (Figures 12 and 13). In relation to the double-layer system without sealing (Figure 12), Bayerite was found as a main component (α-$Al(OH)_3$), results that coincide with the ones reported by Pombo et al. [21], who affirm that if aluminum is thermal-sprayed by the flame process, the formation of this oxide will prevail (formed on the mouthpiece of the pistol) in the surface of the test probe. Additionally, the presence of SiO_2 was detected as a result of the accumulation of environmental dust during the period of exposure in the LMC station.

The obtained spectrum for the double-layer system sealed with wash primer 1 (Figure 13) showed phosphates formation as a product of the reaction of the coating with the sealer. It is important to highlight that the wash primer is a type of coating that is commercially used as a promoter of adherence, made up of two components: a component A which is a resin of polyvinyl butyral with pigment of Zn chromate; and a component B that is an alcoholic solution of phosphoric acid. When the latter is applied on zinc or aluminum, it reacts with them and forms a fine film of crystals of the respective phosphates, with the purpose of sealing the pores left by the application process. It is necessary to highlight that the formed phosphates contribute with a higher resistance to erosion and to the ingress of polluting agents from the environment; in contrast to the unsealed systems, it offers an additional protective barrier to the surface of the probe.

Titanium was also detected in wash primer 1 (Table 4), which is present as TiO_2 in paintings and sealers with the purpose of adding solids to confer them a better capacity to seal the pores. This element was not detected in the wash primer 2, which may be the reason why the wash primer 1 had a much better performance.

4. Conclusions

(1) Thermal-sprayed coatings with Zn/Al (double layer) sealed with wash primer extend the service life of this system, besides improving the physical appearance of the test coupon.

(2) Along the six years of studies, it was found that La Voz Station is a more aggressive area than Lake Maracaibo Crossing Station.

Figure 9: Comparison of different sealed systems used at the LMC Station, after 41 months of exposure (a) Thermal sprayed coating of zinc by flame process (wire), sealed with wash primer 1 (173 μm). (b) Thermal sprayed coating of zinc by the electric arch process, sealed with wash primer 1 (264 μm). (c) Thermal sprayed coating of zinc by electric arch process, sealed with phenolic (261 μm) (66 months of exposure). (d) Thermal sprayed coating of aluminum by electric arch process with phenolic (261 μm). (e) Thermal sprayed coating of aluminum by electric arch process, sealed with wash primer 1 (251 μm). (f) Thermal sprayed coating of Zn/Al by the flame process, sealed with wash primer 1 (345 μm).

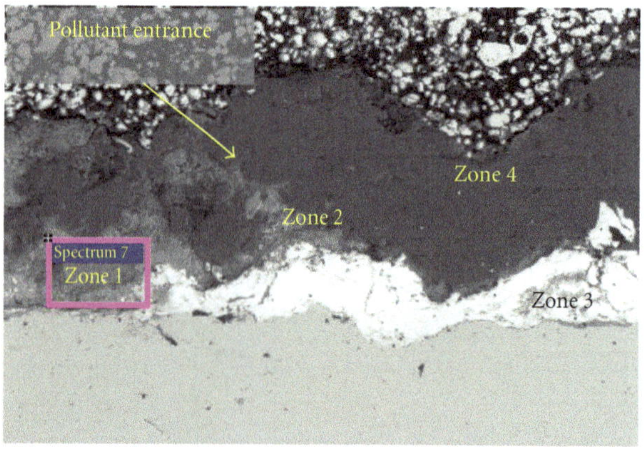

Figure 10: Thermal-sprayed Zn/Al double-layer, without sealer (thickness = 345 μm). View in the SEM. Area near the incidence edge.

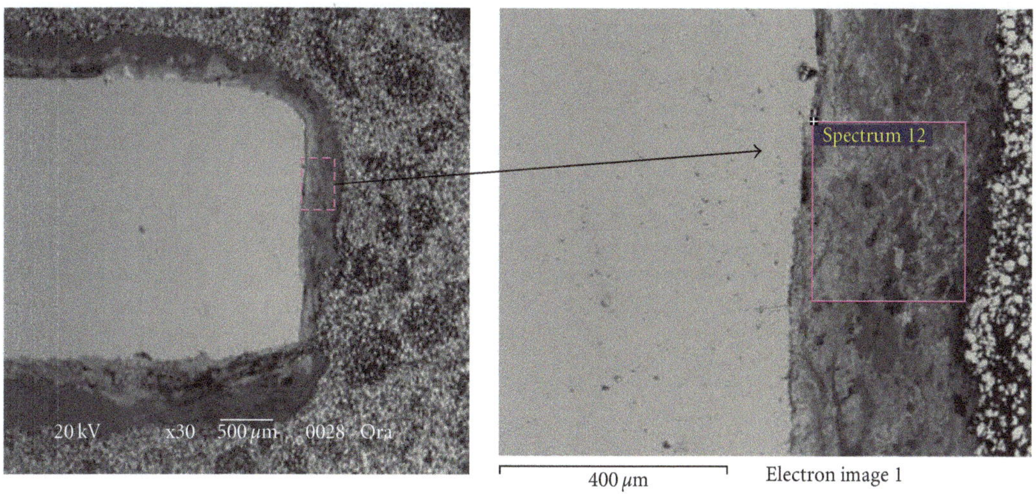

FIGURE 11: Thermal-sprayed Zn/Al double layer, without sealer. Thickness = 345.44 μm. View in the SEM. Right edge.

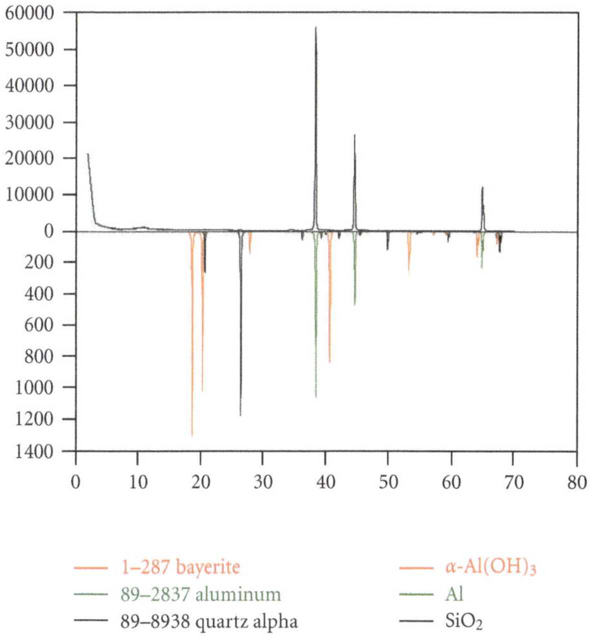

FIGURE 12: Spectrum of the double-layer system without sealer.

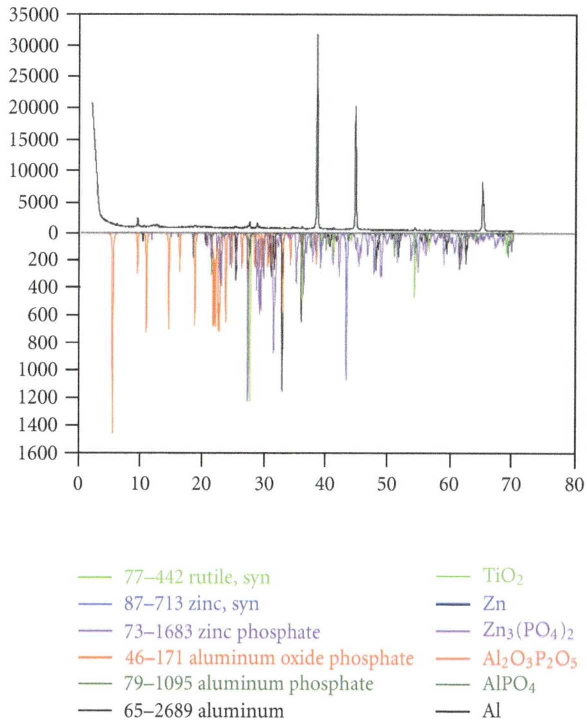

FIGURE 13: Spectrum of the double-layer system with sealer.

(3) The best performance in the sealed systems is due to the reaction of the wash primer with the Al (external layer of coating) of the test coupons, forming phosphates that contribute to a higher resistance to the polluting agents and erosive effects of the exposure environments; this resistance increases in the case of the wash primer 1, with the presence of TiO_2.

(4) The double-layer system with sealer represents a good protection option against the corrosion in marine and coastal marine environments with a high incidence of winds.

Acknowledgment

Thanks are due to CONDES (Consejo de Desarrollo Cientifico y Humanistico), from the Universidad del Zulia, for the economical support.

References

[1] S. Kuroda, J. Kawakita, M. Komatsu, T. Aoyaggi, and H. Saitoh, "Characterization of thermal sprayed Zn and Al coatings after 18 years' exposure in marine environment," Advancing the

Science & Applying the Technology. ASM International USA. pp 343–352, 2003.

[2] R. A. Kogler, D. Brydl, and C. Highsmith, "Recent FHWA experience with metallized coatings for steel bridges," *Materials Performance*, vol. 38, no. 4, pp. 43–45, 1999.

[3] American Welding Society. Corrosion Tests of Flame-Sprayed Coated Steel 19 Year Report, 1974, http://www.intmetl.com/19_year_report.htm.

[4] R. Kain, *Marine Atmospheric Corrosion Museum Report on the Performance of Thermal Spray Coatings on Steel*, American Society for Testing and Materials, Philadelphia, Pa, USA, 1987.

[5] M. Morcillo, M. Almeida, F. Fragata, and Z. Panossian, Corrosión y Protección de Metales en las Atmósferas de Ibero América. Parte II. Protección Anticorrosiva de Metales en las Atmósferas de Ibero América (Red Temática Patina, XV.D /CYTED) Madrid, España, 2002.

[6] O. Salas, *Comportamiento de los Recubrimientos por Termorociado de Aluminio y Cinc en Ambientes Marinos y Costeros Marinos*, Trabajo Especial de Grado Magíster Scientarium. Facultad de Ingeniería. Centro de Estudios de Corrosión. Universidad del Zulia, Maracaibo, Venezuela, 2005.

[7] A. Uzcátegui and A. Valdivieso, *Efecto del Ambiente Tropical en el comportamiento de Recubrimiento Termo-Rociados de Zn, Al, y Zn/Al*, Tesis de Grado. Facultad de Ingeniería. Centro de Estudios de Corrosión. Universidad del Zulia, Maracaibo, Venezuela, 2006.

[8] J. Godoy and N. Sualce, *Corrosión- Erosión de Recubrimientos Termorociados de cinc, aluminio, Cinc/Aluminio en Ambientes Marinos Tropicales*, Tesis de Grado. Facultad de Ingeniería. Centro de Estudios de Corrosión. Universidad del Zulia, Maracaibo, Venezuela, 2006.

[9] J. Atencio and L. Hernández, *Desempeño de Recubrimientos Termo-Rociados en Zinc y/o Aluminio en Ambientes Marinos y Tropicales*, Tesis de Grado. Facultad de Ingeniería. Centro de Estudios de Corrosión. Universidad del Zulia, Maracaibo, Venezuela, 2008.

[10] D. Rojas and A. Tosaya, *Resistencia a la Corrosión de Recubrimientos Termorociados de Zn y/o Al en Ambientes Marinos Tropicales*, Trabajo especial de grado. Facultad de Ingeniería. Centro de Estudios de Corrosión. Universidad del Zulia, Maracaibo, Venezuela, 2009.

[11] O. Salas, O. De Rincón, N. Romero et al., "Behavior of thermal-sprayed Al, Zn, and Zn/Al coatings in tropical marine environments," *Materials Performance*, vol. 47, no. 11, pp. 42–46, 2008.

[12] S. Orlando, T. De Rincón Oladis, R. Álvaro, S. Miguel, and L. Maldonado, "Long-term performance of thermal sprayed coatings in tropical marine environments," in *Proceedings of the International Corrosion Conference Series (NACE '07)*, March 2007.

[13] S. Orlando, T. De Rincón Oladis, R. Álvaro et al., "Thermal sprayed Zn/Al double layer: a protection alternative against steel corrosion in tropical marine atmospheres," in *Proceedings of the International Corrosion Conference Series (NACE '08)*, 2008.

[14] ASTM D 1186, *Standard Test Method for Nondestructive Measurement of Dry Film Thickness of Nonmagnetic Coatings Applied to a Ferrous Base*, ASTM International, West Conshohocken, Pa, USA, 2002.

[15] ASTM E 2109, *Test Methods for Determining Area Porcentage Porosity in Thermal Sprayed Coatings*, ASTM International, West Conshohocken, Pa, USA, 2001.

[16] ISO 4624, *Paints and Varnishes—Pull-Off Test for Adhesion*, ISO, Geneva, Switzerland, 2002.

[17] ASTM D 4541, *Pull-Off Adhesion Testing of Paint, Varnish and other Coatings and Films*, ASTM International, West Conshohocken, Pa, USA, 2002.

[18] M. Morcillo, E. Almeida, B. Rosales, J. Uruchurtu, and M. Marrocos, Corrosión y Protección de los Metales en la Atmósfera de Iberoamérica. Parte I. Mapas de Iberoamerica de Corrosividad atmosferica (Proyecto MICAT, XV. 1 /CYTED), Madrid, España, 1998.

[19] ISO 9225, *Corrosion of Metals, and Alloys—Corrosivity of Atmospheres—Measurement of Pollution*, ISO, Geneva, Switzerland, 1992.

[20] ISO 9223, *Corrosion of Metals, and Alloys—Corrosivity of Atmospheres—Classification*, ISO, Geneva, Switzerland, 1992.

[21] R. R. M. H. Pombo, R. S. C. Paredes, S. H. Wido, and A. Calixto, "Comparison of aluminum coatings deposited by flame spray and by electric arc spray," *Surface and Coatings Technology*, vol. 202, no. 1, pp. 172–179, 2007.

13

Evaluation of Corrosion Behavior of Galvanized Steel Treated with Conventional Conversion Coatings and a Chromate-Free Organic Inhibitor

Laura A. Hernandez-Alvarado,[1] Luis S. Hernandez,[2] and Sandra L. Rodriguez-Reyna[3]

[1] *Faculty of Chemical Sciences, Universidad Autonoma de San Luis Potosi, Avenida Salvador Nava No. 6, 78290 San Luis Potosi, SLP, Mexico*
[2] *Institute of Metallurgy, Universidad Autonoma de San Luis Potosi, Avenida Sierra Leona No. 550, 78210 San Luis Potosi, SLP, Mexico*
[3] *Faculty of Engineering, Universidad Autonoma de San Luis Potosi, Avenida Salvador Nava No. 8, 78290 San Luis Potosi, SLP, Mexico*

Correspondence should be addressed to Luis S. Hernandez, hersal@uaslp.mx

Academic Editor: Peter C. Okafor

Conventional weight loss tests and both DC and AC electrochemical techniques were used to study if an organic inhibitor containing an alkanolamine salt of a polycarboxylic acid can substitute toxic coatings as chromating and certain phosphating procedures in the protection of galvanized steel. The electrolyte used was a 0.5 M aerated NaCl solution. All tests gave concordant results, indicating that the chromate-free organic inhibitor does protect galvanized steel in this environment, even though the provided protection was less than that of the chromate conversion coating. It was observed that, after a moderate initial attack, the corrosion rate diminishes due to the appearance and growth of passivating corrosion products layers, mainly constituted by zinc hydroxychloride ($Zn_5(OH)_8Cl_2 \cdot H_2O$) and two varieties of zinc hydroxide, among other crystalline compounds.

1. Introduction

Conversion coatings are applied to galvanized steel to improve adhesion of additional protective coatings and for corrosion protection of the zinc coating. Phosphate conversion coatings (PCCs) provide adhesion but do not provide substantial corrosion protection. PCCs provide uniform surface texture and increased surface area, and, when used as a base for paint, they promote good adhesion, increase the resistance of the paint to humidity and water soaking, and eventually increase the corrosion resistance of the painted system [1]. Most galvanized steel used in manufacturing industries (car, household appliances, etc.) is phosphate coated and painted. However, some authors [2] have reported the harmful effects of phosphating, mainly those compositions that contain nickel. Chromate conversion coatings (CCCs) for zinc have been the most widely used, as they enhance bare or painted corrosion resistance, improve the adhesion of paint or other organic finishes, and provide the metallic surface with a decorative finish. CCCs are distinguished by their easy application, their applicability to a wide range of alloys and, in many cases, their ability to improve the galvanized corrosion resistance by virtue of a built-in inhibitor reservoir [3]. Although chromate is an excellent corrosion inhibitor, it is highly toxic; it has carcinogenic effects and must be handled and disposed of with extreme care. Therefore, there are severe restrictions on its use.

Much effort has been devoted to replace chromate chemicals with safe, nontoxic alternatives that are environmentally benign, and many environmental friendly coating systems are under development [4–7]. However, preparation and corrosion behavior of these alternatives is not clear and their practical usage is doubtful. The purpose of this paper is to evaluate the corrosion protection provided by a chromate-free organic inhibitor that is suggested as a nontoxic alternative to chromate and phosphate conversion coatings on galvanized steel. The water-soluble organic inhibitor is based on an alkanolamine salt of a polycarboxylic acid.

TABLE 1: Some characteristics of the tested coatings.

Coating	Chemical composition	pH	Application method
chromate	200 g potassium dichromate + 6 cc sulphuric acid in 1 L of water	0.8	dipping for 20 sec at room temperature
phosphate	21.0 g acidic zinc phosphate + 1.2 g sodium nitrite + 1.8 g triethanolamine oleate in 1 L of water	5.0	dipping for 600 sec at room temperature
inhibitor 5.0%	water-soluble corrosion inhibitor containing an alkanolamine salt of a polycarboxylic acid	8.0	dipping for 40 sec at 75°C

2. Experimental Procedure

Test specimens of 5 × 10 cm were cut from a mild steel sheet of 1 mm thickness and subsequently submerged in a molten zinc bath to obtain a galvanized coating of 1450 g/m^2, according to ASTM A 90 standard. The specimens were degreased in trichloroethylene, rinsed in distilled water, coated, rinsed again in distilled water, and then dried for seven days prior to submit them to the tests. The chemical composition and other characteristics of the tested coatings are described in Table 1.

The corrosion resistance and paint adhesion characteristics may generally be determined, in addition to the coating weight, by a characteristic color imparted by the chromium compounds formed on the surface of the zinc. After the chromating, the galvanized surface had a dark orange, almost brown, transparent color that made possible to see the zinc grains on the surface. These colored hexavalent chromium conversion coatings are generally considered to give the best overall corrosion protection, as well as poor electrical conductivity and excellent paint adhesion properties [5]. The samples treated with the phosphating solution got a dull black color with granular appearance. Surfaces treated with the inhibitor did not show any visible change at all. The water-soluble corrosion inhibitor contains an alkanolamine salt of a polycarboxylic acid. The tested concentration, application procedures, and elimination of corrosion products of the samples treated with this commercial product were made in strict accordance with the instructions of the manufacturer. For example, a borax solution (35 gL^{-1} at 65°C) was used for elimination of corrosion products. It is claimed that the inhibitor protects galvanized steel by sealing pores and other discontinuities present in the zinc coating. It is also suggested as a replacement of sodium nitride to protect ferrous metallic materials.

The protective effect of these treatments on galvanized steel was studied by immersion tests conducted in a naturally aerated 0.5 M NaCl solution, pH = 6.0, during 134 days and at room temperature. Panels in triplicate were withdrawn from the solution at different periods of time. In all tests, there were included untreated galvanized panels as blanks. The corrosion weight loss parameter was obtained after cleaning the surface of the panels in an acetic acid solution. Other specimens were examined after the same test periods, without stripping the corrosion products, by X-ray diffraction (*XRD*). The *XRD* patterns of the corroded samples were recorded using a diffractometer with a Cu K_α radiation of 1.54056 Å operated at 36 kV/30 mA. The samples were step-scanned in the 2θ range of 5° to 35° with a step size of 0.02 and a time step of 3 s.

In order to perform the electrochemical techniques, glass tubes were attached to the panel surface with a silicone sealer. The tube, that defined an interior exposed area of 10.3 cm^2, was filled with the 0.5 M NaCl solution. The cell contained a graphite electrode as the counter electrode and a saturated calomel electrode (sce) as reference. The open circuit potential (E_{ocp}) and the polarization resistance (Rp) were determined throughout the immersion time. E_{ocp} measurements were carried out previously to Rp. The Rp values were determined by imposing a pulse of ± 10 mV and recording the current system response. In addition, potentiodynamic polarization measurements were carried out at a scan rate of 120 mV/min, from −250 to +500 mV with respect to E_{ocp}. AC impedance measurements were made at the E_{ocp} by using a frequency response analyzer (FRA) operated under microcomputer control. The FRA was connected to the electrochemical cell through a potentiostat. A sinusoidal signal of 10 mV was applied over the frequency range from 100 kHz to 5 mHz. Impedance data were analyzed using a complex nonlinear least-squares program (CNLS). All electrochemical measurements were carried out in triplicate, at room temperature and with no deaeration or stirring of the electrolyte.

3. Results

3.1. Microstructural Characterization of the Coatings. The zinc coating weight of 1450 g/m^2 corresponds to an approximate thickness of 101 μm per side (Figure 1). Figures 2(a) through 2(c) show the backscattered scanning electron micrographs of the untreated and treated galvanized steel samples. The characteristic aspect of microcracking due to dehydration and the shrinkage of the coating during drying are visible in the chromated samples (Figure 2(a)). The surface of the phosphated samples appeared totally covered by a compact layer of small crystals (hopeite) that favors a firm anchorage of organic coatings (Figure 2(b)). Untreated galvanized samples had a very smooth surface and only appeared the limits of the zinc grains, scratches, and white points that correspond to lead particles (Figure 2(c)). The surfaces of the samples treated with the inhibitor (not included here) did not show any difference in comparison with the untreated galvanized ones.

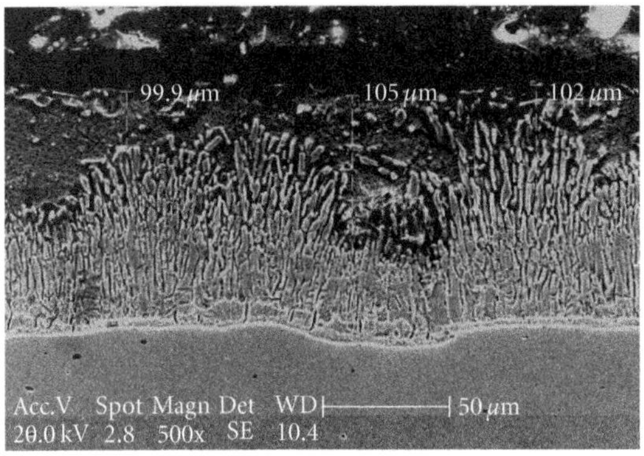

Figure 1: Scanning electron micrograph of the cross section of galvanized steel showing various Zn-Fe alloy layers and the thickness measurements.

3.2. Weight Loss and Electrochemical Measurements. Figure 3 shows the gravimetric results of the considered materials immersed in a 0.5 M NaCl solution up to 134 days. Untreated and treated galvanized steel showed very low corrosion rates, less than 10 mg/cm^2 ($\sim$38 μm/y), which are within a range considered as excellent (25–100 μm/y) according to a widely used classification [8]. Furthermore, the corrosion rate values obtained for untreated galvanized steel are approximate to those previously reported by other authors using similar NaCl solutions [9]. It is clearly observed that the inhibitor and the phosphating protected in a moderate way the galvanized steel, although the highest protection is shown by the chromated samples. It is worth mentioning the less steep curve slope of the chromated samples. Regarding the process kinetics, in the early immersion days the four curves show a moderate corrosion rate that diminished as the exposure time elapsed, indicating that corrosion rate was governed by diffusion processes. As the immersion time increased, a white layer of corrosion products appeared over the sample's surfaces, except for the chromate ones; it restricted the access of the solution to the metallic zinc. The chromated samples did not show any visible deterioration.

Regarding the *Rp* variation as function of time (Figure 4), there is an evident difference between the higher values of the chromated samples and the lower values of the other surface conditions. Even at the end of the immersion period, the chromated samples values are still very high compared to those of the other surface conditions (<2500 $\Omega\cdot$cm^2). In general, all samples depicted a very similar performance. In the case of phosphated samples, the *Rp* values were initially low, later they increased with the immersion time, and finally, after 110 days of immersion, they decreased again. This indicates that corrosion exists at the beginning of the immersion and later the *Rp* values increase due to the plugging of the spaces among the hopeite crystals with corrosion products. Afterwards, the movement of the electrolyte during the measurements uncovers the holes again. This behavior was congruent with the evolution of the E_{ocp} values. The *Rp* data were obtained in triplicate; the tendency shown in Figure 4 is representative of the three samples tested for each surface condition.

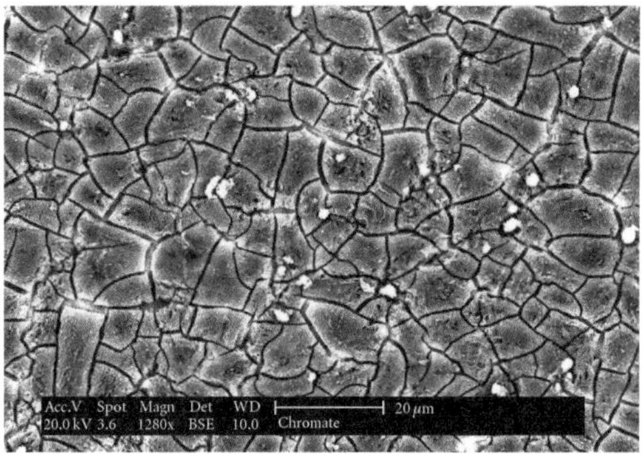

(a)

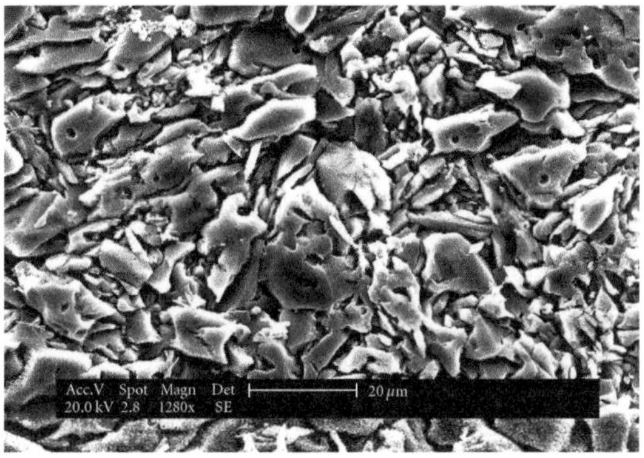

(b)

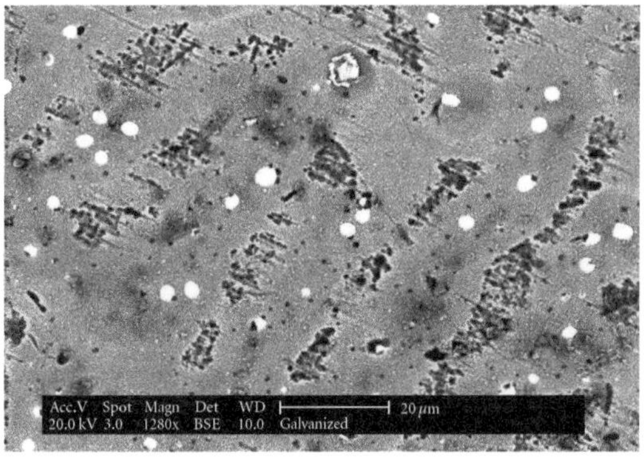

(c)

Figure 2: Backscattered scanning electron (BSE) micrographs of the surface conditions of galvanized steel: (a) chromated 1280x, (b) phosphated 1280x, and (c) untreated 1280x.

The E_{ocp} of the four surface conditions exhibited initial values around the known free open circuit potential of uncoated zinc in aerated saline solutions (-900 to -1100 mV$_{sce}$), presenting afterwards two trends (not shown): a permanency (plateau) in this potential range until the 134th day (chromated and inhibitor-treated samples) or a displacement towards nobler potentials as time elapses (phosphated and untreated samples). For these last two surface conditions, the increase of the E_{ocp} was due to the fact that the system evolved towards the uncoated steel E_{ocp} in this environment, as was demonstrated by the presence of abundant rust on the samples at the end of the immersion test.

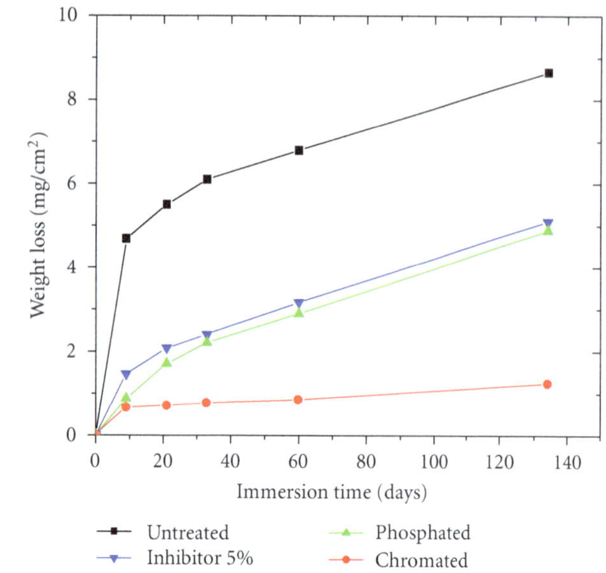

FIGURE 3: Corrosion performance of galvanized steel with various surface conditions immersed in a 0.5 M NaCl solution during 134 days.

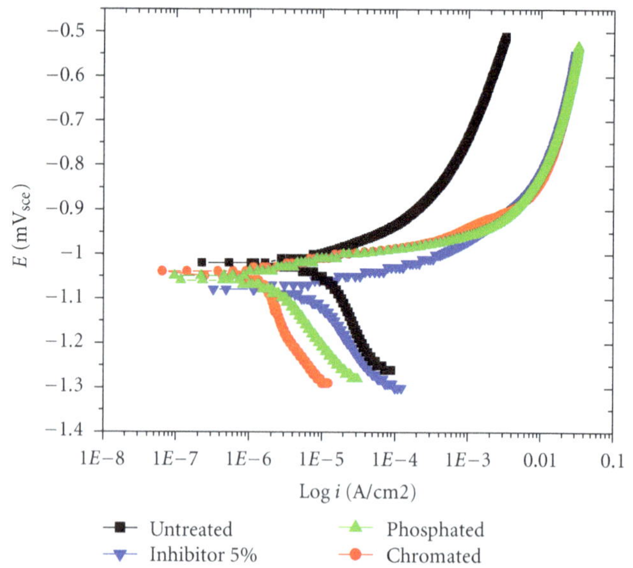

FIGURE 5: Potentiodynamic polarization curves of galvanized steel with various surface conditions immersed in a 0.5 M NaCl solution during 1 h.

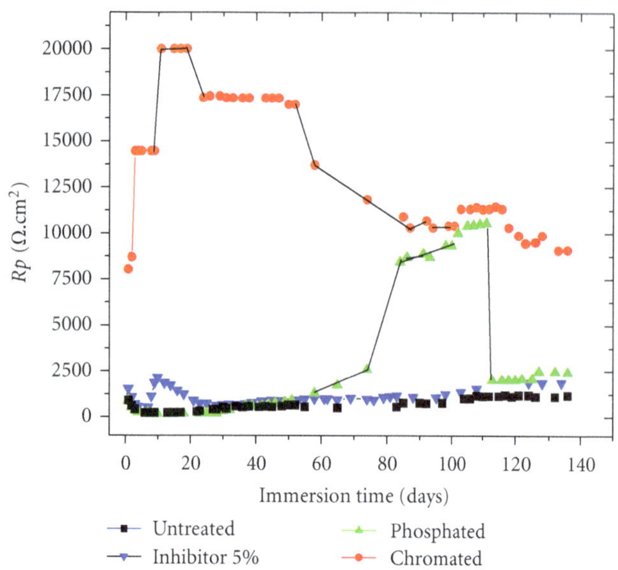

FIGURE 4: Evolution of *Rp* of galvanized steel with the exposure time with various surface conditions immersed in a 0.5 M NaCl solution.

Figure 5 and Table 2 disclose the polarization curves and electrochemical data of the galvanized steel conditions after immersion in the NaCl solution for 1 h. The curves for each surface condition were obtained at least in duplicate. The polarization scanning was made in the vicinities of the Tafel linear regions in order to calculate numeric values of the corrosion rate. The E_{ocp} of all the samples range from −997 to −1050 mV$_{sce}$; the nobler values belong to the phosphate samples and the more negative values correspond to the chromate ones.

All the anodic curves are very similar to each other as well as there is a great similarity of all the cathodic curves. The anodic branches show active zinc dissolution and a monotonically increase of current up to very steep slopes, which indicates a diffusion-controlled reaction rate. These branches also show that the different surface treatments increased the anodic dissolution rate with respect to untreated galvanized. This unexpected behavior may be due to the untreated sample's air drying after immersion in the NaCl solution the day previous to the potentiodynamic polarization measurements. It has been reported that air drying after immersion in a 5% NaCl solution drastically reduces the corrosion current of a zinc material. This decrease is attributed to the formation of a more compact corrosion products film as a result of drying [10]. In the cathodic curves, however, there exist a clear reduction in the values of the current density, from the untreated sample to the chromated one, which is exactly of an order of magnitude and that can be confirmed with the i_{corr} values from Table 2, calculated with the potentiostat software. It is worth mentioning the low i_{corr} values of all surface conditions and the fact that anodic slopes values decrease with corrosion rate.

The inhibitive action of chromating is manifested in this decrease of the oxygen rate reduction and has been previously reported by some authors [11, 12]. Moreover, the position of the extreme surface conditions, corresponding to the fastest and the slowest oxygen rate reduction (untreated and chromating, respcting) coincide with the results of weight loss test (Figure 3).

The impedance measurements were taken 1 and 168 h after the electrolyte had made contact with the sample surface. Typical Nyquist impedance diagrams are shown in Figures 6(a) to 6(c) for the phosphated, chromated and untreated galvanized samples after 1 h, respectively, whereas Figure 6(d) depicted the diagram for an untreated sample after 168 h. In these figures, the experimental diagrams are

TABLE 2: Electrochemical data of the tested materials after immersion in 0.5 M NaCl for 1 h.

Condition	β_a (mV/decade)	β_c (mV/decade)	i_{corr} (μA/cm^2)	E_{oc}/E_{corr} (mV$_{sce}$)	Corrosion rate (mpy)	Ch Sq
Untreated	88	367	11.9	−1033/−1050	7.0	$3.67e^{-6}$
Inhibitor 5%	44	224	6.9	−1046/−1080	4.0	$3.75e^{-6}$
Phosphated	52	201	1.8	−1010/−1020	1.0	$4.07e^{-6}$
Chromated	35	278	1.1	−1050/−1040	0.7	$4.09e^{-6}$

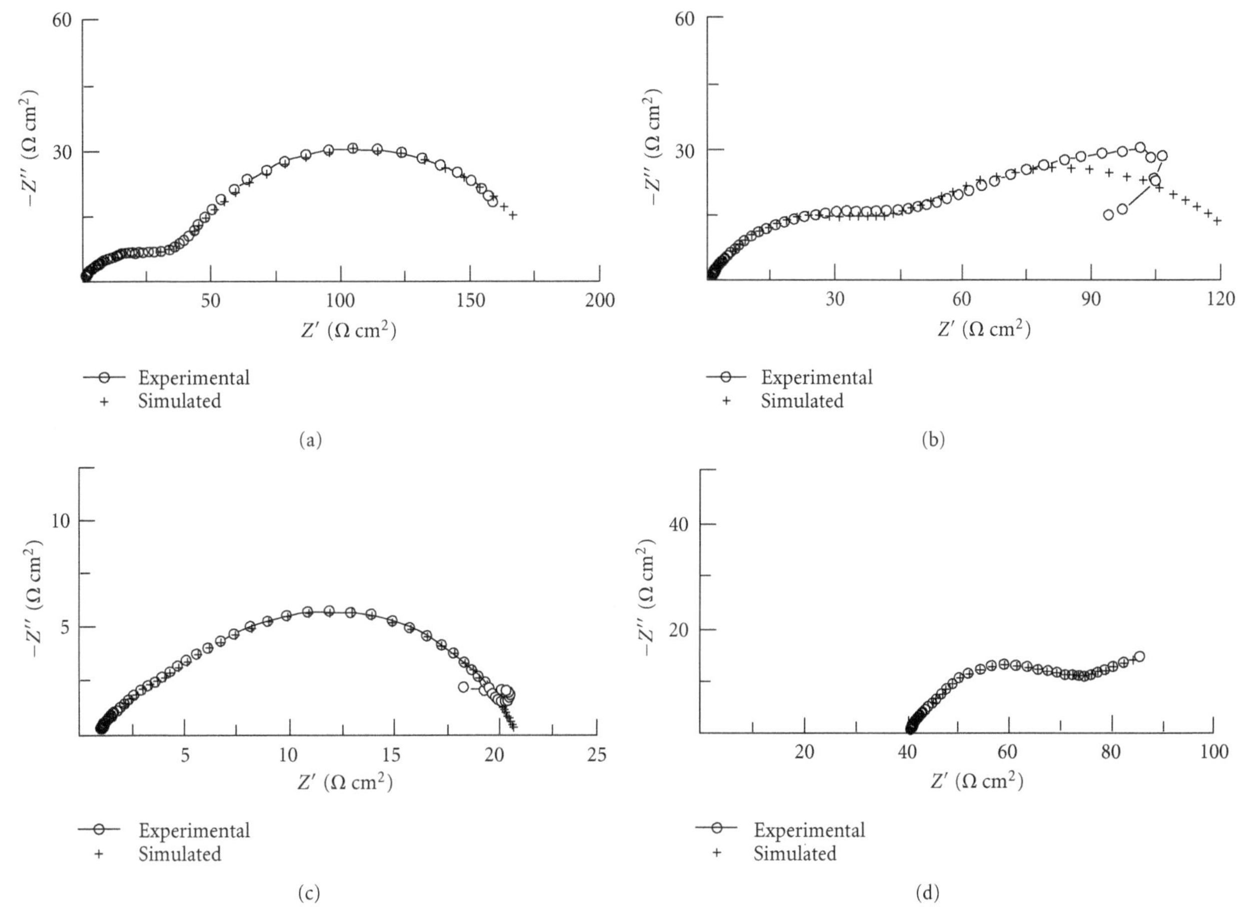

FIGURE 6: Experimental and simulated Nyquist impedance diagrams of galvanized steel with various surface treatments immersed in a 0.5 M NaCl solution during 1 h (a) phosphated, (b) chromated, (c) untreated, and (d) untreated immersed during 168 h.

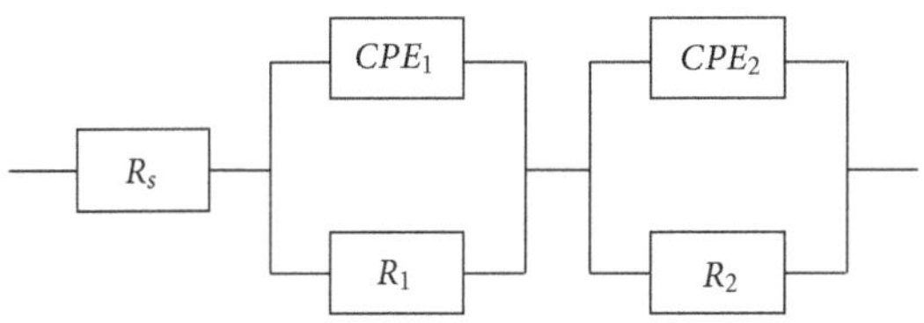

FIGURE 7: Equivalent electric circuit model used to fit the impedance data for treated and untreated galvanized samples.

overlapped with the simulation ones; these were obtained using a nonlinear least-squares fitting analysis procedure with the equivalent electric circuit model shown in the Figure 7. This circuit showed the best adjustment to the coated samples impedance data: R_s is the resistance of the electrolyte, CPE_1 is a constant-phase element related to the capacitance of the coating, R_1 is the coating pore resistance due to electrolyte penetration, CP_2 is a constant-phase element related to the double-layer capacitance and diffusion processes [13], and R_2 is the charge transfer resistance. CPE_1 and R_1 are associated with the zinc coating on steel in the untreated galvanized samples. The diagrams have different shapes and values at the low frequency limit. The phosphated and chromated samples exhibited larger values than the untreated ones. The samples treated with the organic inhibitor showed impedance diagrams very similar to those of untreated samples and are not shown. In general, the impedances measured after 1 h decreased after 168 h of contact with the electrolyte, except for the chromated condition.

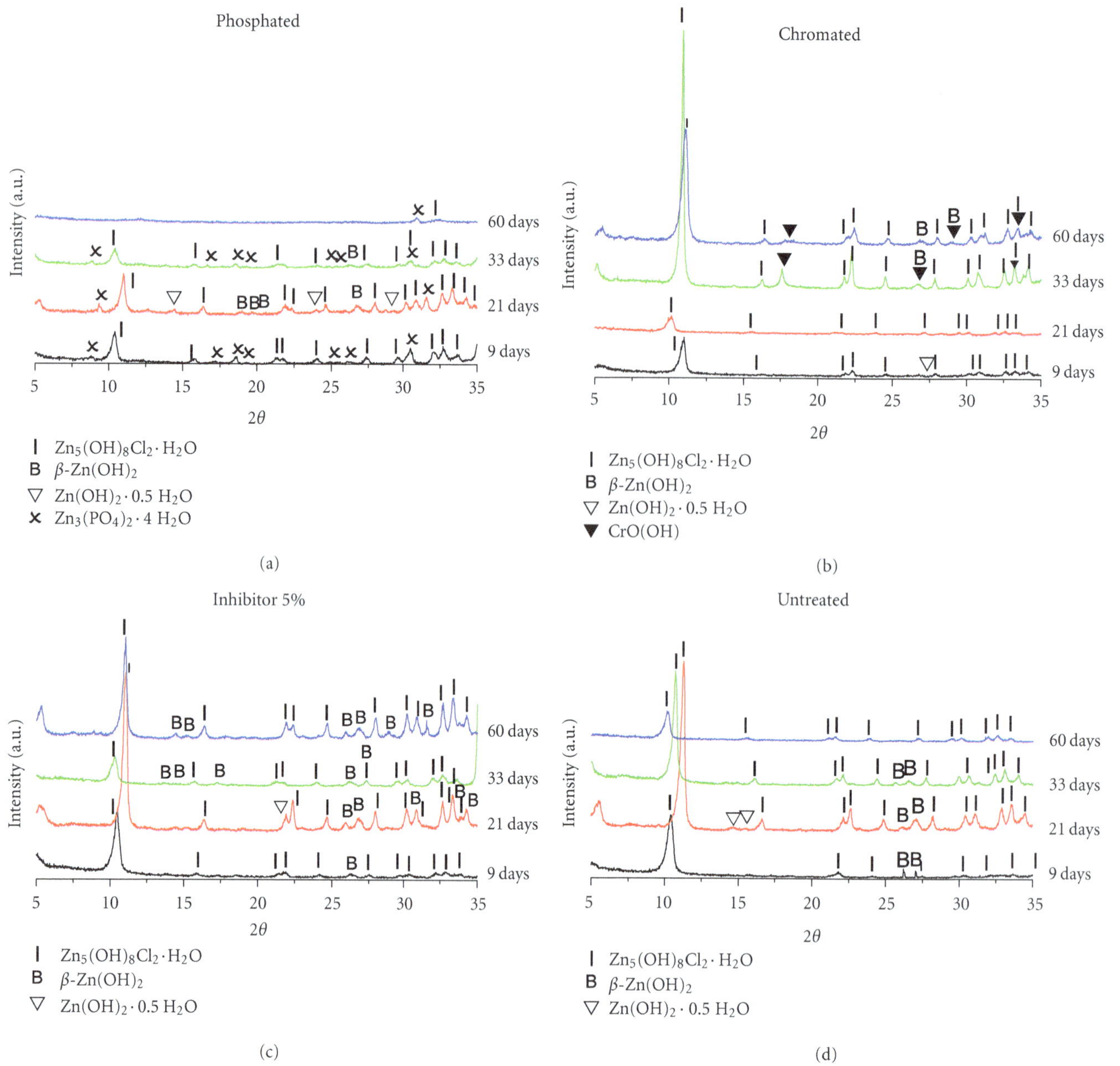

FIGURE 8: XRD spectra of corrosion products obtained after different immersion periods.

According to the interpretation of this type of complex plane plots, the samples after one hour of immersion showed two frequency-dependent components: an initial high frequency semicircle, less defined for the untreated sample (Figure 6(c)), and a second component clearly defined in the low frequency range. This component is a greater second semicircle for phosphated and untreated surfaces and seemingly a diffusion tail (Warburg impedance) for chromate sample. It is generally accepted that both second components (semicircle or diffusion tail) appear as a consequence of a diffusion controlled corrosion process, related mainly with the oxygen reduction. An equivalent electric circuit for plots like these assumes that the diffusion process is much slower than the charge-transfer reaction and that diffusion is rate-controlling [14].

For untreated galvanized steel samples, the size of the second semicircle decreases as the immersion time increases and a capacitive branch appears in the low frequency range. Concerning the low frequency end observed in Figures 6(b) and 6(c), this behavior has been attributed, namely, to a potential drift [15]; this "tail back" is due probably to the effect of changing potential on the impedance response.

3.3. XRD of Corrosion Products and Salt Spray Test. Figure 8 presents the *XRD* spectra of corrosion products, the chemical composition of the detected compounds and the periods when they appeared. In all the samples, was identified zinc hydroxychloride ($Zn_5(OH)_8Cl_2 \cdot H_2O$), known as Simonkolleite, and in most of the samples at least two varieties of $Zn(OH)_2$ were detected.

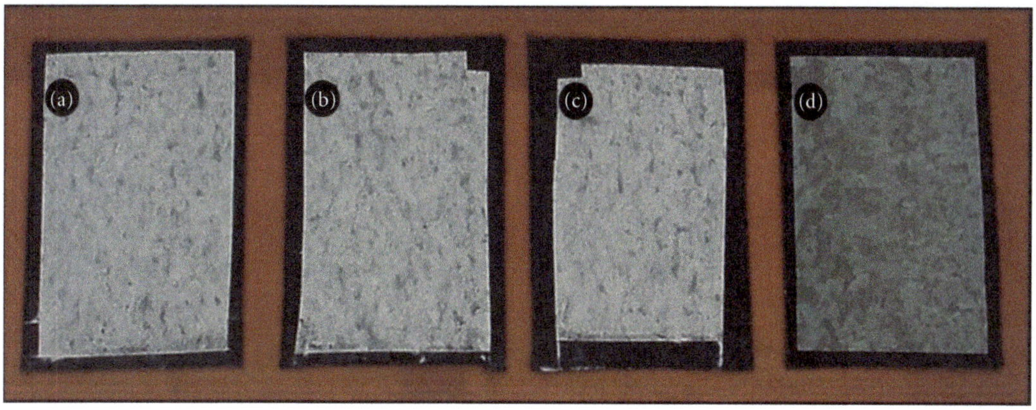

FIGURE 9: Aspects of the surface conditions of the galvanized steel after 168 h in the saline fog chamber: (a) untreated, (b) inhibitor 5%, (c) phosphated, (d) chromated. The whitish aspect of the first three samples is due to the zinc corrosion products.

Hopeite ($Zn_3(PO_4)_2 \cdot 4H_2O$) was detected in all the phosphated samples analyzed. Bracewellite ($CrO(OH)$) appeared only after 33 and 60 days of immersion on the surfaces of the chromated samples. It is a compound supposed to exist in accelerated chromium chromate coatings [16]. The peaks displacement in de *XRD* spectra are due to an inadequate sample alignment.

The presence of $CrO(OH)$ after 33 days of immersion is due to the dissolution of the coating's amorphous hexavalent compounds, remaining an insoluble trivalent chromium compounds layer. Composition depth profiles indicate that the hexavalent chromium compounds exist predominantly in the outer portion of the coating [16, 17]. The ß-$Zn(OH)_2$ seems also to be present only in two specific periods, since the composition of zinc corrosion products in chloride solutions depends on the exposure time, so the hydroxides exists just temporarily, becoming hydroxychlorides as immersion time elapses, as it will be explained later.

Samples of the different surface conditions of galvanized steel were submitted to a salt spray test applying the ASTM B 117 method, which was carried out in triplicate. The appearance of the samples was evaluated at different time intervals with a maximum of 168 hours. By then all samples, except the chromated ones, were totally covered by white corrosion products and also presented incipient red corrosion. The chromated samples, on the other hand, appeared lustrous and only in one of them the dark orange color had disappeared (Figure 9).

4. Discussion

In the immersion test, the high initial corrosion rates, except for the chromated samples, are a consequence of the anodic dissolution of treated or untreated galvanized steel before forming the corrosion products. As long as immersion time increased, a white layer of corrosion products appeared over all samples' surfaces, except on the chromated ones, and the high initial values of corrosion rate decreased. These white films, which resulted in being protector, got thicker with time, causing the corrosion rate to decrease even more. Regarding the chromate samples, it has been reported that, as far as the surface preserves its dark orange, almost brown aspect, Cr^{6+} ions have not been leached and the coating is still protecting the zinc substrate [18].

With regard to this white corrosion products layer, Feitknecht [19] previously reported the presence of zinc hydroxychloride and zinc hydroxide in immersion tests in a NaCl solution of the same concentration as that used in this study, and they also appear in the reaction sequence for zinc exposed in natural marine atmospheres [20]:

$$Zn \longrightarrow Zn(OH)_2 \longrightarrow ZnCl_2 \cdot nZn(OH)_2 \qquad (1)$$

with $n = 4$ for Simonkolleite. Several explanations for the apparent corrosion-inhibiting nature of Simonkolleite have been proposed, trying to justify the fact that the higher the amount of Simonkolleite in the corrosion products, the higher the resistance to corrosion [21]: (1) it has been considered that it is a more compact corrosion product than ZnO because the *c* axis of the hexagonal close-packed structure of Simonkolleite is parallel to the film corrosion products growth direction, (2) it probably decreases the oxygen reduction rate on itself, (3) that is due to its low solubility product (8.2×10^{-76}). On the other hand, crystalline and amorphous varieties of $Zn(OH)_2$ have been detected using cyclic voltammetry in chloride containing electrolytes [22]. It has been reported that the amorphous variety has a higher solubility product and is metastable.

The impedance diagrams, whose low frequency semicircle or tail may indicate a finite thickness layer diffusion process, related mainly with the oxygen reduction, confirmed that the corrosive process is governed by diffusion, as it is also indicated by the weight loss and *DRX* results. According to Barranco et al. [23], it is likely that a complex mass transport mechanism intervenes in the system formed by the corrosive medium/corrosion products/metal, both in the liquid phase and through the corrosion products layer that coats the metallic surface.

Finally, the *Rp* values of the samples, except for the chromated ones, were too low to ensure an acceptable protection. These values were inferior to 2500 $\Omega \cdot cm^2$ during the test period and did not increase at the end of the test in spite of the rust presence in the phosphated and untreated galvanized samples.

5. Conclusions

The organic inhibitor does protect galvanized steel in a 0.5 M NaCl solution, even though the higher protection was provided by the chromate conversion coating. So, if the organic inhibitor will substitute chromating, other considerations should be kept in mind. The quiescent nature of the electrolyte definitely contributed to this protection, because it allowed the corrosion products layers to remain on the surface during the immersion time.

The corrosion rate of galvanized steel, with the different surface conditions, decreased with time. This decrease was associated with the formation and growth of adherent, white corrosion products layers on the metallic surface, mainly composed of zinc hydroxychloride and zinc hydroxide. These layers limited the dissolved oxygen access to the treated or untreated surfaces of galvanized steel, causing the processes in the surface to be controlled by diffusion.

Regarding the chromate conversion coating, even though there was not a great amount of white corrosion products films, they were detected in XRD spectra. The higher protection provided by the this conversion coating to galvanized steel was mainly due to its own nature and it was confirmed by the polarization resistance results, and by the notable inhibition of the oxygen reduction reaction observed in the cathodic potentiodynamic polarization branches.

References

[1] X. G. Zhang, *Corrosion and Electrochemistry of Zinc*, Plenum Press, New York, NY, USA, 1996.

[2] J. F. McIntyre and R. J. Brent, "Tendencies of the pretreatment technology in next millennium," *Pinturas y Acabados Industriales*, vol. 42, no. 261, pp. 36–44, 2000.

[3] R. G. Buchheit, S. B. Mamidipally, P. Schmutz, and H. Guan, "Active corrosion protection in Ce-modified hydrotalcite conversion coatings," *Corrosion*, vol. 58, no. 1, pp. 3–14, 2002.

[4] S. A. Furman, "Novel coating technology replaces chromate in zinc galvanizing process," *Materials Performance*, vol. 49, no. 1, pp. 10–12, 2010.

[5] J. W. Bibber, "Chromium-free conversion coatings for zinc and its alloys," *Journal of Applied Surface Finishing*, vol. 2, no. 4, pp. 273–275, 2009.

[6] T. Peng and R. Man, "Rare earth and silane as chromate replacers for corrosion protection on galvanized steel," *Journal of Rare Earths*, vol. 27, no. 1, pp. 159–163, 2009.

[7] A. D. King and J. R. Scully, "Sacrificial anode-based galvanic and barrier corrosion protection of 2024-T351 by a MG-rich primer and development of test methods for remaining life assessment," *Corrosion*, vol. 67, no. 5, Article ID 055004, 22 pages, 2010.

[8] M. G. Fontana, *Corrosion Engineering*, McGraw-Hill, Singapore, 3rd edition, 2008.

[9] M. Toga, P. Moyo, and D. J. Simbi, "Corrosion of galvanized steel in simulated mine water environments," in *Proceedings 15th International Corrosion Congress*, paper 231, 9 pages, Granada, Spain, 2002.

[10] M. Sagiyama, A. Hiraya, and T. Watanabe, "Electrochemical behavior of electrodeposited zinc-iron alloys in 5%NaCl solution," *Journal of the Iron and Steel Institute of Japan*, vol. 77, no. 2, pp. 244–250, 1991.

[11] R. L. Zeller and R. F. Savinell, "Interpretation of A.C. impedance response of chromated electrogalvanized steel," *Corrosion Science*, vol. 26, no. 5, pp. 389–399, 1986.

[12] A. A. O. Magalhaes, I. C. P. Margarit, and O. R. Mattos, "Electrochemical characterization of chromate coatings on galvanized stell," in *Proceedings of the International Symposium on Electrochemical Impedance Spectroscopy (EIS '98)*, pp. 239–241, Rio de Janeiro, Brazil, 1998.

[13] F. Deflorian, V. B. Miskovic-Stankovic, P. L. Bonora, and L. Fedrizzi, "Degradation of epoxy coatings on phosphatized zinc-electroplated steel," *Corrosion*, vol. 50, no. 6, pp. 438–446, 1994.

[14] C. Lin, T. Nguyen, and M. E. McKnight, "Relation between AC impedance data and degradation of coated steel. 1. Effects of surface roughness and contamination on the corrosion behavior of epoxy-coated steel," *Progress in Organic Coatings*, vol. 20, no. 2, pp. 169–186, 1992.

[15] L. M. Callow and J. D. Scantlebury, "Electrochemical impedance on coated metal electrodes. Part1: Polarization effects," *JOCCA*, vol. 64, no. 2, pp. 83–87, 1981.

[16] R. G. Buchheit and A. E. Hughes, "Chromate and chromate-free conversion coatings," in *Corrosion: Fundamentals, Testing and Protection*, vol. 13A of *ASM Handbook*, pp. 720–35, ASM International, Materials Park, Ohio, USA, 2003.

[17] I. Suzuki, *Corrosion-Resistant Coatings Technology*, Marcel Dekker, New York, NY, USA, 1989.

[18] W. Zhang and R. G. Buchheit, "Effect of ambient aging on inhibition of oxygen reduction by chromate conversion coatings," *Corrosion*, vol. 59, no. 4, pp. 356–362, 2003.

[19] W. Feitknecht, "Studies on the influence of chemical factors on the corrosion of metals," *Chemistry and Industry*, vol. 36, pp. 1101–1109, 1959.

[20] V. Kucera and E. Mattsson, "Atmospheric corrosion," in *Corrosion Mechanisms*, F. Mansfeld, Ed., pp. 211–284, Marcel Dekker, New York, NY, USA, 1987.

[21] X. G. Zhang, *Corrosion and Electrochemistry of Zinc*, Plenum Press, New York, NY, USA, 1996.

[22] C. M. Rangel and L. F. Cruz, "Zinc dissolution in lisbon tap water," *Corrosion Science*, vol. 33, no. 9, pp. 1479–1493, 1992.

[23] V. Barranco, S. Feliu Jr., and S. Feliu, "EIS study of the corrosion behaviour of zinc-based coatings on steel in quiescent 3% NaCl solution. Part 1: directly exposed coatings," *Corrosion Science*, vol. 46, no. 9, pp. 2203–2220, 2004.

14

Green Inhibitors for Corrosion Protection of Metals and Alloys: An Overview

B. E. Amitha Rani and Bharathi Bai J. Basu

Surface Engineering Division, CSIR-National Aerospace Laboratories, Bangalore 560037, India

Correspondence should be addressed to B. E. Amitha Rani, amitharani@nal.res.in

Academic Editor: Ali Y. El-Etre

Corrosion control of metals is of technical, economical, environmental, and aesthetical importance. The use of inhibitors is one of the best options of protecting metals and alloys against corrosion. The environmental toxicity of organic corrosion inhibitors has prompted the search for green corrosion inhibitors as they are biodegradable, do not contain heavy metals or other toxic compounds. As in addition to being environmentally friendly and ecologically acceptable, plant products are inexpensive, readily available and renewable. Investigations of corrosion inhibiting abilities of tannins, alkaloids, organic,amino acids, and organic dyes of plant origin are of interest. In recent years, sol-gel coatings doped with inhibitors show real promise. Although substantial research has been devoted to corrosion inhibition by plant extracts, reports on the detailed mechanisms of the adsorption process and identification of the active ingredient are still scarce. Development of computational modeling backed by wet experimental results would help to fill this void and help understand the mechanism of inhibitor action, their adsorption patterns, the inhibitor-metal surface interface and aid the development of designer inhibitors with an understanding of the time required for the release of self-healing inhibitors. The present paper consciously restricts itself mainly to plant materials as green corrosion inhibitors.

1. Introduction

Corrosion is the deterioration of metal by chemical attack or reaction with its environment. It is a constant and continuous problem, often difficult to eliminate completely. Prevention would be more practical and achievable than complete elimination. Corrosion processes develop fast after disruption of the protective barrier and are accompanied by a number of reactions that change the composition and properties of both the metal surface and the local environment, for example, formation of oxides, diffusion of metal cations into the coating matrix, local pH changes, and electrochemical potential. The study of corrosion of mild steel and iron is a matter of tremendous theoretical and practical concern and as such has received a considerable amount of interest. Acid solutions, widely used in industrial acid cleaning, acid descaling, acid pickling, and oil well acidizing, require the use of corrosion inhibitors in order to restrain their corrosion attack on metallic materials.

2. Corrosion Inhibitors

Over the years, considerable efforts have been deployed to find suitable corrosion inhibitors of organic origin in various corrosive media [1–4]. In acid media, nitrogen-base materials and their derivatives, sulphur-containing compounds, aldehydes, thioaldehydes, acetylenic compounds, and various alkaloids, for example, papaverine, strychnine, quinine, and nicotine are used as inhibitors. In neutral media, benzoate, nitrite, chromate, and phosphate act as good inhibitors. Inhibitors decrease or prevent the reaction of the metal with the media. They reduce the corrosion rate by

(i) adsorption of ions/molecules onto metal surface,

(ii) increasing or decreasing the anodic and/or cathodic reaction,

(iii) decreasing the diffusion rate for reactants to the surface of the metal,

(iv) decreasing the electrical resistance of the metal surface.

(v) inhibitors that are often easy to apply and have *in situ* application advantage.

Several factors including cost and amount, easy availability and most important safety to environment and its species need to be considered when choosing an inhibitor.

2.1. Organic Inhibitors. Organic inhibitors generally have heteroatoms. O, N, and S are found to have higher basicity and electron density and thus act as corrosion inhibitor. O, N, and S are the active centers for the process of adsorption on the metal surface. The inhibition efficiency should follow the sequence O < N < S < P. The use of organic compounds containing oxygen, sulphur, and especially nitrogen to reduce corrosion attack on steel has been studied in some detail. The existing data show that most organic inhibitors adsorbed on the metal surface by displacing water molecules on the surface and forming a compact barrier. Availability of nonbonded (lone pair) and p-electrons in inhibitor molecules facilitate electron transfer from the inhibitor to the metal. A coordinate covalent bond involving transfer of electrons from inhibitor to the metal surface may be formed. The strength of the chemisorption bond depends upon the electron density on the donor atom of the functional group and also the polarizability of the group. When an H atom attached to the C in the ring is replaced by a substituent group ($-NH_2$, $-NO_2$, –CHO, or –COOH) it improves inhibition [4]. The electron density in the metal at the point of attachment changes resulting in the retardation of the cathodic or anodic reactions. Electrons are consumed at the cathode and are furnished at the anode. Thus, corrosion is retarded. Straight chain amines containing between three and fourteen carbons have been examined. Inhibition increases with carbon number in the chain to about 10 carbons, but, with higher members, little increase or decrease in the ability to inhibit corrosion occurs. This is attributed to the decreasing solubility in aqueous solution with increasing length of the hydrocarbon chain. However, the presence of a hydrophilic functional group in the molecule would increase the solubility of the inhibitors.

The performance of an organic inhibitor is related to the chemical structure and physicochemical properties of the compound like functional groups, electron density at the donor atom, p-orbital character, and the electronic structure of the molecule. The inhibition could be due to (i) Adsorption of the molecules or its ions on anodic and/or cathodic sites, (ii) increase in cathodic and/or anodic over voltage, and (iii) the formation of a protective barrier film. Some factors that contribute to the action of inhibitors are

(i) chain length,

(ii) size of the molecule,

(iii) bonding, aromatic/conjugate,

(iv) strength of bonding to the substrate,

(v) cross-linking ability,

(vi) solubility in the environment.

The role of inhibitors is to form a barrier of one or several molecular layers against acid attack. This protective action is often associated with chemical and/or physical adsorption involving a variation in the charge of the adsorbed substance and transfer of charge from one phase to the other. Sulphur and/or nitrogen-containing heterocyclic compounds with various substituents are considered to be effective corrosion inhibitors. Thiophene, hydrazine derivatives offer special affinity to inhibit corrosion of metals in acid solutions. Inorganic substances such as phosphates, chromates, dichromates, silicates, borates, tungstates, molybdates, and arsenates have been found effective as inhibitors of metal corrosion. Pyrrole and derivatives are believed to exhibit good protection against corrosion in acidic media. These inhibitors have also found useful application in the formulation of primers and anticorrosive coatings, but a major disadvantage is their toxicity and as such their use has come under severe criticism. Among the alternative corrosion inhibitors, organic substances containing polar functions with nitrogen, sulphur, and/or oxygen in the conjugated system have been reported to exhibit good inhibiting properties. The inhibitive characteristics of such compounds derive from the adsorption ability of their molecules, with the polar group acting as the reaction center for the adsorption process. The resulting adsorbed film acts as a barrier that separates the metal from the corrodent, and efficiency of inhibition depends on the mechanical, structural, and chemical characteristics of the adsorption layers formed under particular conditions.

Inhibitors are often added in industrial processes to secure metal dissolution from acid solutions. Standard anti corrosion coatings developed till date passively prevent the interaction of corrosion species and the metal. The known hazardous effects of most synthetic organic inhibitors and the need to develop cheap, nontoxic and ecofriendly processes have now urged researchers to focus on the use of natural products. Increasingly, there is a need to develop sophisticated new generation coatings for improved performance, especially in view of Cr VI being banned and labeled as a carcinogen. The use of inhibitors is one of the best options of protecting metals against corrosion. Several inhibitors in use are either synthesized from cheap raw material or chosen from compounds having heteroatoms in their aromatic or long-chain carbon system. However, most of these inhibitors are toxic to the environment. This has prompted the search for green corrosion inhibitors.

3. Green Inhibitors

Green corrosion inhibitors are biodegradable and do not contain heavy metals or other toxic compounds. Some research groups have reported the successful use of naturally occurring substances to inhibit the corrosion of metals in acidic and alkaline environment. *Delonix regia* extracts inhibited the corrosion of aluminum in hydrochloric acid solutions [5], rosemary leaves were studied as corrosion inhibitor for the Al + 2.5Mg alloy in a 3% NaCl solution at 25°C [6], and El-Etre investigated natural honey as a corrosion inhibitor for copper [7] and investigated

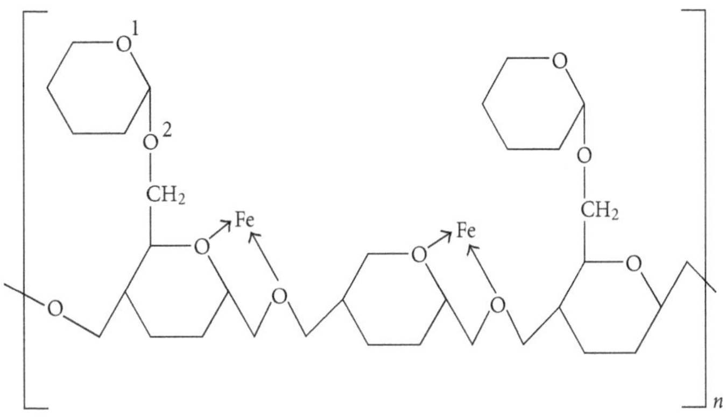

SCHEME 1: Guar gum.

opuntia extract on aluminum [8]. The inhibitive effect of the extract of khillah (*Ammi visnaga*) seeds on the corrosion of SX 316 steel in HCl solution was determined using weight loss measurements as well as potentiostatic technique. The mechanism of action is attributed to the formation of insoluble complexes as a result of interaction between iron cations, and khellin [9] and Ebenso et al. showed the inhibition of corrosion with ethanolic extract of African bush pepper (*Piper guinensis*) on mild steel [10]; *Carica papaya* leaves extract [11]; neem leaves extract (*Azadirachta indica)* on mild steel in H_2SO_4 [12]. Zucchi and Omar investigated plant extracts of *Papaia, Poinciana pulcherrima, Cassia occidentalis,* and *Datura stramonium* seeds and *Papaya, Calotropis procera B, Azadirachta indica,* and *Auforpio turkiale* sap for their corrosion inhibition potential and found that all extracts except those of *Auforpio turkiale* and *Azadirachta indica* reduced the corrosion of steel with an efficiency of 88%–96% in 1 N HCl and with a slightly lower efficiency in 2 N HCl. They attributed the effect to the products of the hydrolysis of the protein content of these plants [13]; Umoren et al. [14], studied the corrosion inhibition of mild steel in H_2SO_4 in the presence of gum arabic (GA) (naturally occurring polymer) and polyethylene glycol (PEG) (synthetic polymer). It was found that PEG was more effective than gum arabic.

Yee [15] determined the inhibitive effects of organic compounds, namely, honey and *Rosmarinus officinalis* L on four different metals—aluminium, copper, iron, and zinc, each polarized in two different solutions, that is, sodium chloride and sodium sulphate. The experimental approach employed potentiodynamic polarization method. The best inhibitive effect was obtained when zinc was polarised in both honey-added sodium chloride and sodium sulphate solutions. Rosemary extracts showed some cathodic inhibition when the metal was polarized in sodium chloride solution. This organic compound, however, displayed less anodic inhibition when compared with honey. The main chemical components of rosemary include borneol, bornyl acetate, camphor, cineole, camphene, and alpha-pinene. Chalchat et al. [16], reported that oils of rosemary were found to be rich in 1,8-cineole, camphor, bornyl acetate, and high amount of hydrocarbons. Recently, work has been emphasized on the use of *Rosmarinus officinalis* L as corrosion inhibitor for Al-Mg corrosion in chloride solution [6]. It is believed that the catechin fraction present in the rosemary extracts contributes to the inhibitive properties that act upon the alloy. Ouariachi et al. [17] also reported the inhibitory action of *Rosmarinus officinalis* oil as green corrosion inhibitors on C38 steel in 0.5 M H_2SO_4.

Odiongenyi et al. [18] reported that the ethanolic extract of *Vernonia amygdalina* appears to be a good inhibitor for the corrosion of mild steel in H_2SO_4 and action is by classical Langmuir adsorption isotherm.

The effect of addition of halides (KCl, KBr, and KI) was also studied, and the results obtained indicated that the increase in efficiency was due to synergism [13]. Umoren et al. also investigated the corrosion properties of *Raphia hookeri* exudates gum—halide mixtures for aluminum corrosion in acidic medium [16]. *Raphia hookeri* exudates gum obeys Freundlich, Langmuir, and Temkin adsorption isotherms. Phenomenon of physical adsorption is proposed. Abdallah also tested the effect of guar gum on carbon steel. It is proposed that it acts as a mixed type inhibitor [14]. The mechanism of action of C-steel by guar gum is due to the adsorption at the electrode/solution interface. Guar gum is a polysaccharide compound containing repeated heterocyclic pyrane moiety as shown in Scheme 1. The presence of heterooxygen atom in the structure makes possible its adsorption by coordinate type linkage through the transfer of lone pairs of electron of oxygen atoms to the steel surface, giving a stable chelate five-membered ring with ferrous ions. The chelation between O1 and O2 with Fe++ seems to be impossible due to proximity factor presented as in Scheme 1:

The simultaneous adsorption of oxygen atoms forces the guar gum molecule to be horizontally oriented at the metal surface, which led to increasing the surface coverage and consequently protection efficiency even in the case of low inhibitor concentrations.

Okafor et al. looked into the extracts of onion (*Allium sativum*), *Carica papaya* extracts, *Garcinia kola,* and *Phyllanthus amarus* [19–22]. El-Etre, Abdallah M used Natural honey as corrosion inhibitor for metals and alloys. II C-steel in high saline water [23]. Jojoba oil has also been evaluated [24]. Artemisia oil has been investigated for it is anticorrosion properties [25]. Oguzie and coworkers evaluated *Telfaria occidentalis*, *Occinum viridis*, *Azadirachta indica*, and *Sanseviera trifasciata* extracts [26–29]. Bendahou et al., studied using the extracts of rosemary in steel [27], and Sethuraman studied *Datura* [30]. Recently, studies on the use of some drugs as corrosion inhibitors have been reported by some researchers [31, 32]. Most of these drugs are heterocyclic compounds and were found to be environmentally friendly, hence, they have great potentials of competing with plant extracts. According to Eddy et al. drugs are environmentally friendly because they do not contain heavy metals or other toxic compound. In view of this adsorption and inhibitive efficiencies of ACPDQC (5-amino-1-cyclopropyl-7-[(3*R*, 5*S*) 3, 5-dimethylpiperazin-1-YL]-6,8-difluoro-4-oxo-uinoline-3-carboxylic acid), on mild steel corrosion have been studied and found to be effective.

Eddy et al. [33] studied inhibition of the corrosion of mild steel by ethanol extract of *Musa species* peel using hydrogen evolution and thermometric methods of monitoring corrosion. Inhibition efficiency of the extract was found to vary with concentration, temperature, period of immersion, pH, and electrode potentials. Adsorption of *Musa* species extract on mild steel surface was spontaneous and occurred according to Langmuir and Frumkin adsorption isotherms and also physical adsorption. Deepa Rani and Selvaraj [34] report the inhibition efficacy of *Punica granatum* extract on the corrosion of Brass in 1 N HCl evaluated by mass loss measurements at various time and temperature. Langmuir and Frumkin adsorption isotherms appear to be the mechanism of adsorption based on the values of activation energy, free energy of adsorption. Few researchers have summarized the effect of plant extracts on corrosion [35–38].

Efforts to find naturally organic substances or biodegradable organic materials to be used as corrosion inhibitors over the years have been intensified. Several reports are available on the various natural products used as green inhibitors as shown in Tables 1 and 2. Low-grade gram flour, natural honey, onion, potato, gelatin, plant roots, leaves, seeds, and flowers gums have been reported as good inhibitors. However, most of them have been tested on steel and nickel sheets. Although some studies have been performed on aluminum sheets, the corrosion effect is seen in very mild acidic or basic solutions (millimolar solutions).

3.1. Mechanism of Action of Green Inhibitors

Many theories to substantiate the mode of action of these green inhibitors have been put forth by several workers. Mann has suggested that organic substances, which form onium ions in acidic solutions, are adsorbed on the cathodic sites of the metal surface and interfere with the cathodic reaction.

Various mechanisms of action have been postulated for the corrosion inhibition property of the natural products.

Argemone mexicana. It is a contaminant of mustard seeds contain an alkaloid berberine which has a long-chain of aromatic rings, an N atom in the ring, and, at several places

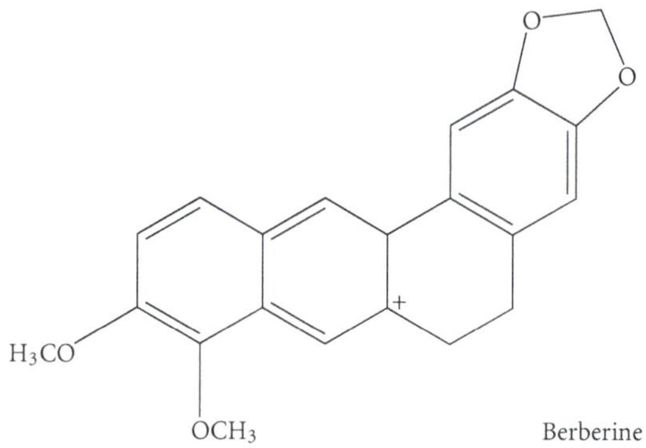

Berberine

Scheme 2: Berberine.

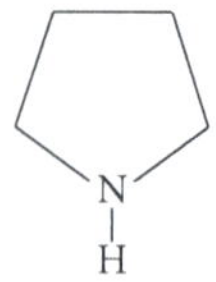

Pyrrolidine

Scheme 3: Pyrrolidine.

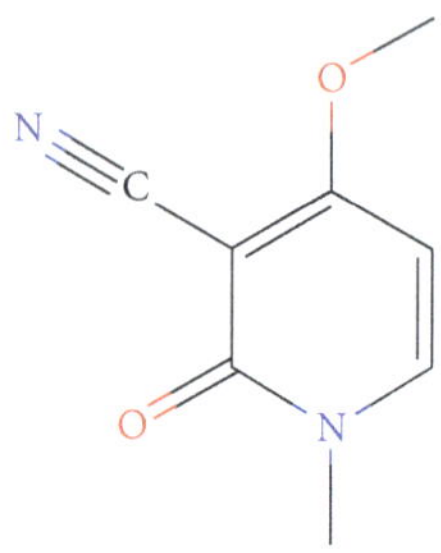

Ricinine

Scheme 4: Ricinine.

H atoms attached to C are replaced by groups, –CH, $–OCH_3$, and –O. The free electrons on the O and N atoms form bonds with the electrons on the metal surface. Berberine in water ionizes to release a proton, thus the now negatively charged O atom helps to free an electron on the N atom and forms a stronger bond with the metallic electrons. These properties confer good inhibition properties to *Argemone mexicana* (Scheme 2).

Garlic. It contains allyl propyl disulphide. Probably, this S-containing unsaturated compounds affects the potential cathodic process of steel.

Carrot. It contains pyrrolidine in aqueous media, pyrrolidine ionizes, and the N atom acquires a negative charge, and the free electrons on N possess still higher charge, resulting in stronger bond formation at N Carrot does not ionize in acidic media and thus does not protect in acids (Scheme 3).

Castor Seed. They contain the alkaloid ricinine. The N atom is in the ring attachment of the $–OCH_3$ (Scheme 4).

Table 1: Green inhibitors used for corrosion inhibition of steel.

Sl. no.	Metal	Inhibitor source	Active ingredient	References
(1)	Steel	Tamarind		[39]
(2)	Steel	Tea leaves		[40]
(3)	Steel	Pomegranate juice and peels		[41]
(4)	Steel	*Emblica officinalis*		[42]
(5)	Steel	*Terminalia bellerica*		[43]
(6)	Steel	Eucalyptus oil	Monomtrene 1,8-cineole	[44]
(7)		Rosemary		[45]
(8)	C-steel, Ni, Zn	Lawsonia extract (Henna)	Lawsone (2-hydroxy-1, 4-napthoquinone resin and tannin, coumarine, Gallic, acid, and sterols)	[46]
(9)	Mild steel	Gum exudate	Hexuronic acid, neutral sugar residues, volatile monoterpenes, canaric and related triterpene acids, reducing and nonreducing sugars	[47]
(10)	Mild steel	*Musa sapientum* peels (Banana peels)		[48]
(11)	Carbon steel	Natural amino acids—alanine, glycine, and leucine		[48]
(12)	Steel	Natural amino acids		[15]
(13)	Mild steel	*Garcinia kola* seed	Primary and secondary amines Unsaturated fatty acids and biflavnone	[49]
(14)	Steel	*Auforpio turkiale*	Protein hydrolysis	[50]
(15)	Steel	*Azydracta indica*	Protein hydrolysis	[51]
(16)	Steel	Aloe leaves		[52]
(17)	Steel	Mango/orange peels		[53]
(18)	Steel	*Hibiscus sabdariffa (Calyx extract) in* 1 M H_2SO_4 and 2 M HCl solutions, Stock 10–50%	Molecular protonated organic species in the extract. Ascorbic acid, amino acids, flavonoids, Pigments and carotene	[54]

Black Pepper. Quraishi et al. [73] studied corrosion inhibition of mild steel in hydrochloric solution by black pepper extract (*Piper nigrum* family: *Piperaceae*) by mass loss measurements, potentiodynamic polarisation, and electrochemical impedance spectroscopy (EIS). Black pepper extract gave maximum inhibition efficiency (98%) at 120 ppm at 35°C for mild steel in hydrochloric acid medium. Electrochemical evaluation revealed it to be a mixed-type inhibitor and that charge transfer controls the corrosion process. The corrosion inhibition property was attributed to an alkaloid "Piperine".

3.1.1. Fennel Seeds. Essential oil from *fennel (Foeniculum vulgare) (FM)* was tested as corrosion inhibitor of carbon steel in 1 M HCl using electrochemical impedance spectroscopy (EIS), Tafel polarisation methods, and weight loss measurements [74]. The increase of the charge-transfer resistance (R_{ct}) with the oil concentration supports the molecules of oil adsorption on the metallic surface. The polarization plots reveal that the addition of natural oil shifts the cathodic and anodic branches towards lower currents, indicative of a mixed-type inhibitor. The analysis of FM oil, obtained by hydrodistillation, using Gas Chromatography (GC) and Gas Chromatography/Mass Spectrometry (GC/MS) showed that the major components were limonene (20.8%) and pinene (17.8%). Interestingly, the composition of FM oil was variable according to the area of harvest and the stage of development. The analysis allowed the identification of 21 components which accounted for 96.6% of the total weight. The main constituents were limonene (20.8%) and pinene (17.8%) followed by myrcene (15%) and fenchone (12.5%). The adsorption of these molecules could take place via interaction with the vacant d-orbitals of iron atoms

Table 2: Green Inhibitors used for corrosion inhibition of aluminum, aluminum alloys, and other metals and alloys.

Sl. no.	Metal	Inhibitor source	Active ingredient	References
(1)	Al	$CeCl_3$ and mercaptobenzothiazole (MBT)		[55]
(2)	Al, steel	Aqueous extract of tobacco plant and its parts	Nicotine	[56]
(3)	Al	Vanillin		[57]
(4)	Al-Mg alloy	Aqueous extract of *Rosmarinus officinalis*—Neutral phenol subfraction of the aqueous extract	Catechin	[58]
(5)	Al	Sulphates/molybdates and dichromates as passivators		[59]
(6)	Al	Amino and polyamino acids—aspartic acid		[6]
(7)	Al	Pyridine and its selected derivatives (symmetric collidine and 2,5-dibrompyridine)		[60]
(8)	Al	Citric acid		[61]
(9)	Fe, Al	Benzoic acid		[62]
(10)	Al	Rutin and quercetin		[63]
(11)	Al			US Patent 5951747
(12)	Al	Polybutadieonic acid		[64]
(13)	Al and Zn	Saccharides—mannose and fructose		[65]
(14)	Al, Al-6061 and Al-Cu	Neutral solutions using sulphates, molybdates, and dichromates		[66]
(15)	Al	*Vernonia amygdalina* (Bitter leaf)		[67]
(16)	Al	Prosopis—cineraria (khejari)		[60]
(17)	Al	Tannin beetroot		[68]
(18)	Al	Saponin		[69]
(19)	Al	Acacia concianna		[70]
(20)	Al and Zn	Saccharides		[71]
(21)	Al	Opuntia (modified stems cladodes)	Polysaccharide (mucilage and pectin)	[72]
(22)	Al-Mg alloy	*Rosmarinus officinalis*		[8]
(23)	Zn	Metal chelates of citric acid		[61]
(24)	Zn	Onion juice	S-containing acids (glutamyl peptides) S-(1-propenyl) L-cysteine sulfoxide, and S-2-carboxypropyl glutathione	[63]
(25)	Sn	Natural honey (acacia chestnut)		[64]
(26)	Sn	Black radish	120	[8]

(a) (b)

Figure 1: (a) Mangostana fruit. (b) Pericarp.

(chemisorption). It is logical to assume that such adsorption is mainly responsible for the good protective properties by a synergistic effect of various molecules [74–76].

3.1.2. Garcinia mangostana. Vinod Kumar et al. [77] studied the corrosion inhibition of acid extract of the pericarp of the fruit of *G. mangostana* on mild steel in hydrochloric acid medium. *G. mangostana,* colloquially known as "the mangosteen", is a tropical evergreen tree. Mangosteen fruit, (Figure 1) on ripening the fruit, turns from green to purple in colour.

The extract of the pericarp of *G. mangostana* contains oxygenated prenylated xanthones, 8-hydroxycudraxanthone G and mangostingone [7-methoxy-2-(3- methyl-2-butenyl)-8-(3-methyl-2-oxo-3-butenyl)-1,3,6-trihydroxyxanthone, along with other xanthones such as cudraxanthone G, 8-deoxygartanin, garcimangosone B, garcinone D, garcinone E, gartanin, 1-isomangostin, $\acute{\alpha}$-mangostin, γ-mangostin, mangostinone, smeathxanthone A, and tovophyllin A [77, 78]. Electrochemical parameters such as Ecorr, β_a, and β_c indicate the mixed mode of inhibition, but predominantly cathodic. IR analysis and impedance studies indicate that the adsorption on the metal surface is due to the heteroatoms present in the organic constituents of the extract of *G. mangostana.*

3.1.3. Ipomea involcrata. Obot et al. [79] studied the corrosion inhibition efficiency of *Ipomoea involcrata* (IP) (family: Convolulaceae) leaf extract on aluminium. It is a common ornamental vine with heart-shaped and bright white pink or purple flowers that has a long history of use in central to southern Mexico. The plant has been shown to contain mainly d-lysergic acid amide (LSA) (Figure 2) and small amounts of other alkaloids, namely, chanoclavine, elymoclavine, and ergometrine, and d-isolysergic acid amide [79]. D-lysergic acid amide (LSA) (Figure 2) contains N and O in their structure including π-electrons which are required for corrosion inhibiting effects. Probably, chanoclavine, elymoclavine, ergometrine, d-isolysergic acid amide, and

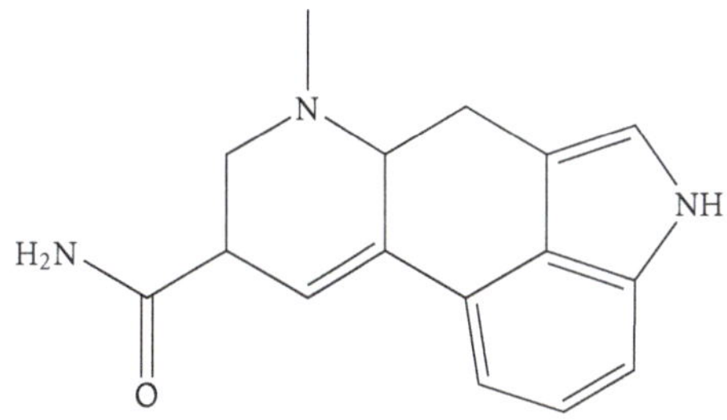

7-Methyl-4, 6.6a, 7, 8, 9-hexahydro-indolo[4, 3-fg]quinoline-9-carboxamide

Figure 2: Structure of lysergic acid.

other ingredients of the plant extracts synergistically increase the strength of the layer formed by the d-lysergic acid amide (LSA). Thus, the formation of a strong physisorbed layer between the metal surface and the phytoconstituents of the plant extract could be the cause of the inhibitive effect. The above authors have also reported that *Chromolaena odorata* as an excellent inhibitor for aluminium corrosion [80]. The environmentally friendly inhibitor could find possible applications in metal surface anodizing and surface coating in industries.

3.1.4. Soya Bean. It is rich in proteins, which are often good inhibitors in acidic media.

Most natural extracts constitute of oxygen- and nitrogen-containing compounds. Most of the oxygen-containing constituents of the extracts is a hydroxy aromatic compound, for example, tannins, pectins, flavonoids, steroids, and glycosides. Tannins are believed to form a passivating layer of tannates on the metallic surface. Similarly, it is postulated that a number of OH groups around the molecule lure them to form strong links with hydrogen and form complexes with metals. The complexes thus formed cause blockage of micro anodes and/or microanodes, which are generated on the metal surfaces when in contact with electrolytes, and, hence, retard subsequent dissolution of the metal.

3.1.5. Terminalia catappa. The inhibitive and adsorption properties of ethanol extract of *Terminalia catappa* for the corrosion of mild steel in H_2SO_4 were investigated using weight loss, hydrogen evolution, and infrared methods of monitoring corrosion. The inhibition potential of ethanol extract of *T. catappa* is attributed to the presence of saponin, tannin, phlobatin, anthraquinone, cardiac glycosides, flavanoid, terpene, and alkaloid in the extract. The adsorption of the inhibitor on mild steel surface is exothermic, spontaneous, and best described by Langmuir adsorption model [81] similar results were reported for *Gnetum Africana* [82].

Caffeic Acid. de Souza and Spinelli [83] studied the inhibitory action of caffeic acid as a green corrosion inhibitor for mild steel. The inhibitor effect of the naturally occurring biological molecule caffeic acid on the corrosion of mild steel in 0.1 M H_2SO_4 was investigated by weight loss, potentiodynamic polarization, electrochemical impedance, and Raman spectroscopy. The different techniques confirmed the adsorption of caffeic acid onto the mild steel surface and consequently the inhibition of the corrosion process. Caffeic acid acts by decreasing the available cathodic reaction area and modifying the activation energy of the anodic reaction.

3.1.6. Gossypium hirsutum. The corrosion inhibition properties of *Gossypium hirsutum* L leave extracts (GLE) and seed extracts (GSE) in 2 M sodium hydroxide (NaOH) solutions were studied using chemical technique. *Gossypium* extracts inhibited the corrosion of aluminium in NaOH solution. The inhibition efficiency increased with increasing concentration of the extracts. The leave extract (GLE) was found to be more effective than the seed extract (GSE). The GLE gave 97% inhibition efficiency while the GSE gave 94% at the highest concentration [83].

It is found that ethanol extract of *M. sapientum* peels (banana) can be used as an inhibitor for mild steel corrosion. The inhibitor acts by being adsorbed on mild steel surface according to classical adsorption models of Langmuir and Frumkin adsorption isotherms. Adsorption characteristics of the inhibitor follow physical adsorption mechanism. It is found that temperature, pH, period of immersion, electrode potential, and concentration of the inhibitor basically control the inhibitive action of *M. sapientum* peels.

3.1.7. Carmine and Fast Green Dyes. The use of dyes such as azo compounds methyl yellow, methyl red, and methyl orange [84] as inhibitors for mild steel has been reported [85–87]. The inhibition action of carmine and fast green dyes on corrosion of mild steel in 0.5 M HCl was investigated using mass loss, polarization, and electrochemical impedance (EIS) methods. Fast green showed inhibition efficiency of 98% and carmine 92%. The inhibitors act as mixed type with predominant cathodic effect.

Corrosion inhibition of mild steel in acidic solution by the dye molecules can be explained on the basis of adsorption on the metal surface, due to the donor-acceptor interaction between π electrons of donor atoms N, O and aromatic rings of inhibitors, and the vacant d-orbitals of iron surface atoms [88, 89]. The fast green molecules possess electroactive nitrogen, oxygen atoms, and aromatic rings, favouring the adsorption while the carmine molecules possess electroactive oxygen atoms and electron rich paraquinanoid aromatic rings. In addition, the large and flat structure of the molecules occupies a large area of the substrate and thereby forming a protective coating. The inhibitors were adsorbed on the mild steel surface according to the Temkin adsorption isotherm (Figure 3).

Torres et al. [90] studied the effects of aqueous extracts of spent coffee grounds on the corrosion of carbon steel in a 1 mol L^{-1} HCl. Two methods of extraction were studied: decoction and infusion. The inhibition efficiency of C-steel in 1 mol L^{-1} HCl increased as the extract concentration and temperature increased. The coffee extracts acted as a mixed-type inhibitor with predominant cathodic effectiveness. In this study, the adsorption process of components of spent coffee grounds extracts obeyed the Langmuir adsorption isotherm. The chlorogenic acids isolated do not appear to be the active ingredient.

3.2. Biocorrosion and Prevention by Green Inhibitors. Biocorrosion relates to the presence of micro organisms that adhere to different industrial surfaces and damage the metal. Bacterial cells encase themselves in a hydrated matrix of polysaccharides and protein and form a slimy layer known as biofilm. The biofilm is a gel consisting of approximately 95% water, microbial metabolic products like enzymes, extracellular polymeric substances, organic and inorganic acids, and also volatile compounds such as ammonia or hydrogen sulphide and inorganic detritus [90–92]. Extracellular polymeric substances play a crucial role in biofilm development. Inhibition of biofilm formation is the simplest way of biocorrosion prevention. Use of naturally produced compounds such as plant extracts could be used as effective biocides [34].

4. Sol-Gel Coatings

In recent years, the sol-gel coatings doped with inhibitors developed to replace chromate conversion coatings show real promise [93]. Results show that the corrosion resistance of the sol-gel coatings containing $CeCl_3$ proves to be better than that of the pure and MBT-added sol-gel coatings by the electrochemical methods. However, unlike chromium, silane-based sol-gel coatings mainly act as physical barrier rather than form chemical bond with substrate. Inhibitors are necessary to release in the coating film to slow the corrosion process through self-healing effect [57, 89, 94–96]. Among the inhibitors, rare-earth elements are generally considered to be effective and nontoxic in sol-gel coatings. Additionally, some organic inhibitors, especially heterocyclic compounds, are effective as slowly released inhibitors in sol-gel coating [97, 98]. Andreeva et al. suggested self-healing anticorrosion coatings based on pH [99, 100]. The approach

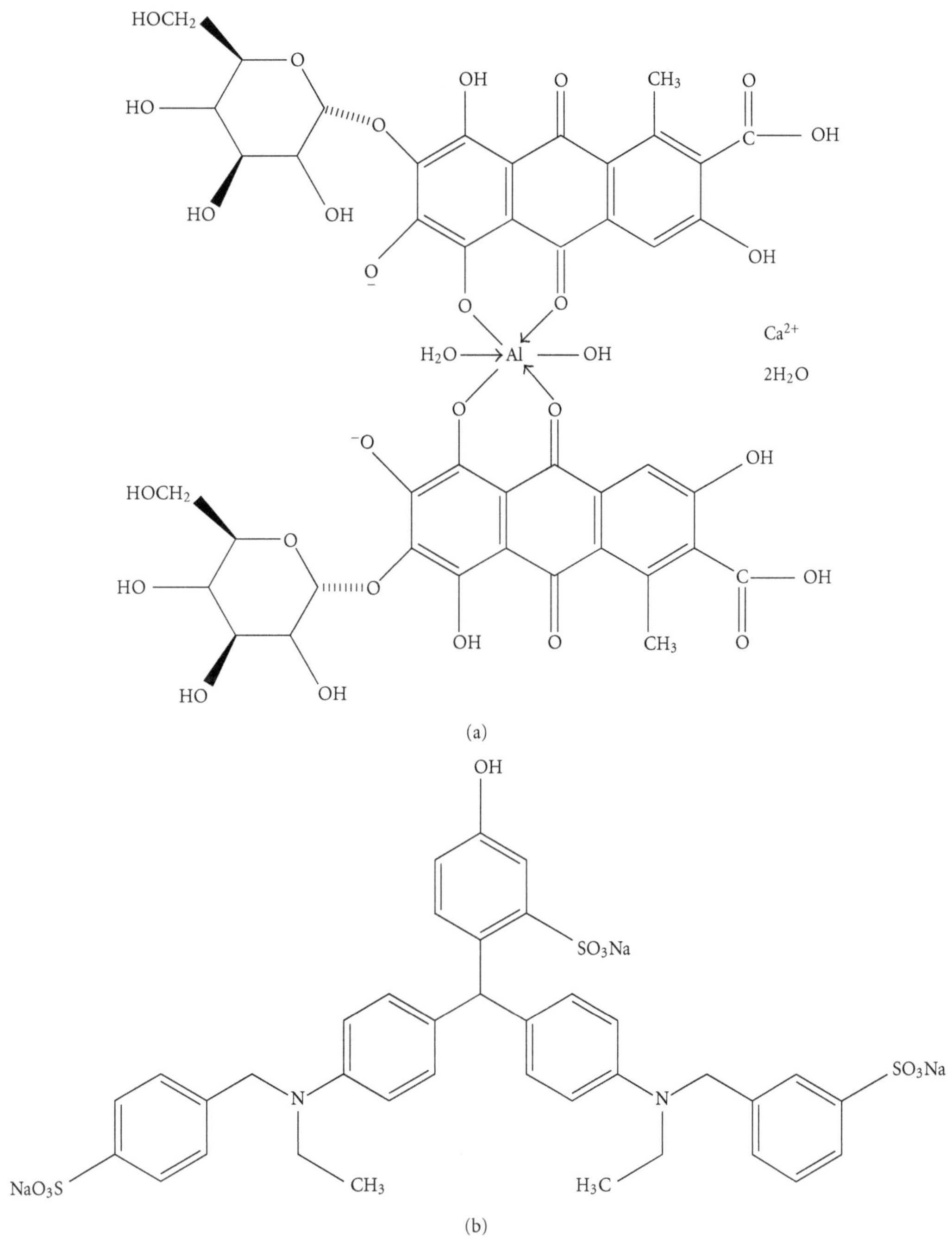

FIGURE 3: Structure of (a) carmine and (b) fast green.

to prevention of corrosion propagation on metal surfaces achieving the self-healing effect is based on suppression of accompanying physicochemical reactions. The corrosion processes are followed by changes of the pH value in the corrosive area and metal degradation. Self-healing or self-curing of the areas damaged by corrosion can be performed by three mechanisms: pH neutralization, passivation of the damaged metal surface by inhibitors entrapped between polyelectrolyte layers, and repair of the coating. The corrosion inhibitor incorporated as a component of the layer-by-layer film into the protective coating is responsible for the most effective mechanism of corrosion suppression. Quinolines are environmentally friendly corrosion inhibitors that are attracting more and more attention as alternatives to the harmful chromates.

Recent awareness of the corrosion inhibiting abilities of tannins, alkaloids, organic and amino acids, as well as organic dyes has resulted in sustained interest on the corrosion inhibiting properties of natural products of plant origin. Such investigation is of much importance because in addition to being environmentally friendly and ecologically acceptable, plant products are inexpensive, readily available, and renewable sources of materials. Although a number of insightful papers have been devoted to corrosion inhibition by plant extracts, reports on the detailed mechanisms of the adsorption process are still scarce. The drawback of most reports on plant extracts as corrosion inhibitors is that the active ingredient has not been identified.

In recent years, sol-gel coatings doped with green inhibitors show real promise for corrosion protection of a variety of metals and alloys.

5. Computational Modeling for Corrosion

Simulation is a prognostic computational tool for complex scientific and engineering problems. The simplest simulation methods have been used for decades, but, with the increase in computational memory and speed simulation, have become the prevalent tool for analysis [101–103]. Simulation turns probability models into statistics problems where the results can be analyzed using standard statistical methods. The challenge of a simulation is to implement a procedure that efficiently captures the desired model characteristics. Often the goal of probability computations is the evaluation of high reliability. In fact, computation of high reliabilities itself is an ongoing research concern. Hence, there is no one way in which to do the computation. Monte Carlo simulation is the traditional and powerful method if computational complexity and time are not limiting. The Box-Muller method is also well known. A variety of techniques have been developed to reduce the number of simulations without compromising accuracy.

The study of corrosion involves the study of the chemical, physical, metallurgical, and mechanical properties of materials as it is a synergistic phenomenon in which the environment is as equally important as the materials involved. Computer modeling techniques can handle the study of complex systems such as corrosion and thus are appropriate and powerful tools to study the mechanism of action of corrosion and its inhibitors.

In the recent past, computer modeling techniques have been successfully applied to corrosion problems as summarized in review articles by Zamani et al. [104] and Munn [105]. The application of computer modeling techniques to corrosion systems requires an understanding of the physical phenomenon of corrosion and the mathematics which govern the corrosion process. In addition, knowledge of the numerical procedures which are the basis of computer modeling techniques is essential for accurate computational analyses. In addition, validation of the computer analysis results with experimental data is mandatory. Without a reasonably accurate description of the damage process at a scale that is pertinent to the desired application, probabilistic computations have minimal value for prognosis and life-cycle assessment.

For corrosion modeling, the materials characterization depends on the orientation of the material. Figure 4 is a composite of three optical micrographs of the perpendicular faces of a typical specimen of 7075-T6 aluminum alloy, where LT, LS, and TS are the rolling, long-transverse, and short-transverse planes, respectively. Visually, there is a difference

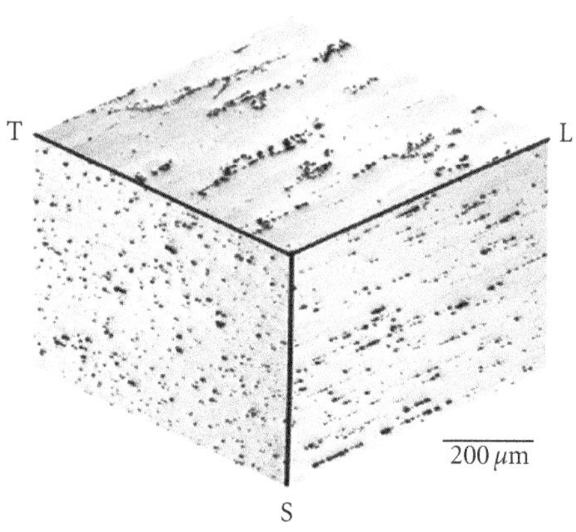

Figure 4: Three optical micrographs of the perpendicular faces of a typical specimen of 7075-T6 aluminum alloy.

in the three surfaces, and the variability in the location, size, and density of the particles is apparent. Thus, for eg when modeling for aircraft wings, the LS surface is the most significant surface to characterize because it is the surface in fastener holes subjected to high-stress loading.

5.1. Some Examples of Computational Modeling in Corrosion Inhibition

5.1.1. Tryptophan. According to the description of frontier orbital theory, HOMO is often associated with the electron donating ability of an inhibitor molecule. High EHOMO values indicate that the molecule has a tendency to donate electrons to the metal with unoccupied molecule orbitals. ELUMO indicates the ability of the molecules to accept electrons. The lower value of ELUMO is the easier acceptance of electrons from metal surface. The gap between the LUMO and HOMO energy levels of the inhibitor molecules is another important index, and the low absolute values of the energy band gap (DE = ELUMO − EHOMO) means good inhibition efficiency. Studies indicated that L-tryptophan has high value of EHOMO and low value of ELUMO with low-energy band gap. Adsorption energy calculated for the adsorption of L-tryptophan on Fe surface in the presence of water molecules equals $-29.5\,\mathrm{kJ\,mol^{-1}}$, which implies that the interaction between L-tryptophan molecule and Fe surface is strong [105, 106]. Molecule dynamics simulation results showed that L-tryptophan molecules assumed a nearly flat orientation with respect to the Fe (1 1 0) surface. The calculated adsorption energy between a L-tryptophan molecule and Fe surface is $-29.5\,\mathrm{kJ\,mol^{-1}}$.

The optimized molecule structure, the highest occupied molecule orbitals, the lowest unoccupied molecule orbital, and the charge distribution of L-tryptophan molecule using DFT functional (B3LYP/6-311*G) are shown in Figure 5. The figure shows that in L-tryptophan molecule, C5, C12, C13, C14, C15, N7, N10, O2, and O4 carry more negative charges, while C8 and C6 carry more positive charges.

This means that C5, C12, C13, C14, C15, N7, N10, O2, and O4 are the negative charge centers, which can offer

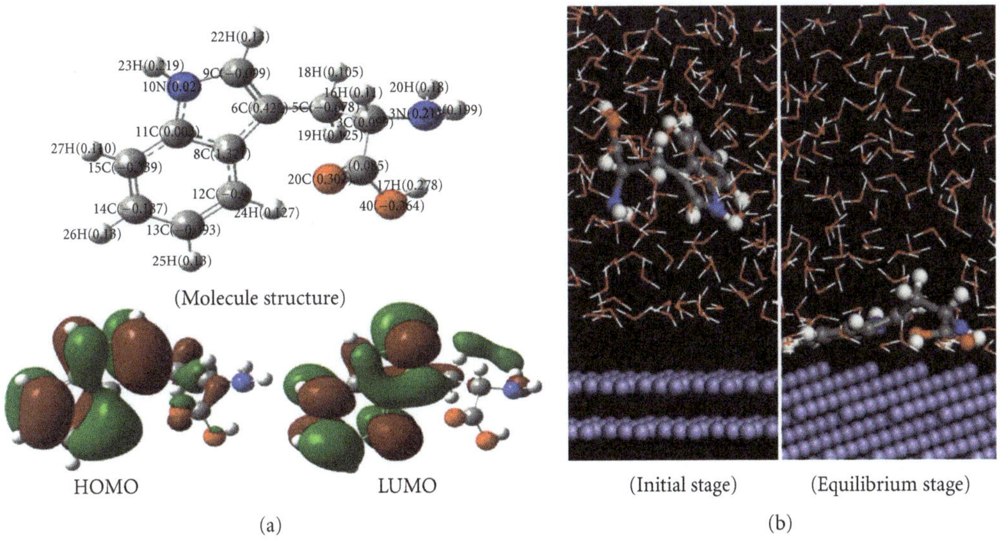

Figure 5: (a) Optimised molecule structure and charge density distribution of L-tryptophan. (b) L-tryptophan adsorbed on Fe surface in water solution.

electrons to the Fe atoms to form coordinate bond, and C8 and C6 are the positive charge centers, which can accept electrons from orbital of Fe atoms to form feedback bond. The optimized structure is in accordance with the fact that excellent corrosion inhibitors cannot only offer electrons to unoccupied orbital of the metal, but also accept free electrons from the metal. Therefore, it can be inferred that indole ring, nitrogen, and oxygen atoms are the possible active adsorption sites.

Presuel-Moreno et al. [107] modeled the chemical throwing power of an Al-Co-Ce metallic coating under thin electrolyte films representative of atmospheric conditions. An Al-Co-Ce alloy coating was developed for an AA2024-T3 substrate that can serve as barrier, sacrificial anode, and reservoir to supply soluble inhibitor ions to protect any defect sites or simulated scratches exposing the substrate. The model calculates the time necessary to accumulate Ce +3 and Co +2 inhibitors over the scratch when released from the Al-Co-Ce coating under different conditions such as the pH-dependent passive dissolution rate of an Al-Co-Ce alloy to define the inhibitor release flux. Transport by both electromigration and diffusion was considered. The effects of scratch size, initial pH, chloride concentration, and electrochemical kinetics of the material involved were studied. Studies indicated that sufficient accumulation of the released inhibitor (i.e., the Ce +3 concentration surpassed the critical inhibitor concentration over AA2024-T3 scratches) was achieved within a few hours (e.g., ~4 h for scratches of $S = 1500\,\mu$m) when the initial solution pH was 6 and the coating was adjacent to the AA2024-T3.

Pradip and Rai [108] modeled design of phosphonic-acid-based corrosion inhibitors using a force field approach.

5.1.2. Piperidine and Derivatives. Khaled and Amin [109] studied the adsorption and corrosion inhibition behaviour of four selected piperidine derivatives, namely, piperidine (pip), 2-methylpiperidine (2mp), 3-methylpiperidine (3mp), and 4-methylpiperidine (4mp) at nickel in 1.0 M HNO_3 solution computationally by the molecular dynamics simulation and quantum chemical calculations and electrochemically by Tafel and impedance methods. The molecular dynamics (MD) simulations were performed using the commercial software MS Modeling from Accelrys using the amorphous cell module to create solvent piperidines cells on the nickel substrate. The behaviour of the inhibitors on the surface was studied using molecular dynamics simulations, and the condensed phase optimized molecular potentials for atomistic simulation studies (COMPASS) force field. COMPASS is an ab initio powerful force field which supports atomistic simulations of condensed phase materials [102]. Molecular simulation studies were applied to optimize the adsorption structures of piperidine derivatives. The nickel/inhibitor/solvent interfaces were simulated, and the charges on the inhibitor molecules as well as their structural parameters were calculated in the presence of solvent effects. Quantum chemical calculations based on the ab initio method were performed to determine the relationship between the molecular structure of piperidines and their inhibition efficiency. Results obtained from Tafel and impedance methods are in good agreement and confirm theoretical studies.

Khaled and Amin [110] also conducted studies on the molecular dynamics simulation on the corrosion inhibition of aluminum in molar hydrochloric acid using some imidazole derivatives. They also adapted Monte Carlo simulations technique incorporating molecular mechanics and dynamics to simulate the adsorption of methionine derivatives, namely, L-methionine, L-methionine sulphoxide, and L-methionine sulphone on iron (110) surface in 0.5 M sulphuric acid. Results show that methionine derivatives have a very good inhibitive effect for corrosion of mild steel in 0.5 M sulphuric acid solution.

5.1.3. Aniline and Its Derivatives. The inhibiting action of aniline and its derivatives on the corrosion of copper in hydrochloric acid has been investigated by Henriquez et al. [39], with emphasis on the role of substituents. With this purpose five different anilines were selected: aniline, p-chloroaniline, p-nitro aniline, p-methoxy, and p-methylaniline. A theoretical study using molecular mechanic and ab initio Hartree Fock methods, to model the adsorption of aniline on copper (100) showed results in good agreement with the experimental data. Aniline adsorbs parallel to the copper surface, showing no preference for a specific adsorption site. On the other hand, from ab initio Hartree Fock calculations, adsorption energy between 2 kcal/mol and 5 kcal/mol is obtained, which is close to the experimental value, confirming that the adsorption of aniline on the metal substrate is rather weak. In view of these results, the orientation of the aniline molecule with respect to the copper surface is considered to be the dominant effect. Mechanic molecular calculations were carried out using the Insight II, a comprehensive graphic molecular modeling program, to obtain configurations of minimum energy.

Acknowledgments

The encouragement and cooperation received from Dr. Upadhya, Director, NAL, Bangalore, Dr. Ranjan Moodithaya, Head, KTMD, and Dr. K.S. Rajam, Head, SED are gratefully acknowledged. Patents used in the paper are: (1) US Patent 5951747—Non-chromate corrosion inhibitors for aluminum alloys; (2) United States Patent 5286357—Corrosion sensors; (3) WO/2002/008345—CORROSION INHIBITORS; and (4) British patent, 2327,1895.

References

[1] M. Bouklah, B. Hammouti, T. Benhadda, and M. Benkadour, "Thiophene derivatives as effective inhibitors for the corrosion of steel in 0.5 M H_2SO_4," *Journal of Applied Electrochemistry*, vol. 35, no. 11, pp. 1095–1101, 2005.

[2] A. S. Fouda, A. A. Al-Sarawy, and E. E. El-Katori, "Pyrazolone derivatives as corrosion inhibitors for C-steel HCl solution," *Desalination*, vol. 201, pp. 1–13, 2006.

[3] A. Fiala, A. Chibani, A. Darchen, A. Boulkamh, and K. Djebbar, "Investigations of the inhibition of copper corrosion in nitric acid solutions by ketene dithioacetal derivatives," *Applied Surface Science*, vol. 253, no. 24, pp. 9347–9356, 2007.

[4] U. R. Evans, *The Corrosion and Oxidation of Metals*, Hodder Arnold, 1976.

[5] O. K. Abiola, N. C. Oforka, E. E. Ebenso, and N. M. Nwinuka, "Eco-friendly corrosion inhibitors: The inhibitive action of Delonix Regia extract for the corrosion of aluminium in acidic media," *Anti-Corrosion Methods and Materials*, vol. 54, no. 4, pp. 219–224, 2007.

[6] M. Kliskic, J. Radoservic, S. Gudic, and V. Katalinic, "Aqueous extract of Rosmarinus officinalis L. as inhibitor of Al-Mg alloy corrosion in chloride solution," *Journal of Applied Electrochemistry*, vol. 30, no. 7, pp. 823–830, 2000.

[7] A. Y. El-Etre, "Natural honey as corrosion inhibitor for metals and alloys. I. Copper in neutral aqueous solution ," *Corrosion Science*, vol. 40, no. 11, pp. 1845–1850, 1998.

[8] A. Y. El-Etre, "Inhibition of aluminum corrosion using Opuntia extract," *Corrosion Science*, vol. 45, no. 11, pp. 2485–2495, 2003.

[9] A. Y. El-Etre, "Khillah extract as inhibitor for acid corrosion of SX 316 steel," *Applied Surface Science*, vol. 252, no. 24, pp. 8521–8525, 2006.

[10] E. E. Ebenso, U. J. Ibok, U. J. Ekpe et al., "Corrosion inhibition studies of some plant extracts on aluminium in acidic medium," *Transactions of the SAEST*, vol. 39, no. 4, pp. 117–123, 2004.

[11] E. E. Ebenso and U. J. Ekpe, "Kinetic study of corrosion and corrosion inhibition of mild steel in H_2SO_4 using *Parica papaya* leaves extract," *West African Journal of Biological and Applied Chemistry*, vol. 41, pp. 21–27, 1996.

[12] U. J. Ekpe, E. E. Ebenso, and U. J. Ibok, "Inhibitory action of *Azadirachta indica* leaves extract on the corrosion of mild steel in H_2SO_4," *West African Journal of Biological and Applied Chemistry*, vol. 37, pp. 13–30, 1994.

[13] F. Zucchi and I. H. Omar, "Plant extracts as corrosion inhibitors of mild steel in HCl solutions," *Surface Technology*, vol. 24, no. 4, pp. 391–399, 1985.

[14] S. A. Umoren, O. Ogbobe, I. O. Igwe, and E. E. Ebenso, "Inhibition of mild steel corrosion in acidic medium using synthetic and naturally occurring polymers and synergistic halide additives," *Corrosion Science*, vol. 50, no. 7, pp. 1998–2006, 2008.

[15] Y. J. Yee, *Green inhibitors for corrosion control: a Study on the inhibitive effects of extracts of honey and rosmarinus officinalis L. (Rosemary)*, M.S. thesis, University of Manchester, Institute of Science and Technology, 2004.

[16] J. C. Chalchat, R. P. Garry, A. Michet, B. Benjilali, and J. L. Chabart, "Essential oils of Rosemary (Rosmarinus officinalis L.). The chemical composition of oils of various origins (Morocco, Spain, France)," *Journal of Essential Oil Research*, vol. 5, no. 6, pp. 613–618, 1993.

[17] E. El Ouariachi, J. Paolini, M. Bouklah et al., "Adsorption properties of Rosmarinus of ficinalis oil as green corrosion inhibitors on C38 steel in 0.5 M H_2SO_4," *Acta Metallurgica Sinica*, vol. 23, no. 1, pp. 13–20, 2010.

[18] A. O. Odiongenyi, S. A. Odoemelam, and N. O. Eddy, "Corrosion inhibition and adsorption properties of ethanol extract of Vernonia Amygdalina for the corrosion of mild steel in H_2SO_4," *Portugaliae Electrochimica Acta*, vol. 27, no. 1, pp. 33–45, 2009.

[19] S. A. Umoren and E. E. Ebenso, "Studies of the anti-corrosive effect of Raphia hookeri exudate gum-halide mixtures for aluminium corrosion in acidic medium," *Pigment and Resin Technology*, vol. 37, no. 3, pp. 173–182, 2008.

[20] M. Abdallah, "Guar gum as corrosion inhibitor for carbon steel in sulphuric acid solutions," *Portugaliae Electrochimica Acta*, vol. 22, pp. 161–175, 2004.

[21] P. C. Okafor, U. J. Ekpe, E. E. Ebenso, E. M. Umoren, and K. E. Leizou, "Inhibition of mild steel corrosion in acidic medium by Allium sativum extracts," *Bulletin of Electrochemistry*, vol. 21, no. 8, pp. 347–352, 2005.

[22] P. C. Okafor and E. E. Ebenso, "Inhibitive action of Carica papaya extracts on the corrosion of mild steel in acidic media and their adsorption characteristics," *Pigment and Resin Technology*, vol. 36, no. 3, pp. 134–140, 2007.

[23] P. C. Okafor, V. I. Osabor, and E. E. Ebenso, "Eco-friendly corrosion inhibitors: Inhibitive action of ethanol extracts of Garcinia kola for the corrosion of mild steel in H_2SO_4 solutions," *Pigment and Resin Technology*, vol. 36, no. 5, pp. 299–305, 2007.

[24] P. C. Okafor, M. E. Ikpi, I. E. Uwah, E. E. Ebenso, U. J. Ekpe, and S. A. Umoren, "Inhibitory action of Phyllanthus amarus extracts on the corrosion of mild steel in acidic media," *Corrosion Science*, vol. 50, no. 8, pp. 2310–2317, 2008.

[25] A. Y. El-Etre and M. Abdallah, "Natural honey as corrosion inhibitor for metals and alloys. II. C-steel in high saline water," *Corrosion Science*, vol. 42, no. 4, pp. 731–738, 2000.

[26] A. Chetouani, B. Hammouti, and M. Benkaddour, "Corrosion inhibition of iron in hydrochloric acid solution by jojoba oil," *Pigment and Resin Technology*, vol. 33, no. 1, pp. 26–31, 2004.

[27] A. Bouyanzer and B. Hammouti, "A study of anti-corrosive effects of Artemisia oil on steel," *Pigment and Resin Technology*, vol. 33, no. 5, pp. 287–292, 2004.

[28] E. E. Oguzie, "Inhibition of acid corrosion of mild steel by Telfaria occidentalis extract," *Pigment and Resin Technology*, vol. 34, no. 6, pp. 321–326, 2005.

[29] E. E. Oguzie, "Studies on the inhibitive effect of *Occimum viridis* extract on the acid corrosion of mild steel," *Materials Chemistry and Physics*, vol. 99, pp. 441–446, 2006.

[30] E. E. Oguzie, "Corrosion inhibition of aluminium in acidic and alkaline media by *Sansevieria trifasciata* extract," *Corrosion Science*, vol. 49, no. 3, pp. 1527–1539, 2007.

[31] M. A. Bendahou, M. B. E. Benadellah, and B. B. Hammouti, "A study of rosemary oil as a green corrosion inhibitor for steel in 2 M H_3PO_4," *Pigment and Resin Technology*, vol. 35, no. 2, pp. 95–100, 2006.

[32] M. G. Sethuraman and P. B. Raja, "Corrosion inhibition of mild steel by Datura metel in acidic medium," *Pigment and Resin Technology*, vol. 34, no. 6, pp. 327–331, 2005.

[33] N. O. Eddy, S. A. Odoemelam, and A. O. Odiongenyi, "Ethanol extract of musa species peels as a green corrosion inhibitor for mild steel: Kinetics, adsorption and thermodynamic considerations," *Electronic Journal of Environmental, Agricultural and Food Chemistry*, vol. 8, no. 4, pp. 243–255, 2009.

[34] P. Deepa Rani and S. Selvaraj, "Inhibitive and adsorption properties of punica granatum extract on brass in acid media," *Journal of Phytology*, vol. 2, no. 11, pp. 58–64, 2010.

[35] S. Rajendran, V. Ganga Sri, J. Arockiaselvi, and A. J. Amalraj, "Corrosion inhibition by plant extracts—an overview," *Bulletin of Electrochemistry*, vol. 21, no. 8, pp. 367–377, 2005.

[36] K. Srivastava and P. Srivastava, "Studies on plant materials as corrosion inhibitors," *British Corrosion Journal*, vol. 16, no. 4, pp. 221–223, 1981.

[37] R. M. Saleh, A. A. Ismail, and A. A. El Hosary, "corrosion inhibition by naturally occurring substances. vii. the effect of aqueous extracts of some leaves and fruit peels on the corrosion of steel, Al, Zn and Cu in acids," *British Corrosion Journal*, vol. 17, no. 3, pp. 131–135, 1982.

[38] P. B. Raja and M. G. Sethuraman, "Natural products as corrosion inhibitor for metals in corrosive media—a review," *Materials Letters*, vol. 62, no. 1, pp. 113–116, 2008.

[39] J. H. Henriquez-Román, M. Sancy, M. A. Páez et al., "The influence of aniline and its derivatives on the corrosion behaviour of copper in acid solution," *Journal of Solid State Electrochemistry*, vol. 9, no. 7, pp. 504–511, 2005.

[40] A. A. El Hosary, R. M. Saleh, and A. M. Shams El Din, "Corrosion inhibition by naturally occurringsubstances-I. The effect of Hibiscus subdariffa (karkade) extract on the dissolution of Al and Zn," *Corrosion Science*, vol. 12, no. 12, pp. 897–904, 1972.

[41] R. M. Saleh and A. M. Shams El Din, "Efficiency of organic acids and their anions in retarding the dissolution of aluminium," *Corrosion Science*, vol. 12, no. 9, pp. 689–697, 1972.

[42] M. J. Sanghvi, S. K. Shuklan, A. N. Misra, M. R. Padh, and G. N. Mehta, "Inhibition of hydrochloric acid corrosion of mild steel by aid extracts of Embilica officianalis, Terminalia bellirica and Terminalia chebula," *Bulletin of Electrochemistry*, vol. 13, no. 8-9, pp. 358–361, 1997.

[43] M. J. Shangvi, S. K. Shukla, A. N. Mishra, M. R. padh, and G. N. Mehta, "Corrosion inhibition of mild steel in hydrochloric acid by acid extracts of Sapindus Trifolianus, Acacia Concian and Trifla," *Transactions of the Metal Finishers Association of India*, vol. 5, no. 3, pp. 143–147, 1996.

[44] A. Chetouani and B. Hammouti, "Corrosion inhibition of iron in hydrochloric acid solutions by naturally henna," *Bulletin of Electrochemistry*, vol. 19, no. 1, pp. 23–25, 2003.

[45] B. Muller, W. Klager, and G. Kubitzki, "Metal chelates of citric acid as corrosion inhibitors for zinc pigment," *Corrosion Science*, vol. 39, no. 8, pp. 1481–1485, 1997.

[46] A. Bouyanzer, L. Majidi, and B. Hammouti, "Effect of eucalyptus oil on the corrosion of steel in 1M HCl," *Bulletin of Electrochemistry*, vol. 22, no. 7, pp. 321–324, 2006.

[47] A. Y. El-Etre, "Natural onion juice as inhibitor for zinc corrosion," *Bulletin of Electrochemistry*, vol. 22, no. 2, pp. 75–80, 2006.

[48] I. Radojcic, K. Berković, S. Kovač, and J. Vorkapić-Furač, "Natural honey and black radish juice as tin corrosion inhibitors," *Corrosion Science*, vol. 50, no. 5, pp. 1498–1504, 2008.

[49] A. Y. El-Etre, M. Abdallah, and Z. E. El-Tantawy, "Corrosion inhibition of some metals using lawsonia extract," *Corrosion Science*, vol. 47, no. 2, pp. 385–395, 2005.

[50] S. A. Umoren, I. B. Obot, and E. E. Ebenso, "Corrosion inhibition of aluminium using exudate gum from *Pachylobus edulis* in the presence of halide ions in HCl," *E-Journal of Chemistry*, vol. 5, no. 2, pp. 355–364, 2008.

[51] N. O. Eddy and E. E. Ebenso, "Adsorption and inhibitive properties of ethanol extracts of *Musa sapientum* peels as a green corrosion inhibitor for mild steel in H_2SO_4," *African Journal of Pure and Applied Chemistry*, vol. 2, no. 6, pp. 046–054, 2008.

[52] S. Lyon, "A natural way to stop corrosion," *Nature*, vol. 427, no. 406, p. 407, 2004.

[53] E. E. Oguzie, K. L. Iyeh, and A. I. Onuchukwu, "Inhibition of mild steel corrosion in acidic media by aqueous extracts from Garcinia kola seed," *Bulletin of Electrochemistry*, vol. 22, no. 2, pp. 63–68, 2006.

[54] R. M. Saldo, A. A. Ismail, and A. A. El Hosary, "Corrosion Inhibition by naturally occurring substances," *British Corrosion Journal*, vol. 17, no. 3, pp. 131–135, 1990.

[55] E. E. Oguzie, "Corrosion inhibitive effect and adsorption bBehaviour of *Hibiscus Sabdariffa* extract on mild steel in acidic media," *Portugaliae Electrochimica Acta*, vol. 26, pp. 303–314, 2008.

[56] E. E. Oguzie, "Corrosion inhibitive effect and adsorption behaviour of Hibiscus sabdariffa extract on mild steel in acidic media," *Portugaliae Electrochimica Acta*, vol. 26, no. 3, pp. 303–314, 2008.

[57] H. W. Shi, F. C. Liu, E. H. Han, and M. C. Sun, "Investigation on a sol-gel coating containing inhibitors on 2024-T3 aluminum alloy," *Chinese Journal of Aeronautics*, vol. 19, pp. S106–S112, 2006.

[58] G. D. Davis, "The Use of Extracts of Tobacco Plants as Corrosion Inhibitors," http://www.electrochem.Org/dl/ma/202/pdfs/0340.PDF.

[59] A. Y. El-Etre, "Inhibition of acid corrosion of aluminum using vanillin," *Corrosion Science*, vol. 43, no. 6, pp. 1031–1039, 2001.

[60] W. A. Badawy, F. M. Allhara, and A. S. ElAzab, "Electrochemical behaviour and corrosion inhibition of Al, Al-6061 and Al-Cu in neutral aqueous solutions," *Corrosion Science*, vol. 41, no. 4, pp. 709–727, 1999.

[61] B. Müller, "Amino and polyamino acids as corrosion inhibitors for aluminium and zinc pigments," *Pigment and Resin Technology*, vol. 31, no. 2, pp. 84–87, 2002.

[62] M. Kliskic, J. Radosevi, and S. Gudic, "Pyridine and its derivatives as inhibitors of aluminium corrosion in chloride solution," *Journal of Applied Electrochemistry*, vol. 27, no. 6, pp. 947–952, 1997.

[63] R. Solmaz, G. Kardaş, B. Yazici, and M. Erbil, "Citric acid as natural corrosion inhibitor for aluminium protection," *Corrosion Engineering Science and Technology*, vol. 43, no. 2, pp. 186–191, 2008.

[64] Z. Sibel, "The effects of benzoic acid in chloride solutions on the corrosion of iron and aluminum," *Turkish Journal of Chemistry*, vol. 26, no. 3, pp. 403–408, 2002.

[65] K. Berkovic, S. Kovac, and J. Vorkapic-Furac, "Natural compounds as environmentally friendly corrosion inhibitors of aluminium," *Acta Alimentaria*, vol. 33, no. 3, pp. 237–247, 2004.

[66] Y. Tao, X. Zhang, and Z. Gu, "The inhibition of corrosion of aluminum in acid environment byin situ electrocoagulation of polybutadienoic acid," *Wuhan University Journal of Natural Sciences*, vol. 3, no. 2, pp. 221–225, 1998.

[67] B. Müller and M. Kurfeß, "Saccharide und deren Derivate als Korrosionsinhibitoren für Aluminiumpigmente in wäßrigen Medien," *Materials and Corrosion*, vol. 44, no. 9, pp. 373–378, 2004.

[68] G. O. Avwiri and F. O. Igho, "Inhibitive action of Vernonia amygdalina on the corrosion of aluminium alloys in acidic media," *Materials Letters*, vol. 57, no. 22-23, pp. 3705–3711, 2003.

[69] S. Manish Kumar, K. Sudesh, R. Ratnani, and S. P. Mathur, "Corrosion inhibition of Aluminium by extracts of Prosopis cineraria in acidic media," *Bulletin of Electrochemistry*, vol. 22, no. 2, pp. 69–74, 2006.

[70] A. A. Hossary, M. M. Gauish, and R. M. Saleh, "Corrosion inhibitor formulations from coal-tar distillation products for acid cleaning of steel in HCl," in *Proceedings of the 2nd International Symposium on Industrial and Oriented Basic Electrochemistry*, pp. 6–18, SAEST, CECRI, Madras, India, 1980.

[71] R. A. Tupikov, Y. G. Dragunov, I. L. Kharina, and D. S. Zmienko, "Protection of carbon steels against atmospheric corrosion in a wet tropical climate using gas-plasma metallization with aluminum," *Protection of Metals*, vol. 44, no. 7, pp. 673–682, 2008.

[72] P. Arora, T. Jain, and S. P. Mathur, *Chemistry*, vol. 1, p. 766, 2005.

[73] M. A. Quraishi, D. K. Yadav, and I. Ahamad, "Green approach to corrosion inhibition by black pepper extract in hydrochloric acid solution," *Open Corrosion Journal*, vol. 2, pp. 56–60, 2009.

[74] N. Lahhit, A. Bouyanzer, J.-M. Desjobert et al., "Fennel *(Foeniculum vulgare)* essential oil as green corrosion Inhibitor of carbon steel in hydrochloric acid solution," *Portugaliae Electrochimica Acta*, vol. 29, no. 2, pp. 127–138, 2011.

[75] E. E. Oguzie, "Corrosion inhibition of mild steel in hydrochloric acid solution by methylene blue dye," *Materials Letters*, vol. 59, no. 8-9, pp. 1076–1079, 2005.

[76] A. Zarrouk, I. Warad, B. Hammouti, A. Dafali, S. S. Al-Deyab, and N. Benchat, "The effect of temperature on the corrosion of Cu/HNO_3 in the Presence of organic inhibitor: Part-2," *International Journal of Electrochemical Science*, vol. 5, no. 10, pp. 1516–1526, 2010.

[77] K. P. Vinod Kumar, M. S. Narayanan Pillai, and G. Rexin Thusnavis, "Pericarp of the fruit of garcinia mangostana as corrosion inhibitor for mild steel in hydrochloric acid medium," *Portugaliae Electrochimica Acta*, vol. 28, no. 6, pp. 373–383, 2010.

[78] H. A. Jung, B. N. Su, W. J. Keller, R. G. Mehta, and A. D. Kinghorn, "Antioxidant xanthones from the pericarp of Garcinia mangostana (Mangosteen)," *Journal of Agricultural and Food Chemistry*, vol. 54, no. 6, pp. 2077–2082, 2006.

[79] I. B. Obot, N. O. Obi-Egbedi, S. A. Umoren, and E. E. Ebenso, "Synergistic and antagonistic effects of anions and ipomoea invulcrata as green corrosion inhibitor for aluminium dissolution in acidic medium," *International Journal of Electrochemical Science*, vol. 5, no. 7, pp. 994–1007, 2010.

[80] I. B. Obot and N. O. Obi-Egbedi, "*Ipomoea involcrata* as an ecofriendly inhibitor for aluminium in alkaline medium," *Portugaliae Electrochimica Acta*, vol. 27, no. 4, pp. 517–524, 2009.

[81] I. B. Obot and N. O. Obi-Egbedi, " An interesting and efficient green corrosion inhibitor for aluminium from extracts of *Chlomolaena odorata* L. in acidic solution," *Journal of Applied Electrochemistry*, vol. 40, no. 11, pp. 1977–1983, 2010.

[82] N. O. Eddy, P. A. Ekwumemgbo, and P. A. P. Mamza, "Ethanol extract of Terminalia catappa as a green inhibitor for the corrosion of mild steel in H_2SO_4," *Green Chemistry Letters and Reviews*, vol. 2, no. 4, pp. 223–231, 2009.

[83] F. A. de Souza and A. Spinelli, "Caffeic acid as a green corrosion inhibitor for mild steel," *Corrosion Science*, vol. 51, no. 3, pp. 642–649, 2008.

[84] O. K. Abiola, J. O. E. Otaigbe, and O. J. Kio, "Gossipium hirsutum L. extracts as green corrosion inhibitor for aluminum in NaOH solution," *Corrosion Science*, vol. 51, no. 8, pp. 1879–1881, 2009.

[85] F. Tirbonod and C. Fiaud, "Inhibition of the corrosion of aluminium alloys by organic dyes," *Corrosion Science*, vol. 18, no. 2, pp. 139–149, 1978.

[86] J. D. Talati and J. M. Daraji, "Inhibition of corrosion of B26S aluminium in phosphoric acid by some azo dyes," *Journal of the Indian Chemical Society*, vol. 68, no. 2, pp. 67–72, 1991.

[87] P. Gupta, R. S. Chaudhary, T. K. G. Namboodhiri, B. Prakash, and B. B. Prasad, "Effect of mixed inhibitors on dezincification and corrosion of 63/37 brass in 1% sulfuric acid," *Corrosion*, vol. 40, no. 1, pp. 33–36, 1984.

[88] P. B. Tandel and B. N. Oza, "Performance of some dyestuffs as inhibitors during corrosion of mild-steel in binary acid mixtures (HCl + HNO^3)," *Journal of the Electrochemical Society of India*, vol. 49, pp. 49–128, 2000.

[89] M. L. Zheludkevich, R. Serra, M. F. Montemor, and M. G. S. Ferreira, "Oxide nanoparticle reservoirs for storage and prolonged release of the corrosion inhibitors," *Electrochemistry Communications*, vol. 7, no. 8, pp. 836–840, 2005.

[90] V. V. Torres, R. S. Amado, C. Faia de Sá et al., "Inhibitory action of aqueous coffee ground extracts on the corrosion of

carbon steel in HCl solution," *Corrosion Science*, vol. 53, no. 7, pp. 2385–2392, 2011.

[91] G. G. Geesey, "Microbial exopolymers: ecological and econimic considerations," *American Society Microbiology News*, vol. 48, pp. 9–14, 1982.

[92] I. B. Beech and C. C. Gaylarde, "Recent advances in the study of biocorrosion—an overview," *Revista de Microbiologia*, vol. 30, no. 3, pp. 177–190, 1999.

[93] P. S. Guiamet and S. G. Gomez De Saravia, "Laboratory studies of biocorrosion control using traditional and environmentally friendly biocides: an overview," *Latin American Applied Research*, vol. 35, no. 4, pp. 295–300, 2005.

[94] M. F. Montemor, W. Trabelsi, M. Zheludevich, and M. G. S. Ferreira, "Modification of bis-silane solutions with rare-earth cations for improved corrosion protection of galvanized steel substrates," *Progress in Organic Coatings*, vol. 57, no. 1, pp. 67–77, 2006.

[95] A. Pepe, M. Aparicio, S. Ceré, and A. Durán, "Preparation and characterization of cerium doped silica sol-gel coatings on glass and aluminum substrates," *Journal of Non-Crystalline Solids*, vol. 348, pp. 162–171, 2004.

[96] X. W. Yu, C. N. Cao, Z .M. Yao, Z. Derui, and Y. Zhongda, "Corrosion behavior of rare earth metal (REM) conversion coatings on aluminum alloy LY12," *Materials Science and Engineering A*, vol. 284, no. 1-2, pp. 56–63, 2000.

[97] A. N. Khramov, N. N. Voevodin, V. N. Balbyshev, and M. S. Donley, "Hybrid organo-ceramic corrosion protection coatings with encapsulated organic corrosion inhibitors," *Thin Solid Films*, vol. 447-448, pp. 549–557, 2004.

[98] E. M. Sherif and S. M. Park, "Effects of 1,4-naphthoquinone on aluminum corrosion in 0.50 M sodium chloride solutions," *Electrochimica Acta*, vol. 51, no. 7, pp. 1313–1321, 2006.

[99] M. S. Donley, R. A. Mantz, A. N. Khramov, V. N. Balbyshev, L. S. Kasten, and D. J. Gaspar, "The self-assembled nanophase particle (SNAP) process: a nanoscience approach to coatings," *Progress in Organic Coatings*, vol. 47, no. 3-4, pp. 401–415, 2003.

[100] D. V. Andreeva, D. Fix, H. Möhwald, and D. G. Shchukin, "Self-healing anticorrosion coatings based on pH-sensitive polyelectrolyte/inhibitor sandwichlike nanostructures," *Advanced Materials*, vol. 20, no. 14, pp. 2789–2794, 2008.

[101] M. L. Zheludkevich, D. G. Shchukin, K. A. Yasakau, H. Möhwald, and M. G. S. Ferreira, "Anticorrosion coatings with self-healing effect based on nanocontainers impregnated with corrosion inhibitor," *Chemistry of Materials*, vol. 19, no. 3, pp. 402–411, 2007.

[102] V. DeGiorgi, "Corrosion basics and computer modeling," in *Industrial Applications of the BEM*, chapter 2, pp. 47–79, 1986.

[103] H. Sun, P. Ren, and J. R. Fried, "The compass force field: parameterization and validation for polyphosphazenes," *Computational and Theoretical Polymer*, vol. 8, pp. 229–246, 1998.

[104] N. G. Zamani, J. F. Porter, and A. A. Mufti, "Survey of computational efforts in the field of corrosion engineering," *International Journal for Numerical Methods in Engineering*, vol. 23, no. 7, pp. 1295–1311, 1986.

[105] R. S. Munn, "A review of the development of Computational Corrosion analysis for special corrosion modelling through its maturity in the Mid 1980's-Computer modelling in Corrosion," ASTM STP 1154 American Study for Testing and Materials, Philadelphia, pp 215-228, 1991.

[106] K. F. Khaled, "Molecular simulation, quantum chemical calculations and electrochemical studies for inhibition of mild steel by triazoles," *Electrochimica Acta*, vol. 53, no. 9, pp. 3484–3492, 2008.

[107] F. J. Presuel-Moreno, H. Wang, M. A. Jakab, R. G. Kelly, and J. R. Scully, "Computational modeling of active corrosion inhibitor release from an Al-Co-Ce metallic coating," *Journal of the Electrochemical Society*, vol. 153, no. 11, Article ID 002611JES, pp. B486–B498, 2006.

[108] Pradip and B. Rai, "Design of tailor-made surfactants for industrial applications using a molecular modelling approach," *Colloids and Surfaces A*, vol. 205, no. 1-2, pp. 139–148, 2002.

[109] K. F. Khaled and M. A. Amin, "Computational and electrochemical investigation for corrosion inhibition of nickel in molar nitric acid by piperidines," *Journal of Applied Electrochemistry*, vol. 38, no. 11, pp. 1609–1621, 2008.

[110] K. F. Khaled and M. A. Amin, "Electrochemical and molecular dynamics simulation studies on the corrosion inhibition of aluminum in molar hydrochloric acid using some imidazole derivatives," *Journal of Applied Electrochemistry*, vol. 39, no. 12, pp. 2553–2568, 2009.

15

Effect of Temperature on the Corrosion Inhibition of Trans-4-Hydroxy-4′-Stilbazole on Mild Steel in HCl Solution

Ayssar Nahlé, Ideisan I. Abu-Abdoun, and Ibrahim Abdel-Rahman

Department of Chemistry, College of Sciences, University of Sharjah, P.O. Box 27272, Sharjah, United Arab Emirates

Correspondence should be addressed to Ayssar Nahlé, anahle@sharjah.ac.ae

Academic Editor: Vesna Mišković-Stanković

The inhibition and the effect of temperature and concentration of trans-4-hydroxy-4′-stilbazole on the corrosion of mild steel in 1 M HCl solution was investigated by weight loss experiments at temperatures ranging from 303 to 343 K. The studied inhibitor concentrations were between 1×10^{-7} M and 1×10^{-3} M. The percentage inhibition increased with the increase of the concentration of the inhibitor. The percentage inhibition reached about 94% at the concentration of 1×10^{-3} M and 303 K. On the other hand, the percentage inhibition decreased with the increase of temperature. Using the Temkin adsorption isotherm, the thermodynamic parameters for the adsorption of this inhibitor on the metal surface were calculated. Trans-4-hydroxy-4′-stilbazole was found to be a potential corrosion inhibitor since it contained not only nitrogen and oxygen, but also phenyl and pyridine rings that are joined together with a double bond (–C=C–) in conjugation with these rings.

1. Introduction

Organic compounds containing polar groups by which the molecule can become strongly or specifically adsorbed on the metal surface constitute most organic inhibitors [1, 2]. These inhibitors, which include the organic N, P, S, and OH groups, are known to be similar to catalytic poisons, as they decrease the reaction rate at the metal/solution interface without, in general, being involved in the reaction considered. It is generally accepted that most organic inhibitors act via adsorption at the metal/solution interface. The mechanism by which an inhibitor decreases the corrosion current is achieved by interfering with some of the steps for the electrochemical process.

The corrosion inhibition of mild steel in aggressive acidic solutions has been widely investigated. In industries, hydrochloric acid solutions are often used in order to remove scale and salts from steel surfaces, and cleaning tanks and pipelines. This treatment may be prerequisite for coating by electroplating, galvanizing, or painting techniques. The acid must be treated to prevent an extensive dissolution of the underlying metal. This treatment involves the addition of some organic inhibitors to the acid solution that adsorb at the metal/solution interface by displacing water molecules on the surface and forming a compact barrier film.

Many authors have used various nitrogen-containing compounds in their corrosion inhibition investigations. These compounds included quaternary ammonium salts [3–10], polyamino-benzoquinone polymers [11], azoles [12–19], substituted aniline-N-salicylidenes [20], amides [21], heterocyclic compounds [22, 23], and cationic surfactants [24, 25]. Other authors worked on sulfur-containing inhibitors [26–31]. Other studies involved the effect of addition of some ions on the inhibition efficiency of some organic compounds. These ions included chromium [32], iodide [33, 34], and chloride [35, 36]. Structural effect of organic compounds as corrosion inhibitors was also studied [17, 34, 37–39]. In all these studies, the nitrogen atom(s) in the compounds showed to be able to absorb very well on the metal surface and form protective layer, which in turn increased the corrosion inhibition with the increase in the concentration of the inhibitor, in some cases reaching 99% inhibition [5, 40, 41].

No studies have been reported on trans-4-hydroxy-4′-stilbazole inhibitor employed in this current work, in terms of studying the effect of temperature on the corrosion

TABLE 1: Effect of concentration of trans-4-hydroxy-4′-stilbazole on the corrosion rate ($mg \cdot cm^{-2} \cdot h^{-1}$) and percentage efficiency of mild steel in 1 M HCl at various temperatures.

	Temperature/k									
	303		313		323		333		343	
Concentration of inhibitor	Corr. Rate	% Efficiency	Corr. Rate	% Efficiency	Corr. Rate	% Efficiency	Corr. Rate	% Efficiency	Corr. Rate	% Efficiency
1 M HCl	0.961	—	1.394	—	4.671	—	12.225	—	26.280	—
1 M HCl + 1×10^{-7} M	0.950	1	1.381	0.9	4.648	0.5	14.057	−15	28.139	−7.1
1 M HCl + 1×10^{-6} M	0.916	5	1.343	3.7	4.570	2.2	12.733	−4	27.061	−3.0
1 M HCl + 1×10^{-5} M	0.714	26	1.203	14	4.192	10.3	12.714	−4	24.901	5.2
1 M HCl + 1×10^{-4} M	0.191	80	0.361	74	2.179	53.4	8.959	27	24.020	8.6
1 M HCl + 1×10^{-3} M	0.060	94	0.112	92	0.543	88.4	2.321	81	8.619	67

inhibition of mild steel in 1 M HCl solution. Mild steel was chosen in our studies since high-temperature aggressive acids are widely used in industries in connection to mild and low alloy steels.

In this work, trans-4-hydroxy-4′-stilbazole was prepared via a synthesis which produced a high percentage yield of this pure compound.

The aim of this work is to study, using weight loss measurements, the effect of temperature on the corrosion inhibition of mild steel in 1 M HCl solution by trans-4-hydroxy-4′-stilbazole, a synthesized compound that was previously studied using electrochemical technique [40] and found to be a very efficient corrosion inhibitor, and to calculate the thermodynamic parameters. Hence, the output of this study is intended to be the building block or the nucleus for a new family or group of stilbazole derivatives in all studies of corrosion inhibitors.

2. Experimental Details

2.1. Instrumentation. The experimental setup consisted of a 250 mL round bottom flask fitted with a reflux condenser and a long glass rod on which the specimen was hooked and in turn immersed in a thermally controlled water bath [28, 42].

Analytical-grade hydrochloric acid (Ajax), 4-iodophenol, triethylamine, palladium acetate, tri(o-tolyl)phosphine, acetonitrile, dichloromethane, $MgSO_4$, hexane, and pentane were obtained from Aldrich Chemical Company and used without further purification. 4-Vinylpyridine was distilled under reduced pressure and stored in cold place.

2.2. Sample Preparation. Rectangular specimens (1 cm × 2.3 cm × 0.3 cm) cut from large sheet of 3 mm thick mild steel (IS 226 containing 0.18% C, 0.6% Mn, and 0.35% Si) supplied by "Reliable Steel Traders," Sharjah, UAE, were used for weight loss measurements. A 2 mm diameter hole was drilled close to the upper edge of the specimen and served to be hooked with a glass rod for immersion purposes. Prior to each experiment, the specimens were polished with 600 grade emery paper, rinsed with distilled water, degreased with acetone, dried, and finally weighed precisely on an accurate analytical balance.

2.3. Synthesis of Trans-4-Hydroxy-4′-Stilbazole. Several methods for the synthesis of trans-4-hydroxy-4′-stilbazole are available including the reaction of 4-picoline with 4-Hydroxy benzaldehyde in acetic anhydride [43, 44], which, under such conditions, gave only tiny yields after long reactions. A better procedure was developed and which used Heck-type vinylic addition reaction [45] catalyzed by palladium. Using this procedure, we were able to obtain a yield of about 80% and quantities reaching up to 20 g in one single reaction, following the modified procedure set by [46] and described in our previous work [40].

2.4. Measuring Procedure. The flask was filled with 100 mL of 1 M HCl solution either with or without trans-4-hydroxy-4′-stilbazole of various concentrations, then placed in water bath. As soon as the required working temperature was reached, the mild steel sample was immersed in the solution, and left there for exactly six hours, after which the sample was removed, rinsed with distilled deionized water, degreased with acetone, dried, and finally weighed precisely on an accurate analytical balance. This procedure was repeated with all the samples with a variety of inhibitor concentrations ranging from 1×10^{-7} M up to 1×10^{-3} M, and at temperatures ranging from 303 K to 343 K.

3. Results

Weight loss corrosion tests were carried out on the mild steel in 1 M HCl in the absence or presence of trans-4-hydroxy-4′-stilbazole over a period of 6 hours. Table 1 represents the corrosion rates [$mg \cdot cm^{-2} \cdot h^{-1}$] and the percentage efficiencies for the studied inhibitor with concentrations varying from 1×10^{-7} M to 1×10^{-3} M at 303, 313, 323, 333, and 343 K, respectively. The percentage efficiency was calculated according to the following expression:

$$\% \text{ Inhibition} = \frac{W_{\text{Uninh.}} - W_{\text{Inh.}}}{W_{\text{Uninh.}}} \times 100, \quad (1)$$

where $W_{\text{Uninh.}}$ is corrosion rate without inhibitor and $W_{\text{Inh.}}$ is corrosion rate with inhibitor.

Figures 1 and 2 show the plots of the corrosion rate of trans-4-hydroxy-4′-stilbazole as a function of concentration

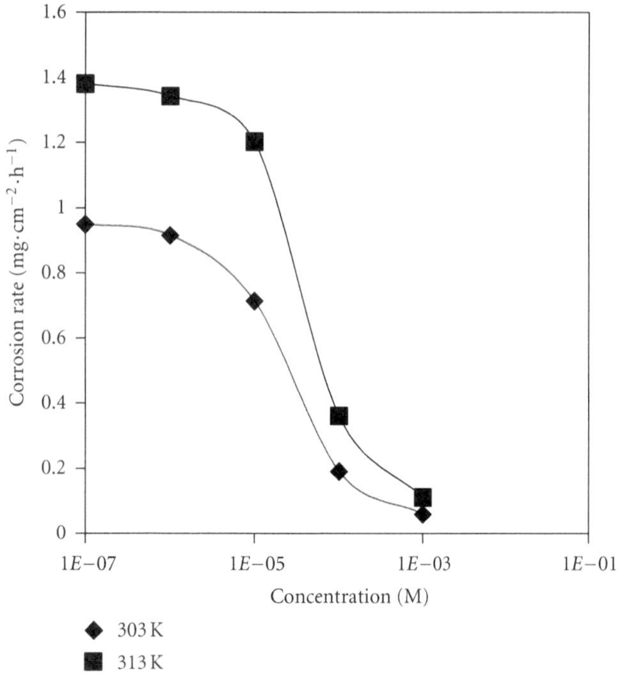

FIGURE 1: Effect of concentration of trans-4-hydroxy-4′-stilbazole on the corrosion rate ($mg\cdot cm^{-2}\cdot h^{-1}$) of mild steel in 1 M HCl at various temperatures.

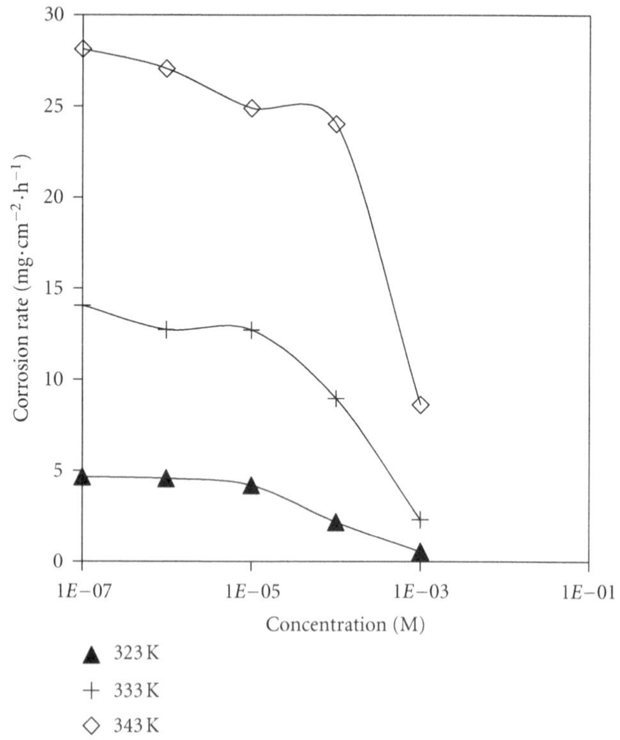

FIGURE 2: Effect of concentration of trans-4-hydroxy-4′-stilbazole on the corrosion rate ($mg\cdot cm^{-2}\cdot h^{-1}$) of mild steel in 1 M HCl at various temperatures.

at 303, 313, 323, 333, and 343 K. At 303 K (Figure 1), the corrosion rate dropped from 0.961 $mg\cdot cm^{-2}\cdot h^{-1}$ (1 M HCl in the absence of the inhibitor) to 0.950 $mg\cdot cm^{-2}\cdot h^{-1}$ when 1×10^{-7} M of trans-4-hydroxy-4′-stilbazole was present in the 1 M HCl. This corrosion rate continued to decrease slightly to reach 0.714 $mg\cdot cm^{-2}\cdot h^{-1}$ (26% inhibition) at a concentration of 1×10^{-5} M, followed by a steep decrease to reach 0.191 $mg\cdot cm^{-2}\cdot h^{-1}$ when the inhibitor's concentration was 1×10^{-4} M, and finally, at higher concentration (1×10^{-3} M) the corrosion rate as initially decreased slightly to reach 0.060 $mg\cdot cm^{-2}\cdot h^{-1}$ (94% inhibition). At 313 K (Figure 1), the curve showed to have similar shape as that obtained at 303 K. At concentrations greater than 1×10^{-5} M, the corrosion rate decreased steeply and reached about 0.112 $mg\cdot cm^{-2}\cdot h^{-1}$ (92%) at 1×10^{-3} M.

At 323 K (Figure 2), when the concentration of the inhibitor was between 1×10^{-7} M and 1×10^{-5} M, it had very slight effect on the corrosion rate; whereas at higher concentrations, the corrosion rate dropped from 4.192 $mg\cdot cm^{-2}\cdot h^{-1}$ (at 1×10^{-5} M) down to 2.179 and 0.543 $mg\cdot cm^{-2}\cdot h^{-1}$ at 1×10^{-4} M and 1×10^{-3} M, respectively.

At 333 K (Table 1), the presence of trans-4-hydroxy-4′-stilbazole acted as an accelerator between 1×10^{-7} M and 1×10^{-5} M, and at concentrations of 1×10^{-4} M and greater, it acted as inhibitor reaching percentage inhibitions of 27% and 81% at 1×10^{-4} M and 1×10^{-3} M, respectively. At 343 K, likewise, trans-4-hydroxy-4′-stilbazole acted as an accelerator between 1×10^{-7} M and 1×10^{-6} M, then inhibited the mild steel at 1×10^{-5} M and greater concentration, reaching a percentage inhibition of about 67% at 1×10^{-3} M. It was noticed that a tenfold increase in the inhibitor's concentration, namely, from 1×10^{-4} M to 1×10^{-3} M, had a very sizeable increase in the percentage inhibition as this increased from 27% to 81% at 333 K, and from 8.6% to 67% at 343 K.

Figure 3 shows the plots of the percent inhibition versus the concentration of the inhibitor at temperatures of 303, 313, 323, 333, and 343 K, respectively. This figure showed that the percent inhibition was not sufficiently affected by the increase of temperature (303 to 323 K), especially when low concentrations (1×10^{-7} to 1×10^{-5} M) of inhibitors were employed; whereas at higher concentrations (above 1×10^{-5} M), the presence of the inhibitor greatly increased the percent inhibition at all temperatures. At higher temperatures, it was noticed that at concentrations between 1×10^{-7} M and 1×10^{-4} M (343 K) and in some cases to 1×10^{-5} M (333 K) trans-4-hydroxy-4′-stilbazole acted as a corrosion accelerator.

The data obtained from the weight loss measurements were plotted in accordance to Arrhenius equation:

$$\ln \text{rate} = -\frac{E_a}{RT} + \text{const.}, \qquad (2)$$

where E_a is activation energy [$kcal\cdot mol^{-1}$], R is gas constant [$kcal\cdot mol^{-1}$], T is absolute temperature [K], and const. is constant.

Figure 4 represents the Arrhenius plot of the corrosion of mild steel in 1 M HCl solution (ln corrosion rate as a function of $1/T$) with or without the presence of trans-4-hydroxy-4′-stilbazole at concentrations ranging from 1×10^{-7} M to 1×10^{-3} M. From this figure, the slope ($-E_a/R$) of each

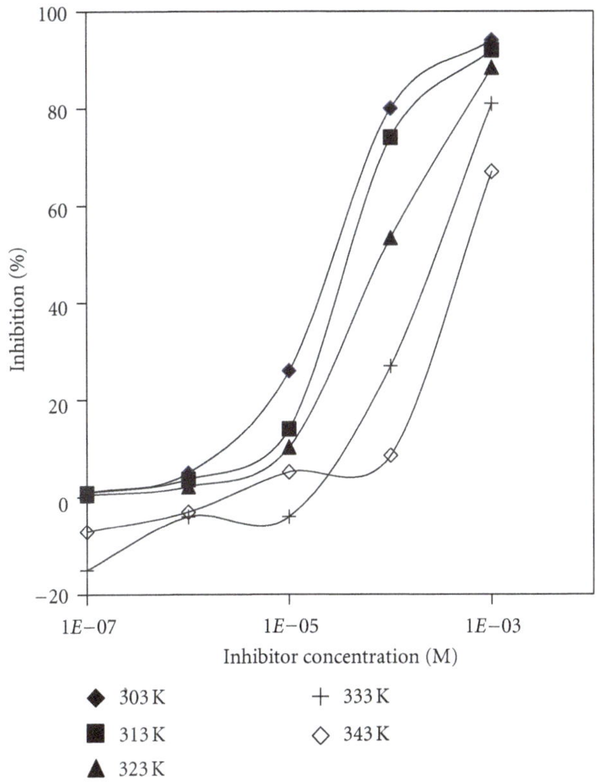

FIGURE 3: Effect of concentration of trans-4-hydroxy-4′-stilbazole on the percent inhibition of mild steel in 1 M HCl at various temperatures.

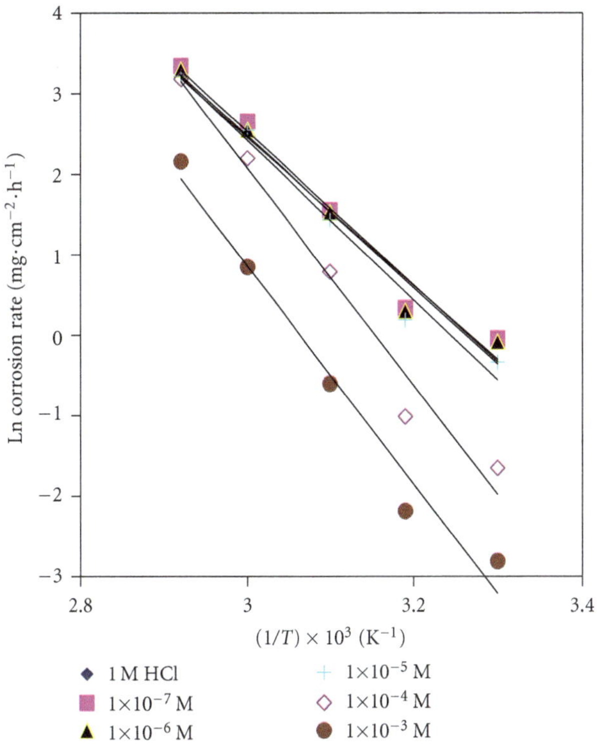

FIGURE 4: Effect of temperature on the corrosion rate of mild steel in 1 M HCl solution with and without the presence of various concentrations of trans-4-hydroxy-4′-stilbazole inhibitor.

individual line was determined and used to calculate the activation energy according to (2), and taking $R = 1.987 \times 10^{-3}$ kcal·mol^{-1} (Table 2). The increase of concentration of trans-4-hydroxy-4′-stilbazole (from 1×10^{-7} M to 1×10^{-3} M) increased the activation energies for the corrosion of mild steel in 1 M HCl (initially 18.27 kcal·mol^{-1}) (Table 2).

Table 3 shows the surface coverage of various concentrations of trans-4-hydroxy-4′-stilbazole (from 1×10^{-7} M to 1×10^{-3} M) on mild steel surface as a function of temperature. These values were extracted from the corresponding percent efficiency values reported earlier in Table 1. The plot of surface coverage, θ, against the natural logarithm of the concentration of the inhibitor, ln C, for mild steel at various inhibitor temperatures is shown in Figure 5. After examining these data and adjusting them to different theoretical adsorption isotherms, it was concluded that the inhibitor was adsorbed on the mild steel surface according to Temkin isotherm:

$$-2a\theta = \ln KC, \tag{3}$$

where a is molecular interaction constant, θ is degree of coverage, K is equilibrium constant for the adsorption reaction, and C is concentration of the inhibitor. The equilibrium constant for the adsorption reaction, K, is related to the standard free energy of adsorption [24] via:

$$K = \frac{1}{55.5}\exp\left(-\frac{\Delta G}{RT}\right), \tag{4}$$

where K is equilibrium constant for the adsorption reaction, 55.5 is concentration of water [mol·L^{-1}], ΔG is standard free energy [kcal·mol^{-1}], R is gas constant [kcal·mol^{-1}], and T is absolute temperature [K].

According to (3), the straight lines shown in Figure 5 will have the following slopes and intercepts:

$$\text{Slope} = -\frac{1}{2a}, \tag{5}$$

$$\text{Intercept} = -\frac{1}{2a}\ln K. \tag{6}$$

Combination of (5) and (6) leads to the following relationship:

$$\text{Intercept} = \text{Slope} \cdot (\ln K), \tag{7}$$

from which the equilibrium constant for the adsorption reaction, K, can be calculated:

$$K = e^{(\text{Intercept/Slope})}. \tag{8}$$

The free energy of adsorption of the inhibitor, ΔG, can be calculated from the results in Figure 5 used to calculate the equilibrium constant, K, and (4) at various temperatures (303 K to 323 K).

The enthalpy of adsorption, ΔH, for the inhibitor can be calculated from the following equation:

$$\Delta H = E_a - RT. \tag{9}$$

Table 2: The activation energy (E_a) for the corrosion of mild steel in 1 M HCl with and without trans-4-hydroxy-4′-stilbazole inhibitor at various concentrations.

System	Activation energy, E_a (kcal·mol^{-1})				
	1×10^{-3} M	1×10^{-4} M	1×10^{-5} M	1×10^{-6} M	1×10^{-7} M
1 M HCl	18.27	18.27	18.27	18.27	18.27
1 M HCl+trans-4-hydroxy-4′-stilbazole	26.93	26.82	19.72	18.76	18.93

Table 3: Effect of concentration of trans-4-hydroxy-4′-stilbazole on surface coverage for mild steel in 1 M HCl at various temperatures.

	Temperature/K		
	303	313	323
Concentration of inhibitor	Surface coverage θ	Surface coverage θ	Surface coverage θ
1 M HCl + 1×10^{-6} M	0.05	0.037	0.022
1 M HCl + 1×10^{-5} M	0.26	0.14	0.103
1 M HCl + 1×10^{-4} M	0.80	0.74	0.534
1 M HCl + 1×10^{-3} M	0.94	0.92	0.884

The entropy, ΔS, can be calculated at various temperatures for the inhibitor using the following equation:

$$\Delta G = \Delta H - T\Delta S. \quad (10)$$

4. Discussion

The results in Table 2 show that the activation energy (E_a) for the corrosion of mild steel in the presence of the inhibitor are higher compared to the activation energy in the absence of inhibitor at both 1×10^{-3} M and 1×10^{-4} M (about 27 versus about 18 kcal·mol^{-1}). Whereas at lower concentration (from 1×10^{-5} to 1×10^{-7} M), the activation energy is very close to that in uninhibited HCl solution. This can be attributed to the fact that higher values of E_a in the presence of inhibitor compared to its absence are generally consistent with a physisorption, while unchanged or lower values of E_a in inhibited solution suggest charge sharing or transfer from the organic inhibitor to the metal surface to form coordinate covalent bonds (chemisorption).

The increase in the activation energies for the corrosion is attributed to a decrease in the adsorption of the inhibitor on the metal surface as the temperature increased; subsequently, an increase in the corrosion rate will result due to the greater exposed area of the metal surface to the acid.

Tables 4, 5, and 6 show the thermodynamic data obtained in the presence of the inhibitor at various temperatures. These thermodynamic quantities represent the algebraic sum of the values for adsorption and desorption. The negative value of ΔG indicates the spontaneous adsorption of inhibitor on the surface of the mild steel. The free energy, ΔG, varies from -11.54 kcal·mol^{-1} at 303 K to -11.12 kcal·mol^{-1} at 323 K. The adsorption process is believed to be exothermic and associated with a decrease in entropy (ΔS) of solute, while the opposite is true for the solvent [31]. The gain in entropy which accompanies the substitutional adsorption process is attributed to the increase in the solvent entropy (Table 6). This agrees with the general suggestion that the values of ΔG increase with the increase of inhibition efficiency [20, 26] as the adsorption of organic compound is accompanied by desorption of water molecules off the surface.

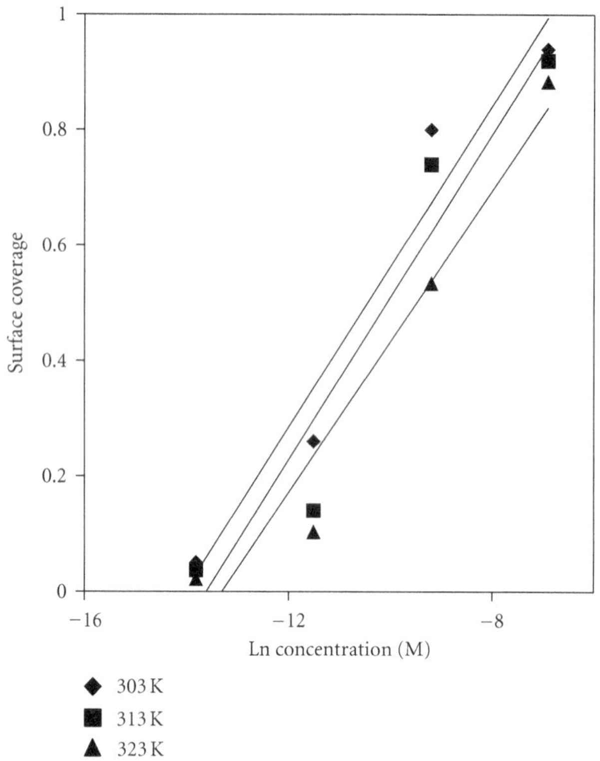

Figure 5: Effect of concentration of trans-4-hydroxy-4′-stilbazole on the surface coverage of mild steel in 1 M HCl at various temperatures.

Table 4: The enthalpy of adsorption (ΔH) for mild steel in 1 M HCl in the presence of 1×10^{-3} M trans-4-hydroxy-4′-stilbazole inhibitor at various temperatures (303 K–343 K).

ΔH, kcal·mol^{-1}				
303 K	313 K	323 K	333 K	343 K
26.33	26.31	26.29	26.27	26.25

The electron density group (–OH) on the phenyl ring gives an extra inhibition effect. The high inhibition efficiency may be attributed to the preferred flat orientation of this compound on the metal surface. The interaction occurs

TABLE 5: The free energy of adsorption (ΔG_{ads}) for mild steel in 1 M HCl in the presence of trans-4-hydroxy-4′-stilbazole inhibitor at various temperatures (303 K–323 K).

ΔG, kcal·mol^{-1}		
303 K	313 K	323 K
−11.54	−11.76	−11.12

TABLE 6: The change in entropy (ΔS) for mild steel in 1 M HCl in the presence of trans-4-hydroxy-4′-stilbazole inhibitor at various temperatures (303 K–323 K).

ΔS, kcal·K^{-1}·mol^{-1}		
303 K	313 K	323 K
0.125	0.122	0.116

between the delocalized π-electrons of the two rings, the double bond (–C=C–), and the lone pair of electrons on N and O atoms with the positively charged metal surface. In addition, (–C=C–) bond plays the major role in the stabilization of both sides of the Stilbazole structure.

These results agree with [27] who suggested that the inhibition efficiency of organic compounds depends on many factors including their charge density, number of adsorption sites, heat of hydrogenation, mode of interaction with the metal surface, and formation of metallic complexes [37].

5. Conclusion

Trans-4-hydroxy-4′-stilbazole was found to be a highly efficient inhibitor for mild steel in 1.0 M HCl solution, reaching about 94% at 1.0×10^{-3} M and 303 K, a concentration considered to be very moderate.

Trans-4-hydroxy-4′-stilbazole is a potential corrosion inhibitor since it contains not only nitrogen and oxygen, but also phenyl and pyridine rings that are joined together with a double bond (–C=C–) in conjugation with these rings. It was apparent from the molecular structure that this compound would be adsorbed onto the metal surface through the lone pair of electron of nitrogen and oxygen and pi electrons of the two aromatic rings (phenyl and pyridine) and the double bond that join them together.

The percentage of inhibition in the presence of this inhibitor was decreased with temperature, which indicates that physical adsorption was the predominant inhibition mechanism because the quantity of adsorbed inhibitor decreases with increasing temperature.

References

[1] B. B. Damaskin, O. A. Pietrij, and W. W. Batrokov, "Adsorpcja organiczeskich sojedinienij na electrodach," Moskva, 1968.

[2] G. Okamoto, M. Nagayama, J. Kato, and T. Baba, "Effect of organic inhibitors on the polarization characteristics of mild steel in acid solution," *Corrosion Science*, vol. 2, no. 1, pp. 21–27, 1962.

[3] S. M. Beloglazov, Z. I. Dzhafarov, V. N. Polyakov, and N. N. Demushia, "Quaternary ammonium salts as corrosion inhibitors of steel in the presence of sulfate-reducing bacteria," *Protection of Metals USSR*, vol. 27, no. 6, pp. 810–813, 1991.

[4] A. V. Fokin, M. V. Pospelov, and A. N. Levichev, "Structural and protective capacities of organic corrosion inhibitors. 2. Alkylethylene diamines and quaternary ammonium salts," *Protection of Metals USSR*, vol. 19, no. 2, pp. 242–244, 1983.

[5] A. H. Nahlé, "Electrochemical studies of corrosion inhibition of a series of quaternary ammonium salts on iron in HCl solution," *Corrosion Prevention and Control*, vol. 44, no. 4, pp. 99–105, 1997.

[6] A. H. Nahlé, "Inhibition of iron in HCl using benzyl trimethyl- and triethyl-ammonium chlorides," *Corrosion Prevention and Control*, vol. 45, no. 4, pp. 124–130, 1998.

[7] A. Nahlé, "Effect of triethanolamine on the electrochemical dissolution of solder in NaOH solution," *Bulletin of Electrochemistry*, vol. 18, no. 3, pp. 105–110, 2002.

[8] A. Nahle and F. C. Walsh, "Electrochemical studies of two corrosion inhibitors for iron in HCl: cetyltrimethyl ammonium bromide and tetraphenyl phosphonium chloride," *Corrosion Prevention and Control*, vol. 42, no. 2, pp. 30–34, 1995.

[9] B. V. Savithri and S. M. Mayanna, "Tetrabutyl ammonium iodide, cetyl pyridinium bromide and cetyl trimethyl ammonium bromide as corrosion inhibitors for mild steel in sulphuric acid," *Indian Journal of Chemical Technology*, vol. 3, no. 5, pp. 256–258, 1996.

[10] T. Vasudevan, S. Muralidharan, S. Alwarappan, and S. V. K. Iyer, "The influence of N-hexadecyl benzyl dimethyl ammonium chloride on the corrosion of mild steel in acids," *Corrosion Science*, vol. 37, no. 8, pp. 1235–1244, 1995.

[11] S. Muralidharan, K. L. N. Phani, S. Pitchumani, S. Ravichandran, and S. V. K. Iyer, "Polyamino-benzoquinone polymers: a new class of corrosion inhibitors for mild steel," *Journal of the Electrochemical Society*, vol. 142, no. 5, pp. 1478–1483, 1995.

[12] O. Benali, L. Larabi, M. Traisnel, L. Gengembre, and Y. Harek, "Electrochemical, theoretical and XPS studies of 2-mercapto-1-methylimidazole adsorption on carbon steel in 1 M $HClO_4$," *Applied Surface Science*, vol. 253, no. 14, pp. 6130–6139, 2007.

[13] E. S. H. E. Ashry, A. E. Nemr, S. A. Essawy, and S. Ragab, "Corrosion inhibitors part V: QSAR of benzimidazole and 2-substituted derivatives as corrosion inhibitors by using the quantum chemical parameters," *Progress in Organic Coatings*, vol. 61, no. 1, pp. 11–20, 2008.

[14] K. F. Khaled, "The inhibition of benzimidazole derivatives on corrosion of iron in 1 M HCl solutions," *Electrochimica Acta*, vol. 48, no. 17, pp. 2493–2503, 2003.

[15] A. Popova, "Temperature effect on mild steel corrosion in acid media in presence of azoles," *Corrosion Science*, vol. 49, no. 5, pp. 2144–2158, 2007.

[16] A. Popova, M. Christov, S. Raicheva, and E. Sokolova, "Adsorption and inhibitive properties of benzimidazole derivatives in acid mild steel corrosion," *Corrosion Science*, vol. 46, no. 6, pp. 1333–1350, 2004.

[17] A. Popova, M. Christov, and A. Zwetanova, "Effect of the molecular structure on the inhibitor properties of azoles on mild steel corrosion in 1 M hydrochloric acid," *Corrosion Science*, vol. 49, no. 5, pp. 2131–2143, 2007.

[18] V. S. Sastri, J. R. Perumareddi, M. Lashgari, and M. Elbouj-daini, "Application of ligand field theory in corrosion inhibition," *Corrosion*, vol. 64, no. 4, pp. 283–288, 2008.

[19] K. Tebbji, A. Aouniti, A. Attayibat et al., "Inhibition efficiency of two bipyrazole derivatives on steel corrosion in

hydrochloric acid media," *Indian Journal of Chemical Technology*, vol. 18, pp. 244–253, 2011.

[20] J. D. Talati, M. N. Desai, and N. K. Shah, "Meta-substituted aniline-N-salicylidenes as corrosion inhibitors of zinc in sulphuric acid," *Materials Chemistry and Physics*, vol. 93, no. 1, pp. 54–64, 2005.

[21] T. Tüken, B. Yazici, and M. Erbil, "The effect of nicotinamide on iron corrosion in chloride solutions," *Turkish Journal of Chemistry*, vol. 26, no. 5, pp. 735–742, 2002.

[22] A. A. A. Fattah, E. M. Mabrouk, R. M. A. Elgalil, and M. M. Ghoneim, "N-heterocyclic compounds as corrosion inhibitors for Zn in HCl acid solutions," *Bulletin de la Societé Chimique de France*, vol. 1, pp. 48–53, 1991.

[23] S. L. Granese, B. M. Rosales, C. Oviedo, and J. O. Zerbino, "The inhibition action of heterocyclic nitrogen organic compounds on Fe and steel in HCl media," *Corrosion Science*, vol. 33, no. 9, pp. 1439–1453, 1992.

[24] H. A. Al-Lohedan, E. Khamis, and Z. A. Issa, "Studies on the influence of temperature on the adsorption of some cationic surfactants on to steel," *Adsorption Science and Technology*, vol. 13, no. 3, pp. 137–152, 1996.

[25] L. G. Qiu, A. J. Xie, and Y. H. Shen, "A novel triazole-based cationic gemini surfactant: synthesis and effect on corrosion inhibition of carbon steel in hydrochloric acid," *Materials Chemistry and Physics*, vol. 91, no. 2-3, pp. 269–273, 2005.

[26] B. G. Ateya, B. E. El-Anadouli, and F. M. El-Nizamy, "The effect of thiourea on the corrosion kinetics of mild steel in H_2SO_4," *Corrosion Science*, vol. 24, no. 6, pp. 497–507, 1984.

[27] A. S. Fouda, M. N. Mousa, F. I. Taha, and A. I. Elneamaa, "The role of some thiosemicarbazide derivatives in the corrosion inhibition of aluminum in HCl," *Corrosion Science*, vol. 26, no. 9, pp. 719–726, 1986.

[28] A. Nahlé, "Effect of temperature on the corrosion inhibition of carbon steel in HCl solutions," *Bulletin of Electrochemistry*, vol. 17, no. 5, pp. 221–226, 2001.

[29] A. Nahlé, "Inhibition of corrosion of iron in HCl solution by semicarbazides and thiosemicarbazides," *Bulletin of Electrochemistry*, vol. 21, no. 6, pp. 275–281, 2005.

[30] S. N. Raicheva, B. V. Aleksiev, and E. I. Sokolova, "The effect of the chemical structure of some nitrogen- and sulphur-containing organic compounds on their corrosion inhibiting action," *Corrosion Science*, vol. 34, no. 2, pp. 343–350, 1993.

[31] S. H. Sanad, A. A. Ismail, and A. A. El-Meligi, "The effect of temperature on the corrosion and corrosion inhibition of steel alloys in hydrochloric acid solutions," *Bulletin of Electrochemistry*, vol. 11, no. 10, pp. 462–469, 1995.

[32] F. Zucchi, G. Trabanelli, and G. Brunoro, "The influence of the chromium content on the inhibitive efficiency of some organic compounds," *Corrosion Science*, vol. 33, no. 7, pp. 1135–1139, 1992.

[33] Y. L. Huang, C. N. Cao, M. Lu, and H. C. Lin, "Inhibition effects of I^- and I_2 on stress corrosion cracking of stainless steel in acidic chloride solutions," *Corrosion*, vol. 49, no. 8, pp. 644–649, 1993.

[34] A. Popova, M. Christov, and T. Deligeorgiev, "Influence of the molecular structure on the inhibitor properties of benzimidazole derivatives on mild steel corrosion in 1 M hydrochloric acid," *Corrosion*, vol. 59, no. 9, pp. 756–764, 2003.

[35] M. Scendo and M. Hepel, "Inhibiting properties of benzimidazole films for Cu(II)/Cu(I) reduction in chloride media studied by RDE and EQCN techniques," *Journal of Electroanalytical Chemistry*, vol. 613, no. 1, pp. 35–50, 2008.

[36] M. Yamaguchi, H. Nishihara, K. Aramaki, M. Tomihari, and S. Furuya, "The inhibitive effect of organic cations on passive film breakdown of iron in a chloride-containing borate buffer solution," *Corrosion Science*, vol. 36, no. 7, pp. 1133–1141, 1994.

[37] A. S. Fouda, A. Abd El-Aal, and A. B. Kandil, "The effect of some phthalimide derivatives on the corrosion behaviour of copper in nitric acid," *Anti-Corrosion Methods and Materials*, vol. 52, no. 2, pp. 96–101, 2005.

[38] K. Kobayashi, K. Shimizu, and M. Iida, "Structural effects of organic compounds as corrosion inhibitors for hydrogen entry into iron in sulphuric acid," *Corrosion Science*, vol. 35, no. 5-8, pp. 1431–1435, 1993.

[39] L. D. Skryler, E. A. Streltsova, and T. L. Skryleva, "Hydrocarbon chain length and their effect on corrosion inhibition by alkylammonium chlorides," *Protection of Metals USSR*, vol. 27, no. 6, pp. 755–758, 1991.

[40] A. Nahlé, I. Abu-Abdoun, and I. Abdel-Rahman, "Electrochemical studies of the effect of trans-4-hydroxy-4′-stilbazole on the corrosion inhibition of mild steel in HCl solution," *Anti-Corrosion Methods and Materials*, vol. 54, no. 4, pp. 244–248, 2007.

[41] A. Nahlé, I. Abu-Abdoun, and I. Abdel-Rahman, "Inhibition of carbon steel corrosion by 4-vinylbenzyl triphenyl phosphonium chloride in HCl solution," *Anti-Corrosion Methods and Materials*, vol. 55, no. 4, pp. 217–224, 2008.

[42] A. Nahlé, I. Abdel-Rahman, and M. Alfarouk, "Effect of temperature on the inhibition of corrosion of carbon steels by semicarbazides and thiosemicarbazides," *Bulletin of Electrochemistry*, vol. 21, no. 8, pp. 353–361, 2005.

[43] U. Kumar, T. Kato, and J. M. J. Fréchet, "Use of intermolecular hydrogen bonding for the induction of liquid crystallinity in the side chain of polysiloxanes," *Journal of the American Chemical Society*, vol. 114, no. 17, pp. 6630–6639, 1992.

[44] B. D. Shaw and E. A. Wagstaff, "26. The reaction between 2-picoline and aromatic aldehydes," *Journal of the Chemical Society (Resumed)*, pp. 77–79, 1933.

[45] W. C. Frank, Y. C. Kim, and R. F. Heck, "Palladium-catalyzed vinylic substitution reactions with heterocyclic bromides," *Journal of Organic Chemistry*, vol. 43, no. 15, pp. 2947–2949, 1978.

[46] D. W. Bruce, S. C. Davis, D. A. Dunmur, S. A. Hudson, P. M. Maitlis, and P. Styring, "Mesomorphic metal complexes derived from 4-alkyloxystilbazoles," *Molecular Crystals and Liquid Crystals Science and Technology Section A*, vol. 215, no. 2, pp. 1–11, 1992.

16

Environmental and Material Influences on the Stress-Corrosion Cracking of Steel in H_2O–CO–CO_2 Solutions

J. W. van der Merwe

School of Chemical and Metallurgical Engineering, University of the Witwatersrand, Johannesburg, Private Bag 3, Wits 2050, South Africa

Correspondence should be addressed to J. W. van der Merwe, josias.vandermerwe@wits.ac.za

Academic Editor: Rokuro Nishimura

The stress-corrosion cracking of A516 pressure vessel steel was investigated by the use of slow strain-rate tests. The orientation of samples to the rolling direction was investigated, and it was found that samples machined longitudinal to the rolling direction showed a slightly increased sensitivity to stress corrosion. The temperature variation showed that for different gas mixtures, the maximum sensitivity to stress corrosion was in the region of 45° to 55°C for the 25% CO gas mixture, whereas with higher CO concentrations, this temperature region of maximum sensitivity moved to higher temperatures. Surface finish showed a slight increase in sensitivity to cracking with increased surface roughness. The most significant increase was found with increased total gas pressures and when samples have been exposed to the environment for an extended period. This was as a result of the inhibition of the corrosion reaction by the passivation of the carbon monoxide, which is a time-dependent process.

1. Introduction

During coal gasification processes, a significant volume of water comes in contact with the combustion gas mixture, and the result of this is a contaminated, black water by-product that is transported and processed. Due to water shortages, the by-product water is used as a cooling agent. This water by-product then comes in contact with steel piping which leads to embrittlement thereof during exposure to the appropriate conditions. The water by-product consists of water with dissolved gasses mainly as carbon monoxide and carbon dioxide. Steel is susceptible to stress-corrosion cracking (SCC) in CO/CO_2/H_2O environments. Extensive research [1, 2] has indicated that SCC in this system is probably due to inhibition of iron corrosion in the H_2O-CO_2 system by the time-dependent adsorption of CO. Rupture of the inhibited passive surface by emerging slip steps then creates the highly localised active region on a passive surface necessary for the development of a sharp crack. Previous work done by various researchers [1–3] using a wide spectrum of mechanical tests and potentio-dynamic polarisation has indicated that SCC of steel in this system occurs over a wide range of CO partial pressures and temperatures which include the typical operating conditions for coal gasification plants. This indicated that high susceptibility to SCC has not always been found in the plant, from personal experience, where SCC is indeed limited to certain areas of the plant. Malik [4] performed an investigation on steel exposed to the vapour phase of H_2O–CO–CO_2 mixtures and found that in the presence of both carbon monoxide and carbon dioxide, the steel was more sensitive to cracking.

This investigation concentrated on the effect of temperature, carbon monoxide concentration, pressure, sample orientation, and surface finish on the susceptibility to cracking of carbon-manganese steels. The slow strain-rate test technique was used since it is a simple test to discriminate between environments and material conditions to reveal the relative sensitivities to cracking. With the appropriate strain rate, it is possible to induce stress-corrosion cracking. The most important characteristic of the test is the relatively slow strain rate generated at the region of crack initiation and growth. A principle advantage of the test is the rapidity with which the stress-corrosion susceptibility may be assessed. For many systems, a tensile strain rate of 10^{-5} to $10^{-6}\,s^{-1}$

Table 1: The A516 pressure vessel steel composition.

Element	Al	B	C	Co	Cr	Cu	V	Mn
wt%	0.009	0.002	0.217	0.010	0.010	0.010	0.010	1.030
Element	Mo	N_2	Nb	Ni	P	S	Ti	Si
wt%	0.010	0.007	0.004	0.020	0.009	0.003	0.001	0.290

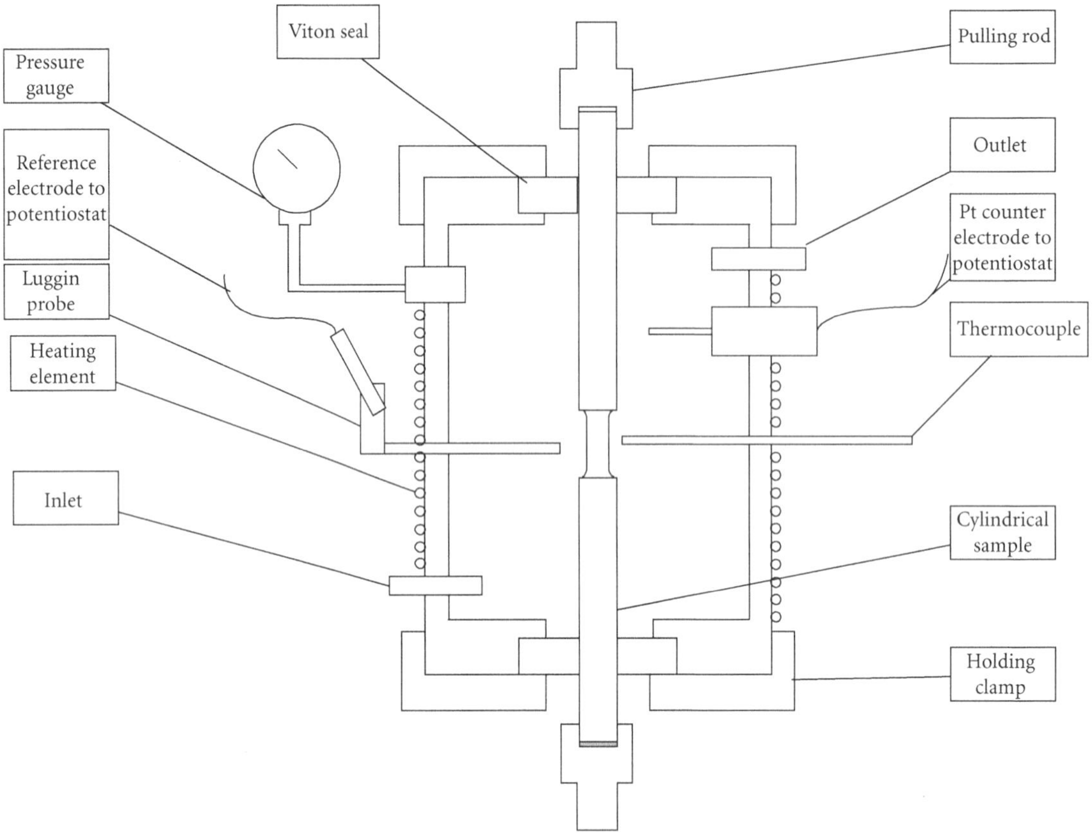

Figure 1: A schematic presentation of the tensile specimen in the autoclave that was used for the slow strain-rate test.

promotes cracking, but the absence of embrittling at these rates would not be indicative of immunity to cracking if other strain rates have not been investigated. Generally, the strain rate is constant for most of the test, but as soon as the necking starts, the localised strain rate increases dramatically, and the sensitivity to cracking diminishes. However, at this stage, the cracks have formed already, and the ductile failure induced by the higher strain rates is reduced, because of the existing cracks that formed during the uniform strain-rate application. The effective strain rate once cracks have initiated cannot be measured. With the slow strain-rate test it is possible to control the rate-determining step which is the strain rate of the deforming specimen.

Comparison between identical specimens exposed to the embrittling environment and inert conditions can be used to assess the susceptibility to cracking. In this study, the parameters used were the percentage elongation, as well as the percentage reduction in area, as these are easily measured.

The stress-corrosion cracking characteristics of A516 pressure vessel steel were investigated in terms of the influence of the gas composition, rolling direction, temperature, total gas pressure, applied electrochemical potential, time of exposure before testing, surface finish, and strain rate. The interrelationship of these factors was investigated to determine conditions of maximum susceptibility to cracking.

2. Experimental Procedure

2.1. Material. The steel used throughout the study was A516 pressure vessel steel, with the composition shown in Table 1. The composition was determined by using Leco Carbon/Sulphur Analysers, 3460 Emission Spectrometer, and 8680 + 72RET XRF Spectrometers.

Stress-corrosion cracking was evaluated by using the slow strain-rate method according to ISO 7539-7 (ASTM E8), with the initial slow strain rate being $10^{-6}\,s^{-1}$. A schematic drawing of the test rig used is shown in Figure 1.

The environment consisted of distilled water saturated with a CO/CO_2 gas mixture. Small-sized specimens proportional to the standard specimen according to ASTM standard E8 for cylindrical tensile test specimens were used, with a diameter of 4.0 mm and a reduced section length of 22.5 mm, slightly longer than the standard which is at 19.0 mm a schematic drawing of the tensile specimen is shown in Figure 2.

2.2. Experimental Procedure. The water-gas mixture was prepared in a stainless steel pressure vessel of approximately 160 litres. This setup was also used for the other laboratory tests performed during this study. The pressure vessel was filled with distilled water and purged with nitrogen for long enough to ensure that the dissolved oxygen concentration was below 0.1 ppm. Following the nitrogen purge, there was a purge with the appropriate gas mixture for approximately 20 minutes, and thereafter, the vessel was sealed and pressurised to the required level. Most of the experiments for the slow strain-rate tests were performed at a pressure of 800 kPa and a gas composition of 25% carbon monoxide and 75% carbon dioxide. The gas mixture was prepared beforehand by a commercial gas company according to the required concentrations.

The prepared solution was then transferred to the cell with the slow strain-rate tensile specimen as shown in Figure 1. The solution was allowed to flow through this unit to ensure that the same conditions were achieved in the environment of the tensile specimen. The temperature was measured with a K-type thermocouple that was introduced into the autoclave with a Swagelok fitting, and the temperature was controlled by a temperature controller that regulated a heating element which was wrapped around the autoclave. The pressure in the slow strain-rate autoclave was measured with a pressure gauge that was also introduced into the autoclave with a 1/8″ stainless steel tube and Swagelok fitting. Two Viton seals were used to seal and insulate tensile specimen from the autoclave. The autoclave was manufactured from stainless steel which would have shown insignificant corrosion with very little contamination. Each tensile specimen was further insulated from the testing rig at the pulling rod ends, to ensure that no galvanic cell was formed between the stainless steel autoclave and the carbon steel specimen.

With most of the tests, unless otherwise noted, the specimen was installed in the autoclave, after which the autoclave was filled with the solution, and the test was started shortly afterwards. The tensile specimen was also slightly prestressed to ensure that no time was wasted during the tensioning of the frame and pulling rods, since the strain-rate was low and this would have increased the duration of the test considerably.

The specimen was measured beforehand with a vernier calliper in terms of its gauge length and original diameter. The accuracy of these measurements was to the nearest 0.1 mm, according to the accuracy of the vernier calliper. The gauge length measurements were made from shoulder to shoulder on the tensile specimen, which included the radii where no elongation would be expected, but these positions were the best defined and consistent on all the specimens. The original and final diameters were measured, and the final diameter, after the test, was measured by carefully rejoining the two fractured pieces. This measurement was difficult and the accuracy disputable with a definite higher margin of error.

3. Results

The results from the slow strain-rate tests for samples machined parallel to the rolling direction are shown in Table 2. Here, the lengths of the two fractured sections were given as l_1 and l_2; therefore, l_f the final length of the specimen, was

$$l_{\text{final}} = l_1 + l_2. \quad (1)$$

The elongation of the specimen was calculated as follows:

$$\text{Elongation} = \frac{l_{\text{final}} - l_0}{l_0} \times 100, \quad (2)$$

where l_0 was the original length of the specimen, measured from shoulder to shoulder. The reduction in area was given by

$$\text{Reduction in area} = \frac{d_0^2 - d_{\text{final}}^2}{d_0^2} \times 100, \quad (3)$$

where d_0 was the original diameter, and d_{final} was the final diameter of the gauge length of the sample. Table 3 gives the results of the samples tested perpendicular to the rolling direction. These results are further presented in Figure 3 to Figure 7. Figures 3 and 4 show the influence of the orientation of the specimens to the rolling direction of the plate, as well as the environment temperature in terms of the reduction in area and elongation, respectively. The stress-corrosion cracking susceptibility was demonstrated by the loss in the reduction in area, as well as elongation. The errors were calculated by considering the accuracy of the measuring equipment, and especially on the reduction in area the error was more significant. These results did not show considerable repeatability; however, the reason for this was established during the course of the investigation. The influence of rolling direction on the susceptibility to cracking was not significant, and it would appear that stress-corrosion was affected to a greater extent by the environmental parameters, such as temperature, rather than material parameters.

Figure 5 illustrates the influence of the total pressure of the medium on the susceptibility to cracking in terms of the elongation and reduction in area. These results were for samples machined parallel to the rolling direction at a temperature of 45°C.

The effect of the total gas pressure on the cracking sensitivity exposed to a 25% carbon monoxide environment is shown in Figures 5 and 6. From these two figures, it appears that the increase in pressure has a significant effect on the crack sensitivity; with higher overall gas pressures the steel becomes more embrittled than for similar conditions at lower pressures. In Figure 5, it is shown that at 20°C an increase in the total gas pressure causes a drop in the reduction in area measured on the samples. When the

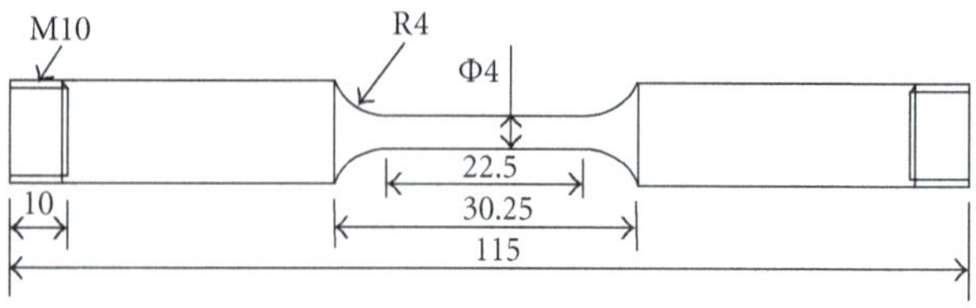

FIGURE 2: The configuration of the tensile specimen used.

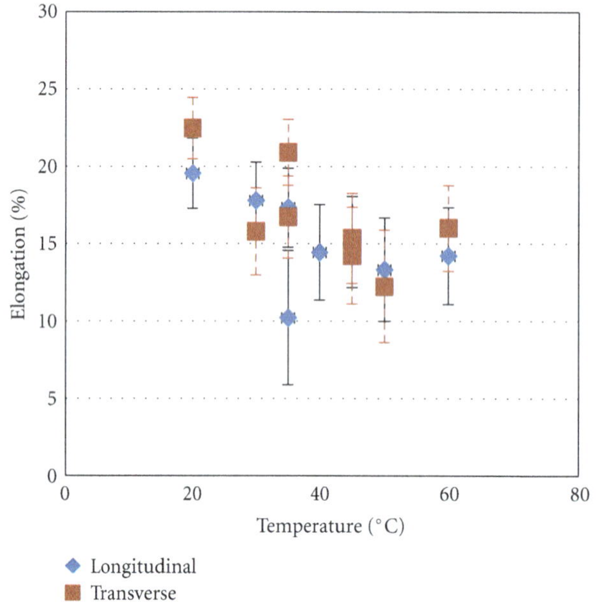

FIGURE 3: Elongation as a function of temperature for specimens machined longitudinal and transverse to the rolling direction, at 800 kPa with a gas mixture of 25% CO and 75% CO_2.

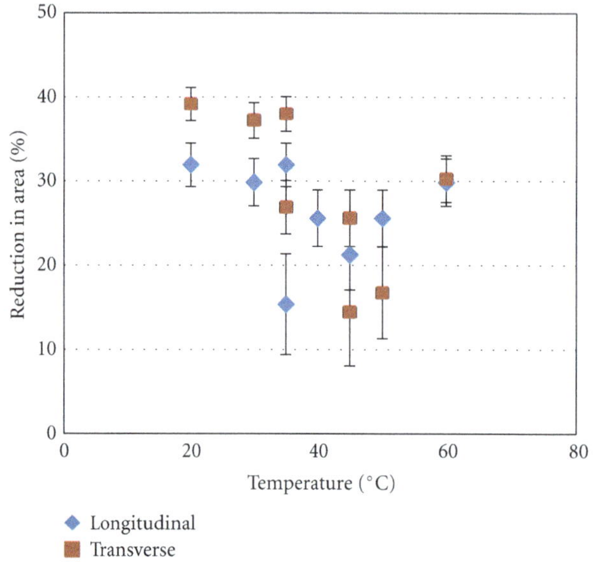

FIGURE 4: Reduction in area as a function of temperature for specimens machined longitudinal and transverse to the rolling direction, at 800 kPa with a gas mixture of 25% CO and 75% CO_2.

Table 2: Temperature-dependent results of samples machined parallel to the rolling direction.

Sample	Temperature (°C)	Total pressure (kPa)	l_1 (mm)	l_2 (mm)	l_{final} (mm)	Elongation (%)	d_{final} (mm)	Reduction in area (%)
L1	20	350	10.3	23.7	34.0	17.8	3.05	41.8
L3	20	500	23.8	10.35	34.15	18.4	3.20	36.0
L15	20	800	14.1	20.3	34.4	19.6	3.30	31.9
L13	20	1000	23.6	10.1	33.7	16.4	3.55	21.2
L10	20	Air	11.0	25.5	36.5	28.9	2.20	69.8
L16	30	800	11.2	22.8	34.0	17.8	3.35	30.3
L7	35	800	10.2	22.1	32.3	10.2	3.68	15.6
L4	35	800	12.5	21.4	33.9	17.3	3.30	31.9
L9	40	200	23.2	11.7	34.9	21.8	3.00	43.8
L6	40	800	11.1	22.15	33.25	14.4	3.45	25.6
L8	45	800	23.2	10.2	33.4	15.1	3.55	21.2
L14	50	800	20.5	12.5	32.0	13.5	3.45	25.6
L17	60	800	24.7	8.5	33.2	14.4	3.35	29.4

* This total length includes the shoulders of the specimen and does not refer to the gauge length.

Table 3: Temperature-dependent results of samples machined perpendicular to the rolling direction.

Sample	Temperature (°C)	Total pressure (kPa)	l_1 (mm)	l_2 (mm)	l_{total} (mm)	Elongation (%)	d_{final} (mm)	Reduction in area (%)
K7	20	800	24.55	10.50	35.05	22.4	3.12	39.2
K4	20	1000	15.10	20.00	35.10	22.7	3.1	39.9
K9	30	800	23.15	10.40	33.55	15.8	3.17	37.2
K2	35	800	17.20	17.50	34.70	20.8	3.15	38.0
K6	35	1000	15.50	18.25	33.75	16.7	3.42	26.9
K1	45	1000	20.55	12.45	33.0	13.3	3.55	21.2
K5	45	800	15.70	17.50	33.20	14.2	3.7	14.4
K8	50	800	24.95	7.80	32.75	12.2	3.65	16.7
K11	60	1000	24.40	9.20	33.6	16.0	3.34	30.3

influence of the total gas pressure was evaluated at various temperatures, as shown in Figure 6, the effect of the total gas pressure was not as significant.

The influence of the temperature on transverse samples exposed to 50% carbon monoxide and 50% carbon dioxide was investigated, and the detailed results are shown in Table 4. The tests were performed at 800 kPa.

The test results shown in Table 4 were similar to the results of the tests performed in 25% CO and 75% CO_2, where there existed a range of greater stress-corrosion susceptibility for the intermediate temperatures, such as from 40°C to 60°C.

Figure 7 shows the reduction in area as well as the elongation for samples that were tested in 50% carbon monoxide and 50% carbon dioxide evaluated over a range of temperatures, between 20°C and 80°C.

The test results are compared later in Figure 8 for the two gas compositions evaluated, namely, 25% and 50% carbon monoxide; the maximum sensitivity increased to higher temperature for the 50% carbon monoxide.

The electrochemical potential dependence was investigated by using a *Voltamograph* CV-27 potentiostat to keep the sample at a certain potential for the whole duration of the slow strain-rate test. These tests were performed at 45°C, 800 kPa, and a carbon monoxide concentration of 25% and 75% carbon dioxide. The potential was measured with an Ag/AgCl reference electrode that was introduced into the autoclave by a Luggin probe. The results are shown in Table 5, where the embrittling factor is plotted as a function of the applied electrochemical potential in terms of the Ag/AgCl reference electrode.

The results shown above in Table 5 are illustrated in Figures 9 and 10, demonstrating the effect of applied potential on elongation and reduction in area, respectively.

From both figures, it is clear that at a particular applied potential range, the cracking sensitivity of the steel was more enhanced.

The effect of surface roughness was investigated when scanning electron micrographs showed crack initiation on grinding marks as seen in Figures 11 and 12. Here, the cracks formed on what seems to be a 45° angle, which turned out to be the grinding angle for the specific sample. Slow strain-rate tests were performed on samples with different surface finishes. Grinding papers with the following ISO/FEPA grit designation grit sizes: P180, P200, P400, P600, P800, and P1200 were used on separate samples, where the P180 is the

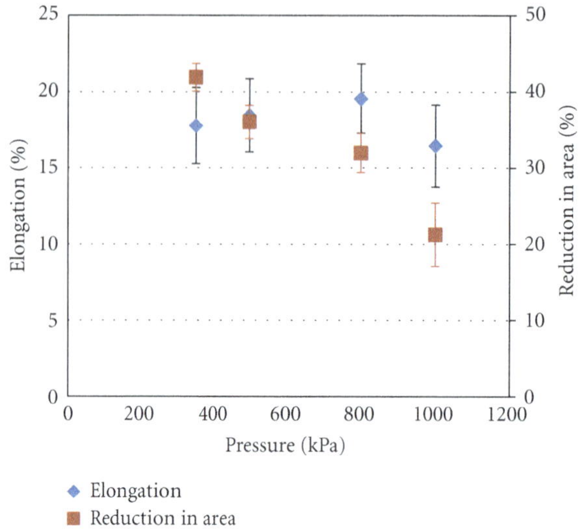

Figure 5: Reduction in area and elongation as a function of pressure of slow strain-rate tests performed at 20°C with a gas mixture of 25% CO and 75% CO_2.

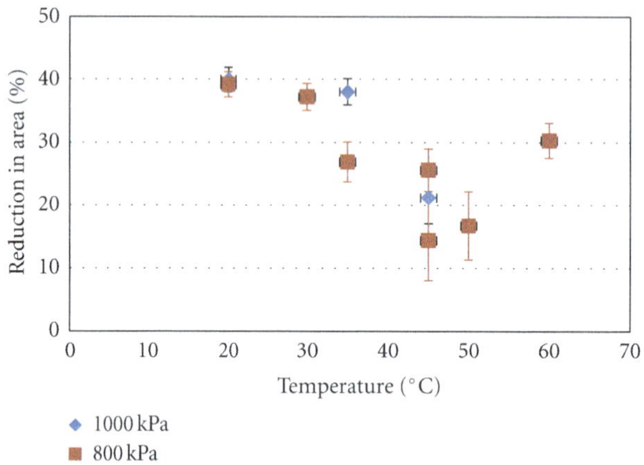

Figure 6: Comparison of two pressures over a temperature range with regard to reduction in area for a 25% CO and 75% CO_2 gas mixture.

roughest and P1200 is the smoothest paper, and the average particle diameters are approximately 82, 75, 35, 26, 22, and 15 μm. Table 6 shows the results of these tests performed on samples exposed to 25% carbon monoxide and 75% carbon dioxide, at 45°C and 800 kPa. Figures 13 and 14 show the results of the slow strain-rate tests in terms of the reduction in area and elongation; here, it can be seen that elongation did not show a marked difference in the influence of the machining marks on the surface of the samples. However, on the reduction in area results, a decrease was found, a greater susceptibility, with increasing surface roughness. This trend is slightly masked by scatter in the results.

The time of exposure of the specimen to the solution before test was started was investigated when it accidently became evident that a specimen showed increased brittleness when the test was not performed immediately. In Figure 15, the embrittling factor index is plotted against the time of exposure prior to testing. For these exposure tests, the experimental parameters were a total gas pressure of 800 kPa, a temperature of 45°C, and a strain rate of $10^{-6}\,s^{-1}$. Here, the time of exposure before the test had a significant effect on the embrittling of the sample, which is unexpected, especially considering the degree of embrittling that was found. The embrittling indices for reduction in area and elongation both dropped in the order of 50% when the samples were not exposed to the environment before the test, and when the samples were exposed for 48 hours before testing. This trend was confirmed by the intermediate sample that was only exposed for 24 hours.

The influence of the strain-rate was investigated and in Figure 16 the embrittling factor index is plotted as a function of the strain rate. These tests were also performed at 45°C, 800 kPa, and 25% CO–75% CO_2. The strain-rate varied from $10^{-5}\,s^{-1}$ down to $10^{-8}\,s^{-1}$ unfortunately only one test was performed at $10^{-8}\,s^{-1}$. The reduction in area index for embrittling increased from 0.2% to 0.4% when the strain rate decreased from $10^{-7}\,s^{-1}$ to $10^{-8}\,s^{-1}$, and the same trend was seen on the elongation of the samples it went from slightly above 0.2 to 0.7 when the strain rate was decreased.

In Figure 17 the fracture surface of a specimen tested at 45°C and 800 kPa indicated the typical transgranular cleavage cracking for these tests. Some pearlite colonies are

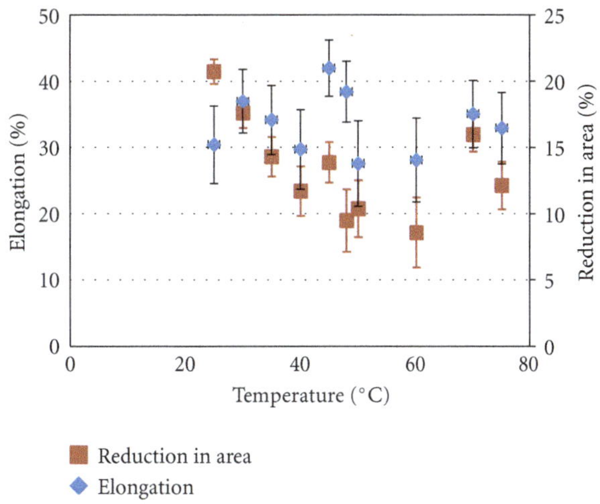

FIGURE 7: Elongation and reduction in area as a function of temperature for specimens machined transverse to the rolling direction, at 800 kPa with a gas mixture of 50% CO and 50% CO_2.

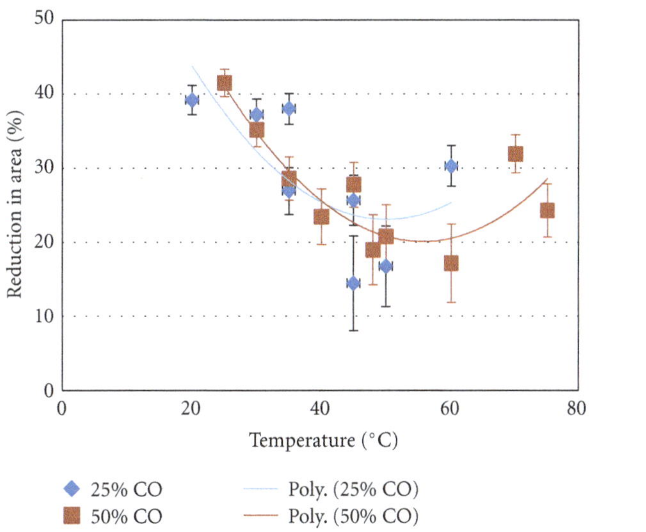

FIGURE 8: Comparison of results with two different gas mixtures, 25% CO and the other 50% CO.

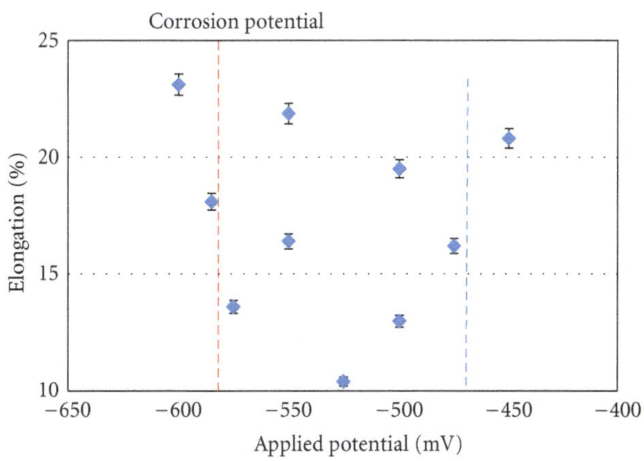

FIGURE 9: Elongation as a function of the applied electrochemical potential at 800 kPa and 45°C with a gas mixture of 25% CO and 75% CO.

TABLE 4: Temperature-dependent results of samples machined perpendicular to the rolling direction and tested in 50% CO–50% CO_2.

Temperature (°C)	l_1 (mm)	l_2 (mm)	l_{total} (mm)	Elongation (%)	d_{final} (mm)	Reduction in area(%)
25	8.18	25.24	33.42	15	3.06	41
30	10.90	23.26	34.16	18	3.22	35
35	7.84	26.00	33.84	17	3.38	29
40	14.82	18.52	33.34	15	3.50	23
45	13.80	20.92	34.72	21	3.40	28
48	12.42	21.90	34.32	19	3.60	19
50	20.92	12.18	33.10	14	3.56	21
60	12.16	21.00	33.16	14	3.64	17
70	8.94	25.00	33.94	18	3.30	32
75	19.50	14.20	33.70	16	3.48	24

TABLE 5: Potential-dependent results of samples tested in 25% CO–75% CO_2 at 45°C.

Potential (mV)	l_1 (mm)	l_2 (mm)	l_{final} (mm)	Elongation (%)	d_{final} (mm)	Reduction in area (%)
−600	8.9	26.3	35.2	23.11	3.05	41.85
−585	13.2	20.9	34.1	18.1	3.34	30.27
−575	17.8	15.3	33.1	13.6	3.64	17.19
−550	15.8	17.9	33.7	16.4	3.56	21.68
−550	8.0	26.9	34.9	21.86	3.19	36.4
−525	12.9	19.45	32.35	10.4	3.45	25.6
−500	24.9	8.0	32.9	12.98	3.57	20.34
−500	26.15	8.2	34.35	19.5	3.05	41.86
−475	10.95	22.7	33.65	16.2	3.16	37.59
−450	23.8	10.8	34.6	20.8	2.9	47.43

TABLE 6: Results of the slow strain-rate tests performed at 45°C, total pressure of 800 kPa, and 25% CO for different surface finishes.

Sample	Grit finish	Elongation	Reduction in area
L52	P1200	24.0	41.9
L70	P1200	21.7	31.9
L68	P800	22.1	38.4
L59	P600	26.1	36.0
L57	P400	22.8	34.4
L67	P400	16.0	26.0
L63	P220	22.2	31.9
L66	P180	26.2	31.9

seen as the crack propagated through the colony and was possibly etched by the environment. A cross section was also made to show the transgranular nature.

A test was performed in 100% CO, 800 kPa, at a strain rate of $10^{-7}\,s^{-1}$ and 45°C. These specimens did show cracking with the elongation at 21.8%.

4. Summary of Results

Here the results of the material and surface characteristics are compared with the environmental influences on the susceptibility of the steel to stress-corrosion cracking.

4.1. Rolling Direction. The results in Figure 4 did not show a significant difference between the two orientations of testing specimens being aligned parallel or perpendicular to the rolling direction, although it did appear that the samples machined longitudinal to the rolling direction showed a slightly higher susceptibility to stress corrosion. At the lower temperature range the samples longitudinal to the rolling direction showed a definitely higher susceptibility. However, at the higher temperatures, between 50°C to 60°C the difference was negligible and in some instances the opposite to the lower temperatures. From both the elongation and reduction in area data, a similar tendency was found.

4.2. Total Pressure. The results in Figure 6 show that according to the % reduction in area, higher pressures increase the susceptibility to stress-corrosion cracking. However, this is in the range of 200 to 1000 kPa. According to the % elongation a similar trend is followed with a slight increase at 800 kPa. It is also seen (Figures 4 and 5) that with an increase in the temperature, the effect of the total pressure is even greater, although this was based on two samples, and more information is needed for statistical certainty. These results are similar to those of Hannah et al. [3] where the reduction in area of the samples decreased with increasing pressures.

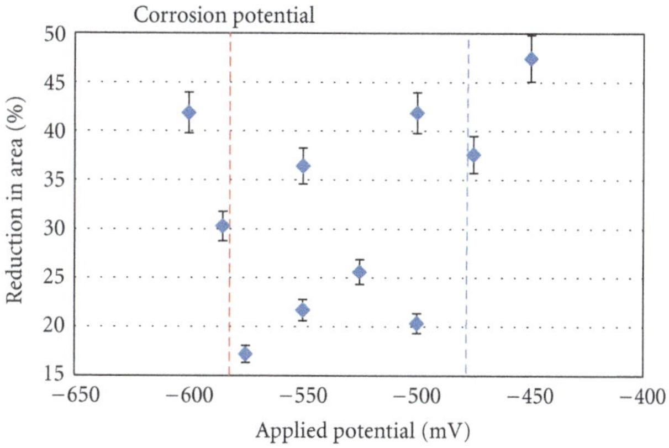

FIGURE 10: Reduction in area as a function of the applied electrochemical potential at 800 kPa and 45°C with a gas mixture of 25% CO and 75% CO.

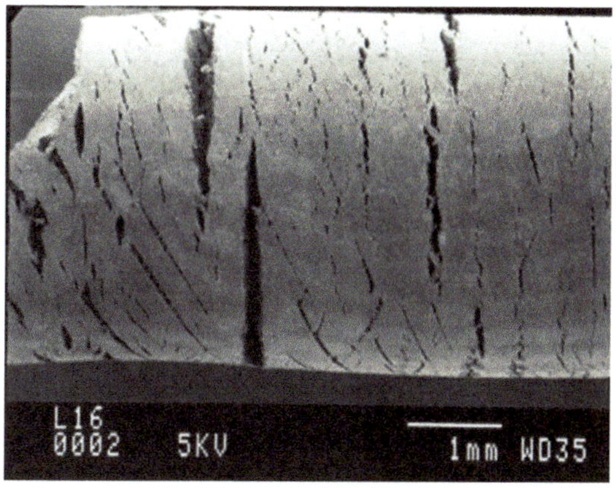

FIGURE 11: The influence of grinding marks on crack initiation for a slow strain-rate sample tested at 45°C and 800 kPa 25% CO–75% CO_2.

4.3. Gas Composition. It is clear that at higher carbon monoxide concentrations, the susceptibility to cracking increases dramatically.

4.4. Temperature. The influence of temperature was examined for two situations, samples parallel and perpendicular to the rolling direction.

4.4.1. Samples Perpendicular to the Rolling Direction. When looking at the reduction in area, these samples show a minimum resistance to stress corrosion in the order of 45°C to 55°C. This means that there is an increase in both sides—the region close to room temperature as well as higher temperatures. The increase in the resistance to stress-corrosion cracking at higher temperatures (60°C) was expected with the lower solubility of gases at the higher temperatures.

4.4.2. Samples Parallel to the Rolling Direction. The results show that a similar tendency in sensitivity to temperature for both the samples machined perpendicularly and parallel to the rolling direction. The reduction in area decreased in the temperature range 40°C to 50°C, and this seems to be in agreement with the work of Kowaka [1] as well as Itoh [5], on the general trend of the effect of temperature. The elongations for these samples gave the same region of greater embrittlement. This behaviour is expected due to the influence of the temperature upon the kinetics of the corrosion reaction and the passivity of the steel. Here, decreased sensitivity to cracking is an indirect effect, due to the decreasing solubility of the carbon monoxide in water at increasing temperatures, which would affect the passivation of the steel. Schmitt and Rothmann [6] also showed a decrease in ductility of various carbon steel with the increase in temperature, from 25°C to 50°C. Although Schmitt et al. [7] found a maximum susceptibility around 40°C for 37 Mn5 steel grades loaded to 90% of yield strength at 10-bar CO_2.

4.5. Applied Electrochemical Potential. In Figure 9, the region where the testing was performed, the cracking resistance increased towards the higher and lower limits. Embrittling was prevalent from around −575 mV to −475 mV, which is very similar to results obtained by Brown et al. [2]. The fractured specimens at the more noble potentials were characterised by general corrosion and little cracking. Towards the more active limit, hydrogen formation could have induced cracking, but here the resistance to cracking also increased, thus indicating that cracking is not enhanced

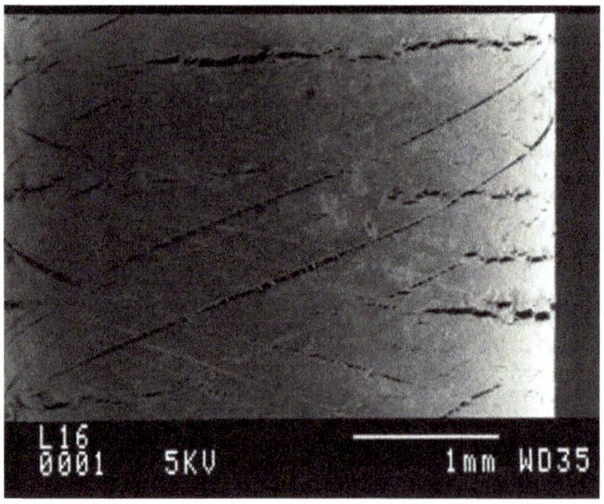

FIGURE 12: The same sample as shown above at a slightly higher magnification revealing how the cracks followed the grinding marks on the sample after a slow strain-rate test at 45°C and 800 kPa 25% CO–75% CO_2.

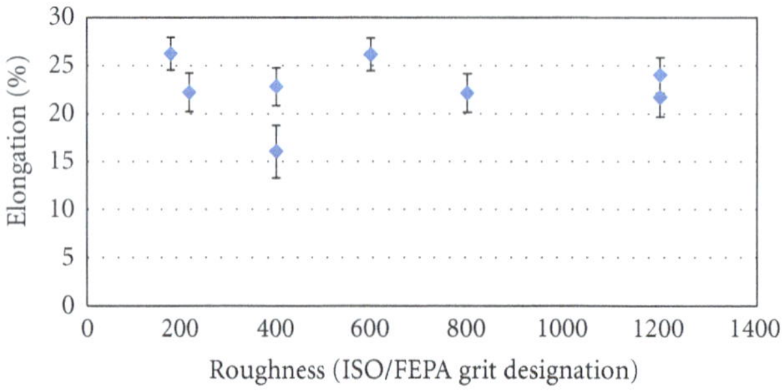

FIGURE 13: Slow strain-rate test results showing the influence of the surface roughness on the stress-corrosion resistance of steel at 45°C and 800 kPa.

by an excess of hydrogen. It could be argued that cracking could still be promoted by the presence of a small amount of hydrogen, but if this was the case, higher concentrations would have decreased the resistance to cracking. The situation is convoluted by the shifting of the corrosion potential with time, but this is evaluated in the next section.

4.6. Time of Exposure. The results in Figure 15 indicated the importance of the adsorbed CO species on the surface of the specimen and the effectiveness of this layer before the sample is stressed. Although only three samples were tested, the trend shows very clearly a greater sensitivity towards cracking. The time of exposure shows a very clear effect on the susceptibility and is a parameter that would cause significant errors if it is not kept constant during testing.

4.7. Surface Finish. The effect of the surface finish was first noted when one of the specimens was studied with the scanning electron microscope, and it was found that cracking occurred along most of the scratches left from the machining and grinding process (Figures 11 and 12). The grinding marks served as crack initiation points. This indicated the sensitivity of the cracking process to crack initiation. Although the influence of grinding grooves was not severe, there was a tendency towards a greater resistance to cracking when the surfaces had finer grinding grooves. Therefore, it would seem that the crack initiation was more probable with the larger grooves and might be one of the reasons why cracking would occur in one area and not another. The grooving is expected to be similar on the surface, although it is possible that larger grooves remained from previously applied, coarser grinding papers.

4.8. Strain Rate. Strain rate is a very important variable and this was evaluated to a greater extent with other test methods. However, as shown by Kim and Wilde [8], with slow strain-rate testing, it is possible to differentiate between the hydrogen-induced cracking and stress-corrosion cracking. This is because of the repassivation reaction of the steel that prevents the corrosion reaction of the bare steel, since the repassivation occurs at a faster rate than the exposure of the steel due to the slip step emergence. From the results, it is clear that the minimum resistance to cracking is in the order of $10^{-7}\,s^{-1}$. The resistance to cracking increased at $10^{-8}\,s^{-1}$, which means that the passivation rate at the crack tip could keep up with the bare surface that is formed due to the crack tip strain rate. This is an important result that indicates that the cracking is dependent upon the passivation of the steel, and again not the hydrogen embrittlement. Therefore, if the passivation of the steel can be influenced, either by enhancing or breaking it down, the stress-corrosion cracking can be mitigated. Unfortunately, the test at $10^{-8}\,s^{-1}$ strain rate was

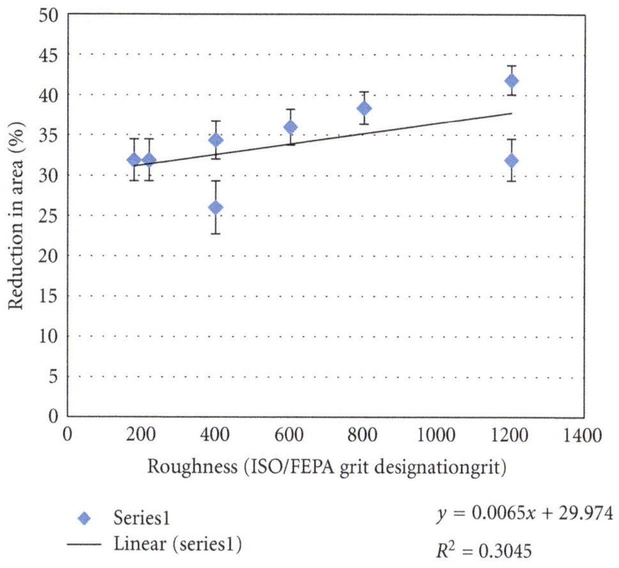

FIGURE 14: Slow strain-rate test results showing the influence of the surface roughness on the stress-corrosion resistance of steel at 45°C and 800 kPa.

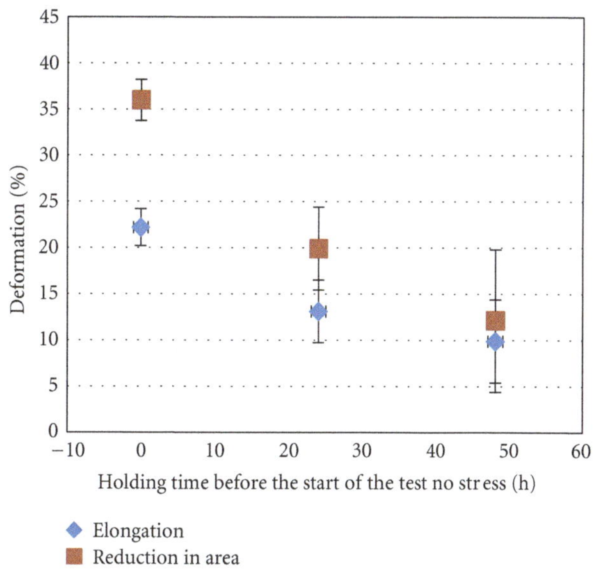

FIGURE 15: The embrittling of the steel as a function of the time of exposure prior to testing at 45°C and 800 kPa.

performed over a very long period, and only one of these tests was performed; therefore, this is not completely conclusive evidence but suggests that strain rate is a very important component in the cracking process.

5. Discussion

Stress-corrosion cracking of steel exposed to CO–CO_2–H_2O appears to be very dependent upon the carbon monoxide adsorption on the steel surface. With the increase of temperature, the solubility of both carbon monoxide and carbon dioxide decreases and therefore, decreasing sensitivity to cracking was found at higher temperatures. A similar decrease in sensitivity to cracking was found at the lower temperatures. However, the effect of temperature was moved to higher temperatures for the higher carbon monoxide concentration, therefore indicating the importance of the higher carbon monoxide concentration. The influence of the gas is further highlighted by the effect of higher total gas pressures. Furthermore, the effect of exposure time before the test illustrates the dependence upon the adsorption of the carbon monoxide on the steel interface. This time-dependent adsorption was demonstrated by Heaver [9], illustrating the effect of the passivation of the steel as a result of exposure time to a carbon-monoxide-containing environment. Therefore, when the time dependency of the embrittlement is considered, it has to be ascribed to the adsorption reaction of the carbon monoxide. The other

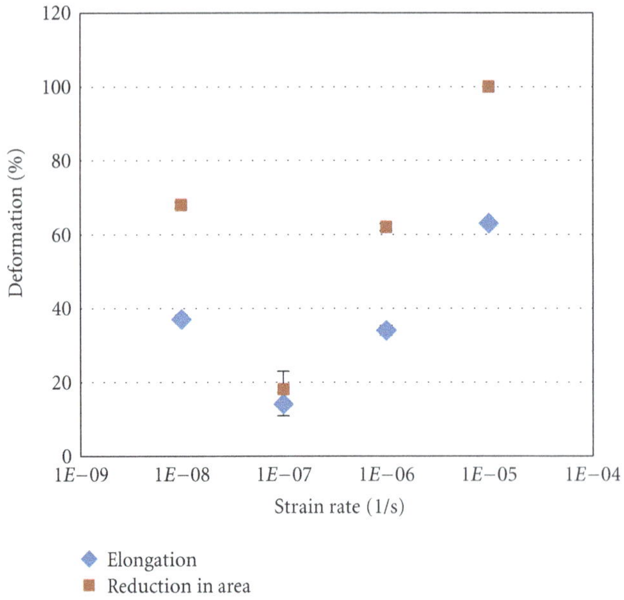

FIGURE 16: The embrittling factor index as a function of the strain rate of the slow strain-rate test at 45°C and 800 kPa.

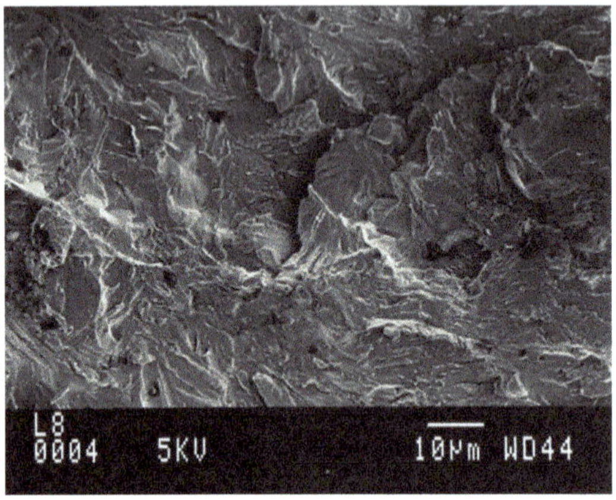

FIGURE 17: A scanning electron micrograph of the fractured surface of a slow strain-rate specimen tested in 25% CO at 45°C and 800 kPa total pressure.

parameters that were investigated did not show similar significant embrittling effects. The orientation of the samples and grinding marks on the surface of the samples both did not have a very significant effect, although these parameters showed some contribution to cracking. However, applied potential showed a significant effect on cracking with a definite potential range where cracking was enhanced. The minimum potential to this crack sensitive range corresponded to the corrosion potential and shows insensitivity towards hydrogen charging, whereas the maximum probably corresponded to the transpassive potential. This confirms the importance of the passivation and the passivation kinetics on the crack propagation.

6. Conclusions

(1) An increase in the total pressure of the gas increases the susceptibility of the steel to stress-corrosion cracking.

(2) Samples machined longitudinal to the rolling direction showed a slightly increased sensitivity to stress corrosion as compared to those machined transverse to the rolling direction.

(3) Maximum susceptibility to stress-corrosion cracking is between 40°C and 50°C.

(4) Cracking is dependent on the electrochemical potential, and cracking decreases on the cathodic side of the polarisation diagram.

(5) A waiting time before the test is started reduces the resistance to cracking.

(6) The surface finish does not have a large effect on the cracking resistance, although with a finer finish, the cracking is less.

(7) Strain-rate dependence shows a maximum susceptibility at around $10^{-7}\,s^{-1}$.

Acknowledgments

The author would like to acknowledge the guidance of Professor R. F. Sandenbergh and Professor G. T. Van Rooyen during the project.

References

[1] M. Kowaka, *Metal Corrosion Damage and Protection Technology*, Allerton Press, 1991.

[2] A. Brown, J. T. Harrison, and R. Wilkins, "Trans-granular stress corrosion cracking (S.C.C.) of ferritic steels," *Corrosion Science*, vol. 10, no. 7, pp. 547–548, 1970.

[3] I. M. Hannah, R. C. Newman, and R. P. M. Procter, "Environmental cracking of C-Mn steels in aqueous CO-CO_2 environments," in *Proceedings of the 4th interntional conference on the effect of hydrogen on the behaviour of materials*, N. R. Moody and A. W. Thompson, Eds., p. 965, 1990.

[4] H. Malik and F. Nawaz, "Stress corrosion cracking and electrochemistry of C-Mn steels in CO-CO_2-H_2O environments," *Anti-Corrosion Methods and Materials*, vol. 52, no. 5, pp. 259–265, 2005.

[5] G. Itoh, Ko-atsu gas, Journal of the High Pressure Gas Safety Institute of Japan, vol. 14, p. 19, 1977.

[6] G. Schmitt and B. Rothmann, "Corrosion of unalloyed and low alloyed steels in carbonic acid solutions," *Werkstoffe und Korrosion*, vol. 29, no. 4, pp. 237–245, 1978.

[7] G. Schmitt, H. Schlerkmann, and Aachen, "Corrosion cracking of steel in the system CO_2/H_2O," in *Proceedings of the 8th International Congress on Metallic Corrosion*, pp. 426–431, 1981.

[8] C. D. Kim and B. E. Wilde, "A review of constant strain-rate stress corrosion cracking test," in *Stress Corrosion Cracking The Slow Strain Rate Technique, STP665*, G. M. Ugianski and J. H. Payer, Eds., pp. 97–112, American Society for Testing and Materials, 1979.

[9] E. E. Heaver, *Stress corrosion cracking of steels in industrial process environments [Ph.D. thesis]*, 1994.

17

Atmospheric Corrosion of Painted Galvanized and 55%Al-Zn Steel Sheets: Results of 12 Years of Exposure

C. I. Elsner,[1,2] P. R. Seré,[1] and A. R. Di Sarli[1]

[1] *CIDEPINT, Centro de Investigación y Desarrollo en Tecnología de Pinturas (CICPBA-CCT CONICET-La Plata), Avenida 52 s/n entre 121 y 122, B1900AYB La Plata, Argentina*
[2] *Facultad de Ingeniería, Universidad Nacional de La Plata, Avenida 1 esq. 47, B1900TAG La Plata, Argentina*

Correspondence should be addressed to A. R. Di Sarli, ardisarli@gmail.com

Academic Editor: Blanca M. Rosales

Zinc or 55%Al-Zn alloy-coated steel sheets, either bare or covered by different painting systems, have been exposed for 12 years to the action of the urban atmosphere at the CIDEPINT station located in La Plata (34° 50′ South, 57° 53′, West), province of Buenos Aires, Argentina. The samples exposed surface was evaluated through periodical visual inspections, standardized adhesion tests, and electrochemical impedance measurements. The ambient variables monitored were average annual rains and temperatures, time of wetness, sulphur and chloride concentration, relative humidity, and speed and direction of the winds. It was found that in this atmosphere, the corrosion resistance of the bare 55% Al-Zn/steel sheets was higher than of the galvanized steel, and the polyurethane painting system was more protective than the alkyd and epoxy ones, which degraded after 6-7 years of exposure.

1. Introduction

Exposed to specific aggressive media, metal or alloy stability depends upon the protective properties of the surface film formed, because its chemical composition, conductivity, adherence, solubility, hygroscopicity, and morphological characteristics determine the film capacity to work as a controlling barrier [1]. In such a sense, steel galvanic protection by means of zinc or zinc alloys is a common example, owing not only to the fact that the zinc, being electrochemically more active than the steel, corrodes preferentially, but also to the barrier effect of the corrosion products precipitated on the metallic surface. In particular, the coatings based on zinc are widely used to protect steel structures against atmospheric corrosion [2], because of the protective properties afforded by an insoluble film of basic carbonate. However, if the exposure conditions are such that there is changes of the ambient variables like atmospheric conditions, UV radiation intensity, type and level of pollutants, wet-dry cycles, depletion of air but high humidity, or a medium containing strongly aggressive species like chloride or sulphate ions, the zinc could dissolve forming soluble, less dense, and scarcely protective corrosion products, which sometimes lead to localized corrosion [2–5]. This condition can be reached during the storage and transportation of galvanized steel sheets or when they are exposed to marine and/or industrial environments [6]. Aluminium coatings have overcome these two factors. Nevertheless, as they cannot provide cathodic protection to exposed steel in most environments, early rusting occurs at coating defects and cut edges; besides, these coatings are also subjected to crevice corrosion in marine environmental [7].

For years, many attempts to improve the corrosion resistance of zinc and aluminium coatings through alloying were carried out. Although the protective effect of combinations of these two elements was known, they were not used until the discovery that silicon inhibits the fast alloying reaction with steel [8]. Thus, the alloy commercially known as Galvalume or Zincalum arose, and its composition: 55% Al, 1.6% Si, the rest zinc, was selected from a systematic study, providing an excellent combination of galvanic protection and low corrosion rate.

When a higher degree of protection of these metallic surfaces is of concern, properly chosen painting systems can provide a more effective corrosion-inhibiting barrier and also a better aesthetic appearance [9]. Some exposure conditions are so aggressive that both protective systems (metallic + organic coatings) must be applied to get longer effectiveness. Such a combination, referred to as a duplex system, has demonstrated a synergistic effect when compared to the individual coating systems. This better corrosion protection is attributed to the double action afforded by the zinc or 55%Al-Zn layer (cathodic protection + blockage of its defects by the corrosion products), and also by the pigmented paint system (barrier effect + steel corrosion-inhibition) [10]. Besides, this duplex system requires less reconditioning and repairs of coating systems after transportation and assembly on site.

The mechanism responsible for the protective action of paint coatings is highly complex because it depends upon the simultaneous action of different factors. Irrespective of their intended function (functional, decorative, or protective), the paint must adhere satisfactorily to the underlying substrate [11, 12].

The organic film permeability is important in metallic substrate corrosion, since this property is directly connected to the permeation of environmental corrosion-inducing chemicals through the polymeric matrix, the chemical composition of the latter, and the presence of pores, voids, or other defects in the coating. It is important to note that particularly water and oxygen can permeate the film, at least to some extent, even if none of the intrinsic structural defects are present. For these reasons, the painted metal's resistance to degradation produced by weathering is a very important variable, since it defines the material's durability. Within this concept, the corrosion and resistance to weathering (degradation due to UV radiation, oxygen, humidity, etc.) are separately evaluated. The corrosion resistance depends upon the permeability (barrier effect) of the primer and galvanic layer as well as of the inhibitive capacity of the contained anticorrosive pigments. The weathering degradation (loss of gloss and/or adhesion, chalking, cracking, blistering, etc.) takes place at shorter times and depends mainly on the topcoat paint properties.

The still unsolved paint delamination or blistering problem, due to a bad bond at the substrate/paint interface, depends upon the chemical nature and crosslinking degree of the polymer as well as the metal substrate and its surface treatment [13]. In principle, paint adhesion can be improved by providing the substrate with a pretreatment layer, followed by applying a corrosion inhibiting primer + intermediate and/or topcoat paints. In line with this definition, the primer is considered the critical element in most paint systems because it is mainly responsible for preserving the metallic state of the substrate, and it must also anchor the total paint coating to the steel. Most coatings adhere to the metal via purely physical attractions (e.g., hydrogen bonds) that develop when two surfaces are brought closely together [14, 15]. Paint vehicles with polar groups (–OH, –COOH, etc.) have good wetting properties and show excellent physical adhesion characteristics (epoxies, alkyds, oil paints, etc.). Much stronger chemically bonded adhesion is possible when the primer can actually react with the metal, as is the case of several pretreatments [16–18].

Paint life depends on several factors such as the metallic substrate, the selected paint system, and the paint-substrate interface [19]. Paint selection is generally based on the aggressive medium properties, while the metal treatment before painting has a substantial impact on the useful life of the selected system.

The susceptibility to degradation of painted metals is estimated by accelerated laboratory tests and natural atmospheric exposure for several years [20–31]. Although the extrapolation of accelerated test results do not compare linearly to the actual performance of the coatings in their service life, it can supply useful information related to the rate and form of the corrosion-inhibiting system degradation. In most cases, such information can help to improve the paint formulation and/or the painting scheme design. Consequently, a comparative evaluation of the protective performance of either bare or covered with three different painting systems is reported in this paper. The corrosion resistance of these samples was tested by exposure for 12 years to the action of the urban atmosphere at the CIDEPINT station located in La Plata (34° 50′ South, 57° 53′, West), province of Buenos Aires, Argentina.

The evolution of the samples exposed surface was evaluated through periodical visual inspections, adhesion tests according to the ASTM D-3359/09 standard, and electrochemical impedance measurements applied to samples immersed for 1 h in 0.5 M Na_2SO_4 solution. The ambient variables monitored were average annual rains and temperatures, time of wetness, sulphur and chloride concentration, relative humidity, and speed and direction of the winds.

2. Experimental Details

A total of 280 commercial-grade steel sheets (15 × 8 × 0.2 cm) hot-dip coated with zinc or 55%Al-Zn plates were used as the metallic substrate. They were degreased by immersion in 5% Na_2CO_3 solution and then rinsed with distilled water to eliminate any possible surface contamination.

The commercial-grade protective painting systems (Table 1) were applied by brushing to maintain the same conditions for all the samples. After applying the painting system, the painted plates were placed in a dessicator cabinet at controlled temperature (30 ± 2°C) until completely dry. Next, measurements of dry film thickness (Table 1) were taken with an Elcometer 300 coating thickness gauge, using a bare sanded plate and standards of known thickness as references. Plates were exposed at 45° from the horizontal and they were oriented East allowing a maximum insolation on them. The test site was formed by wooden structures, where the flat plates were set on. Location coordinates, meteorological data, atmospheric pollutants, time of wetness and corrosion category of the place of location according to ISO 9223 are given in Figure 1 and Tables 2 and 3, respectively. All environmentally data were available from the Seismology and Meteorological Department of the Astronomic and Geophysics Science Faculty of the National University of La Plata, placed very near CIDEPINT Station.

To check reproducibility, a total of 280 samples including bare or painted steel/zinc or steel/55%Al-Zn sheets were exposed to natural weathering in La Plata station for 12 years.

Table 1: Mean thicknesses (μm).

Metal/paint system	Metallic coating	Primer	Topcoat	Total thickness
S/Z/AS	18 ± 0.9	22 ± 0.9	52 ± 2.6	92 ± 4.5
S/ZA/AS	20 ± 0.9	22 ± 0.9	52 ± 2.6	94 ± 4.5
S/Z/ES	18 ± 0.9	4 ± 0.2	87 ± 2.6	109 ± 5.1
S/ZA/ES	20 ± 0.9	4 ± 0.2	87 ± 2.6	111 ± 5.1
S/Z/PS	18 ± 0.9	4 ± 0.2	48 ± 2.6	70 ± 3.5
S/ZA/PS	20 ± 0.9	4 ± 0.2	48 ± 2.6	72 ± 3.5

Note: S/Z/AS: galvanized steel/alkyd-based paint system; S/ZA/AS: steel/55%Al-Zn alloy/alkyd-based paint system; S/Z/ES: galvanized steel/epoxy-based paint system; S/ZA/ES: steel/55%Al-Zn alloy/epoxy-based paint system; S/Z/PS: galvanized steel/polyurethane-based paint system; S/ZA/PS: steel/55%Al-Zn alloy/polyurethane-based paint system.

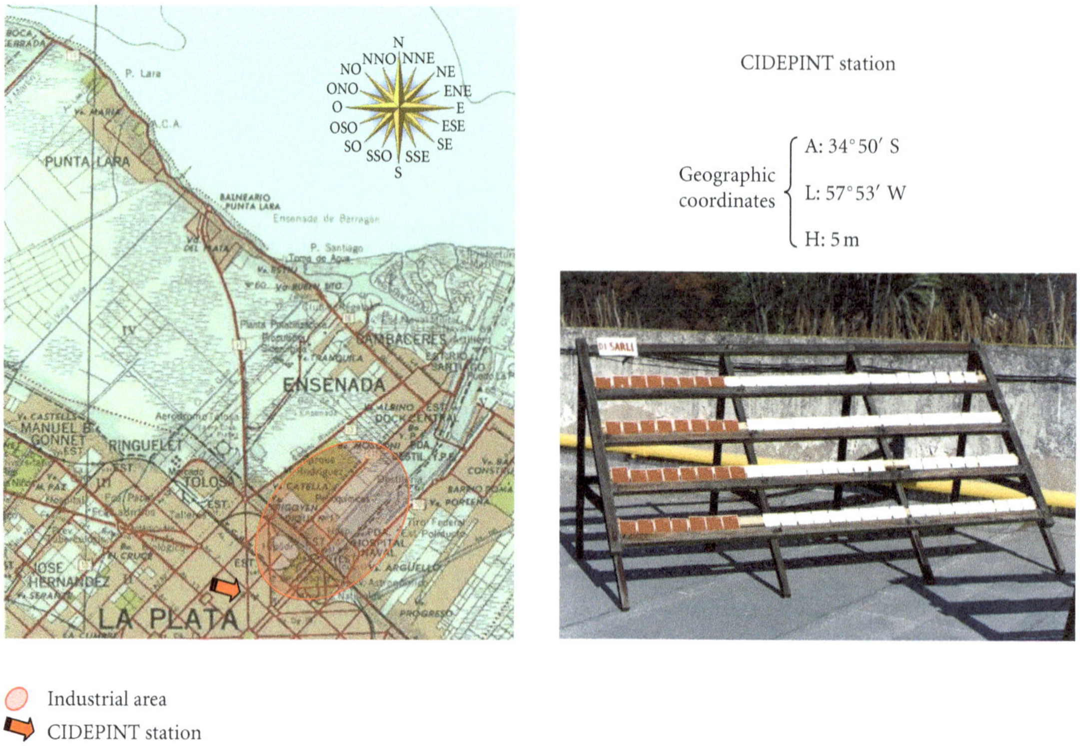

Figure 1: Location of the CIDEPINT station in La Plata, Buenos Aires, Argentina.

The painted samples were build-up with the three painting systems mentioned in Table 1, and their edges were masked with a thick wax base coating to avoid edge effect.

Gravimetric determinations for measuring weight-loss of bare steel/metallic coating samples were carried out in triplicate for each material tested.

The visual inspections and samplings took place according to the following program: during the first year each 1st, 3rd, 6th, 9th, 12th month, and then each 2nd, 4th, 8th, and 12th year. At the same times, adhesion tests according to the Test Tape ASTM D-3359/09 Standard on replicates of each type of painted samples were also performed.

2.1. Electrochemical Measurements. For the impedance measurements periodically carried out on other replicates of each type of samples, a cylindrical clamp-on acrylic (polymethyl methacrylate) cell was positioned on the painted panel by an O-ring defining a surface area of 15.9 cm^2. An aperture in the top of this three electrode electrochemical cell contained a Pt-Rh mesh counter-electrode with negligible impedance, oriented parallel to the working electrode (painted metal surface). A glass-linear Saturated Hg/$HgSO_4$ tipped Reference Electrode was positioned, together with the counter-electrode, close to the exposed painted steel surface panel. For further easy comparison with previous information, all the potential data in the text and figures were referred to the Saturated Calomel Electrode (SCE). Before the electrochemical impedance spectrum of each replicate was obtained, the sample was subjected to 1 hour of wetting in 0.5 M Na_2SO_4 solution. Impedance spectra were obtained from a Solartron 1255 FRA coupled to a Solartron EI 1286 and a PC, all controlled by the Zplot software. Impedance spectra collected in the frequency range $10^{-2} \leq f(Hz) \leq 10^6$ were analyzed and interpreted on the basis of equivalent electrical

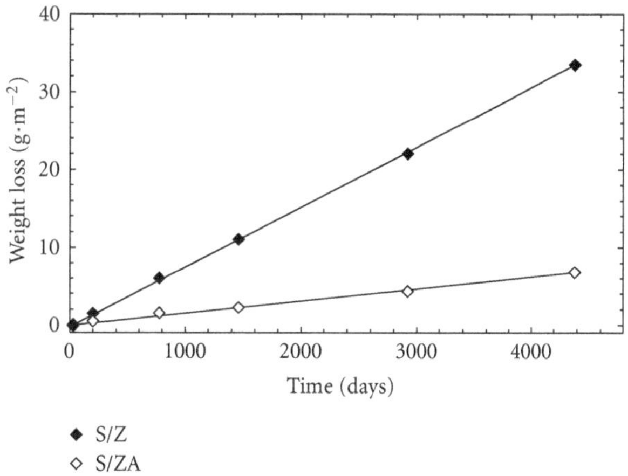

FIGURE 2: Plots showing the time dependence of the weight loss suffered by the bare S/Z and S/ZA sheets.

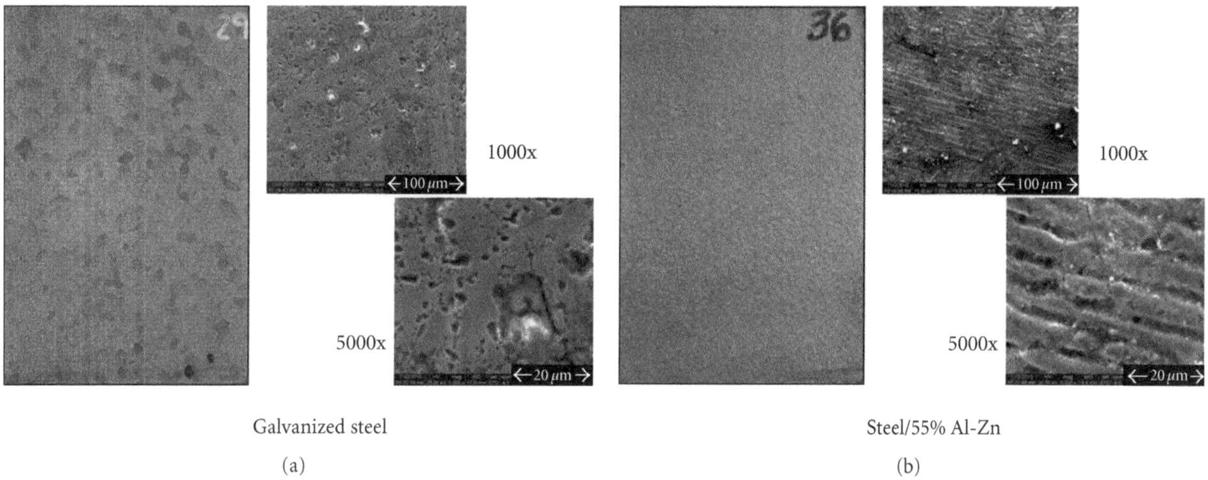

FIGURE 3: Photographs and SEM images of the bare metallic coatings after 12 years of exposure.

circuits, optimizing the values of the circuit parameters by using Boukamp' program [32].

All the electrochemical experiments were carried out at laboratory temperature (23 ± 2°C) and with the electrochemical cell in a Faraday cage to reduce external interferences as much as possible.

To improve the experimental data reliability, three replicates of each sample type were measured in all the tests.

3. Experimental Results and Discussion

3.1. Atmospheric Exposure Test

3.1.1. Corrosion of the Metallic Coatings. It is known that all the materials degrade under the influence of atmospheric factors such as oxygen, humidity, and/or pollutants (SO_2, NaCl, NO_x, etc.). Another important degradation source is the sun radiation, particularly its UV rays. All these influences compose the so-called "Macroclimate" of a determined zone [33]. In change, "Microclimate" is defined as the specific climate formed around an object and it results of vital importance to understand the atmospheric mechanisms causing the materials degradation. Among the parameters used to define it are the surface time of wetness (TOW), the heating by sun radiation, mainly the infrared, and the acidic nature ions (SO_3^{2-}, NO_2^-, Cl^-) gathering within the aqueous layer deposited on the object. On the other hand, the atmospheric corrosion process is the sum of partial corrosion processes taking place each time an electrolyte layer deposits on the surface metal. The rain, snow, fog, and/or humidity condensation produced by temperature changes are the main promoters of atmospheric corrosion. In such sense, the value of some climatological variables characterizing the average exposure conditions corresponding to the station used in the present work are shown in Tables 2 and 3. The aggressiveness of La Plata station was attributed to its high relative humidity, severe and lengthy TOW as well as surface runoff supported by the tested replicates.

Hot-dip zinc is widely used as a coating for carbon steel because of its good corrosion resistance and relatively low price [34]. Due to its practical use, zinc atmospheric corrosion has been studied in field exposures as well as in laboratory with controlled environments [35, 36]. Zincite, ZnO, is the first product formed when the naked metal is

Table 2: Meteorological data for the 12 years of exposure in the CIDEPINT Station, La Plata, Argentina.

Year	Mean temperature (°C)	Mean relative humidity (%)	Precipitation (mm)	Days of rain
1	16.5	80.8	943.1	90.0
2	16.8	80.7	1042.8	94.0
3	16.3	80.4	927.0	93.0
4	15.8	81.8	1342.1	90.0
5	16.4	84.4	1316.6	98.0
6	16.0	82.7	1611.8	107.0
7	15.8	81.1	924.5	98.0
8	16.1	80.8	881.7	88.0
9	15.9	79.4	926.6	91.0
10	16.0	79.0	1083.4	81.0
11	15.4	77.5	1153.2	85.0
12	16.3	78.3	775.1	70.0

Table 3: Average levels of chemical agents, time of wetness and corrosion category of CIDEPINT Station.

Year	Deposition rate of SO_2 ($mg \cdot m^{-2} \cdot d^{-1}$)	Deposition rate of chloride ($mg \cdot m^{-2} \cdot d^{-1}$)	Time of wetness fraction	Corrosion category according to ISO 9223
1	6.22	Negligible	0.61	$P_0S_0\tau_4/C_2$
2	7.25	Negligible	0.62	$P_0S_0\tau_4/C_2$
3	6.81	Negligible	0.61	$P_0S_0\tau_4/C_2$
4	6.53	Negligible	0.65	$P_0S_0\tau_5/C_2$
5	7.42	Negligible	0.63	$P_0S_0\tau_5/C_2$
6	6.94	Negligible	0.69	$P_0S_0\tau_5/C_2$
7	7.60	Negligible	0.59	$P_0S_0\tau_4/C_2$
8	8.12	Negligible	0.57	$P_0S_0\tau_4/C_2$
9	7.93	Negligible	0.61	$P_0S_0\tau_4/C_2$
10	6.82	Negligible	0.62	$P_0S_0\tau_4/C_2$
11	7.45	Negligible	0.62	$P_0S_0\tau_4/C_2$
12	8.01	Negligible	0.55	$P_0S_0\tau_4/C_2$

exposed to the air, creating a protective film that inhibits corrosion process. Under humidity conditions higher than 80%, zinc is oxidized forming zinc hydroxide. If the pH on the surface is high enough, this hydroxide can react with atmospheric components such as CO_2, SO_x, and Cl^-, forming, in the hydroxide/air interface, the corresponding zinc basic salts [37]. Some of these products form a compact film that protects the metal against later corrosive attacks [38]. An important intermediate in the subsequent formation of other corrosion products is hydrozincite $Zn_4CO_3(OH)_6 \cdot H_2O$ [39, 40]. If the pH of the humid surface is low, neither hydroxide nor basic salts are formed [37]. In presence of SO_2 polluted air, the main corrosion product is hydroxysulfate $Zn_4SO_4(OH)_6 \cdot 4H_2O$, and in presence of Cl^- contamination, the precipitation of insoluble hydroxychloride $Zn_5(OH)_8Cl_2 \cdot H_2O$ is possible.

The hot-dip aluminium-zinc alloy, known as Zincalume, actually contains about 55% aluminium, 1.5% silicon, and the balance zinc. A microstructure of the alloy-coated steel which forms on cooling is essentially two phase, comprising about 80% by volume of a dendritic aluminium-rich phase and the remainder an interdendritic zinc-rich phase with a thin intermetallic layer next to the steel substrate. When the coating corrodes initially, the zinc phase corrodes preferentially until the formation of corrosion products reduces further activity in these areas. During the initial stage of corrosion, the coating behaves like zinc coating. In the later stages of corrosion when the coating is essentially comprised of zinc corrosion products carried in an aluminium-rich matrix, the corrosion becomes more characteristic of the aluminium-rich phase, resulting in a lower corrosion rate, more typical of aluminium [41, 42].

Weight-loss measurements provide the most reliable figure concerning the aggressiveness of a given atmosphere, so that the corresponding corrosion data approach to the service conditions more than any other test. In the present case, as it is shown in Figure 2 both materials presented a linear relationship between its weight-loss and the weathering time and, considering the 12 years exposure, the galvanized coating exhibited a degree of corrosion 4.94 times greater than that of the 55%Al-Zn coating.

In general, the zinc coating suffered uniform corrosion with the development of a layer of corrosion products, mainly ZnO and $Zn_4CO_3(OH)_6 \cdot H_2O$, but in particular

Galvanized steel

(a)

Steel/55% Al-Zn

(b)

FIGURE 4: Photographs of the scribed area of bare metallic coatings after 12 years of exposure.

places of the surface, the accumulation of atmospheric dust produced localized corrosion as it is possible to see in the SEM images shown in Figure 3. On the other hand, on the 55%Al-Zn coating and due to the complex structure of the alloy, the development of localized corrosion was observed as a consequence of the preferential dissolution of the interdendritic Zn-rich phase, which provoked loss of surface' brightness and the development of thin dark lines related to the aluminium corrosion process.

During all the weathering period, both metallic coatings were able to afford cathodic protection to the substrate as it is shown in Figure 4.

These results can be explained by considering the climatic conditions prevailing during the outdoor exposure prior to samples removal, that is, level of sulphur compounds, wet/dry cycles, high TOW, and pluvial precipitations dissolving the zinc corrosion products and releasing zinc ions from the corroded surface, which are dispersed to the environment. This phenomenon is known today as a metal runoff process [43–50].

3.1.2. Duplex Systems. As a direct way of evaluating the anticorrosive performance of organic coatings, the exposure test to natural atmospheres either for intact or scribed painted surfaces is, without any doubt, the best.

Experimental results coming from exposure tests to natural weather conditions are further representative of the protective and aesthetic properties provided by the topcoat paint. In such a sense, the results obtained from visual inspections periodically carried out for 12 years of exposure in La Plata station confirmed that weathering of the considered duplex systems proceeds very slowly. The periodical visual inspection put in evidence that no sample presented underrusting, peeling, cracking, or checking. From the 4th year weathering and due to the effect of the UV radiation, the Epoxy paint systems showed significant deterioration by chalking, and the topcoat Alkyd paint began to show significant changes of color, brightness, and chalking that led to the exposure of the primer from the 8th year of test. In any case, blistering or filiform corrosion near the scribed cross was observed. Examples of this behavior are presented on Figures 5 and 6. These results were attributed to the highest resistance of the polyurethane topcoat paint to the UV radiation, rain and temperature changes due to the strong interaction between the reactive components of the polymer and the chemically stable pigment (TiO_2) added to the effective anticorrosive protection offered by the primer.

At this point, it is noteworthy to remark that besides knowing the behavior of painting systems when they are intact, it is important to evaluate how they performed when mechanical damage occurs. For that purpose, some of the samples were scribed in an X-shape before being exposed to the natural atmosphere.

The main characteristic of duplex systems is to get and maintain good adhesion to the metallic-coated steel surface during its weathering period. As it is shown in Figure 7, independent of the metallic substrate the polyurethane coating presented the largest adhesion loss during the weathering. Due to the previously mentioned behavior of Epoxy and Alkyd paint systems from the 4th year of weathering, is important to mention that in the case of both Alkyd systems the adhesion test considered mainly the primer and in the case of the Epoxy-based samples the evaluation was interrupted as a consequence of the observed chalking degree.

3.2. Electrochemical Tests. During atmospheric corrosion, in general, the metal is not immersed in large quantities of electrolyte but in contact with thin layers or monolayers of moisture, due to that the corrosion process develops as localized corrosion cells. In that situation, the measurement of the corrosion potential as well as of the resistive and capacitive parameters governing the electrochemical behavior of the metal/coating interface is not always possible during the atmospheric corrosion [51, 52].

According to Zhang and Lyon [52], the cathodic process for metals like steel, zinc, and copper coated with thin ($<100\,\mu m$) water films reveals a diffusional limiting current whose value depends on the water film thickness. For thinner thicknesses, like in most of the atmospheric corrosion cases, the main cathodic process is controlled by activation. In the case of zinc, due to its high electronegativity, the cathodic process is not sensitive to the water film thickness present on

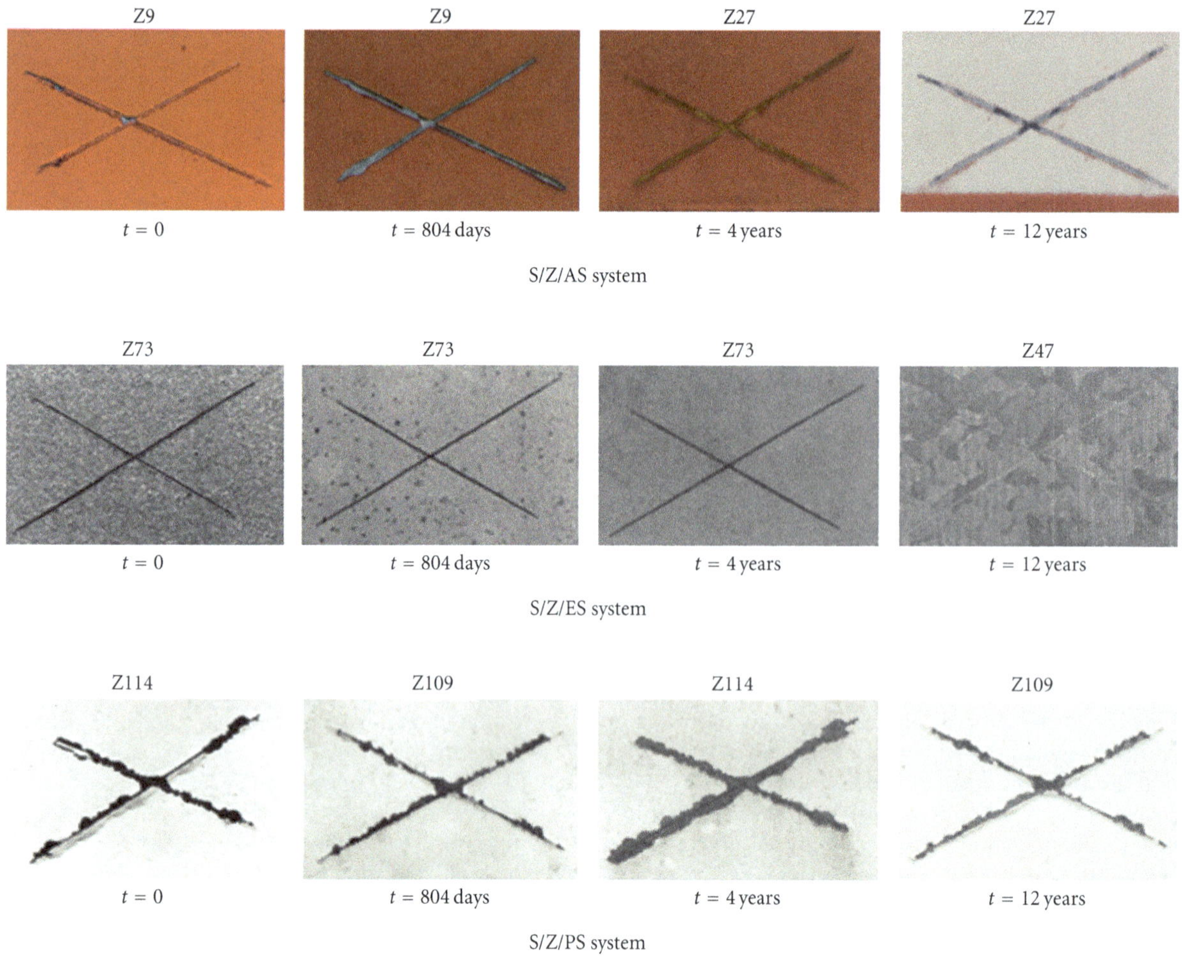

FIGURE 5: Photographs showing the evolution of S/Z/painting systems as a function of the exposure time.

the surface. On the other hand, as it was mentioned earlier, the protective capacity of the corrosion products formed during the atmospheric exposure depends on a variety of properties, which are also dependent on the composition and metallurgical history of the metal as well as on the atmospheric variables [1, 5, 53].

3.2.1. Corrosion Potential and Impedance Results for Bare S/Z and S/ZA Sheets. The measured corrosion potential values (E_{corr}) point out the metal susceptibility to be corroded. In general, when a value for a given medium is nobler (positive), it will result more resistant to corrosion. As it is possible to see in Figure 8, the corrosion potential evolution for the bare coating materials (Z and ZA) along the 12 years of exposure was quite similar. In this case, the surface of both the bare S/Z and the S/ZA sheets remained active up to the end of the test exposure with potential values ranging between −1.04 V/SCE and −1.00 V/SCE, which are characteristics of these metals under free corrosion processes. [54]. The continuous electrochemical reactivity was attributed to the rain runoff effect on the corrosion products, which avoided the formation of an oxide, hydroxide, and/or passive protective layer on the bare surfaces.

These data are in accordance not only with the visual inspection but also with the results obtained elsewhere [55] by optic and electronic microscopy techniques, which put in evidence a developed corrosion process. The surface had hollows and corrosion products characterized as oxides and basic carbonates. On the other hand, the 55%Al-Zn alloy had a big cathodic area where the metals exposed to the atmosphere will corrode by coupling with the oxygen cathodic reduction reaction:

$$O_2 + 2H_2O + 4e^- \longrightarrow 4OH^- \quad (1)$$

although when the level of contamination with acid products is high, the hydrogen evolution gets importance as cathodic reaction:

$$2H^+ + 2e^- \longrightarrow H_2 \quad (2)$$

regardless of which reaction prevails, the pH on the cathodic region increases. From a certain level of acidity, it is possible that the SO_2 of a polluted atmosphere acts as an oxidant able to impart a great acceleration to the cathodic process.

The impedance spectra of the coating Zn (Figure 9) may be interpreted in terms of the corrosion products film

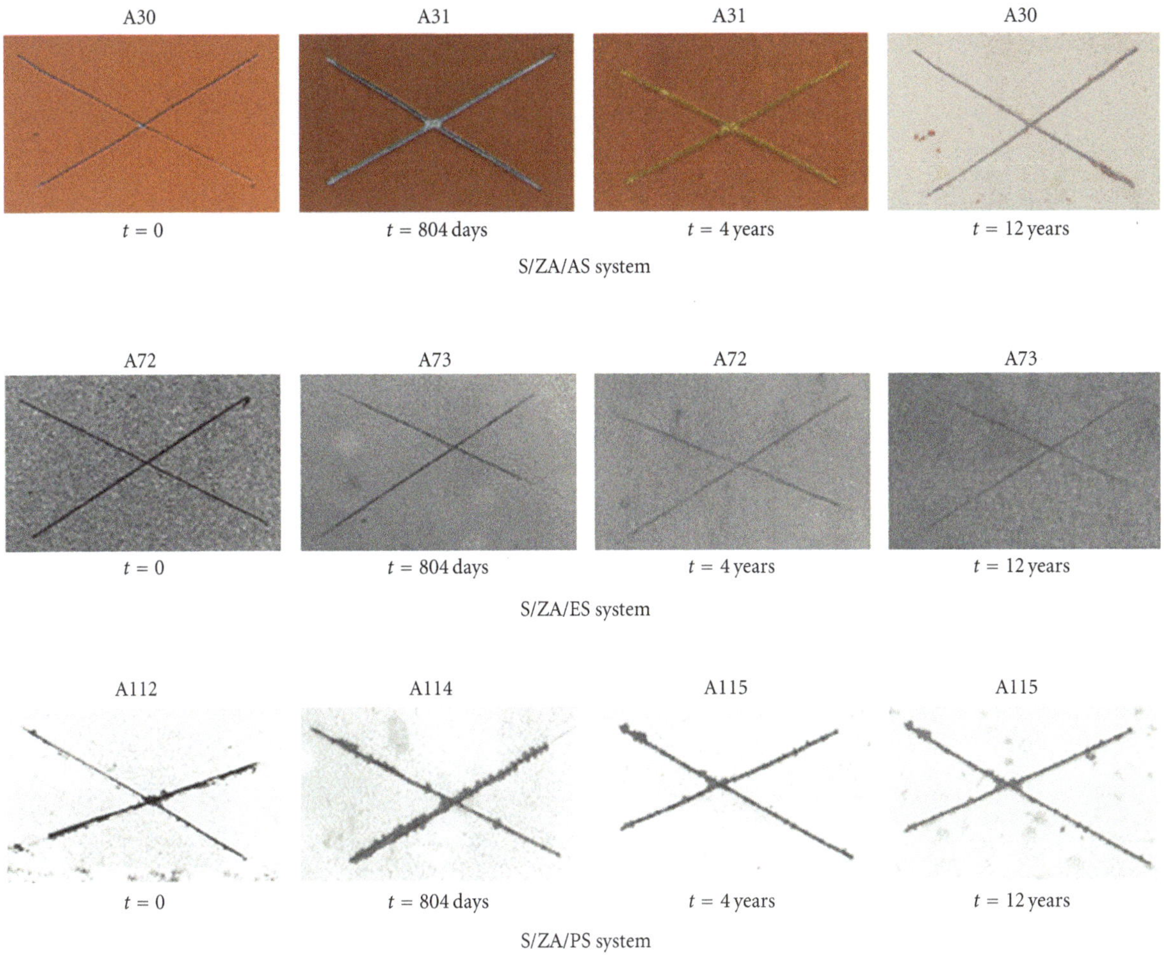

FIGURE 6: Photographs showing the evolution of S/ZA/painting systems as a function of the exposure time.

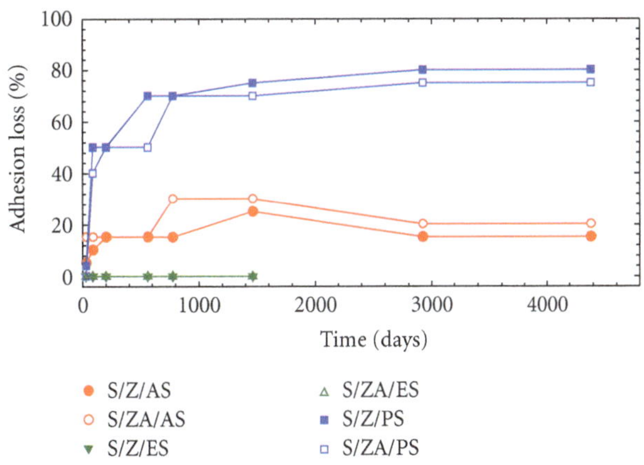

FIGURE 7: Plots showing the time dependence of the adhesion loss suffered by the painted S/Z and S/ZA sheets.

structure that is usually formed on the surface. The first time constant (R_1C_1) may be linked with the compact inner layer of ZnO and the second one (R_2C_2) with the external and porous layer of $Zn_4CO_3(OH)_6 \cdot H_2O$ [55, 56]. This surface film seems to inhibit further metal dissolution, although the environmental conditions determine the extent of corrosion progress due to a competition between film formation and film removal reactions. It was found that data of zinc corrosion measurements correlate with the air pollution levels given as a function of the SO_2 and Cl^- concentrations [57].

For similar exposure conditions, the influence of the coating composition on the bare sheet impedance values is shown in Figure 10. In it can be seen that the charge transfer

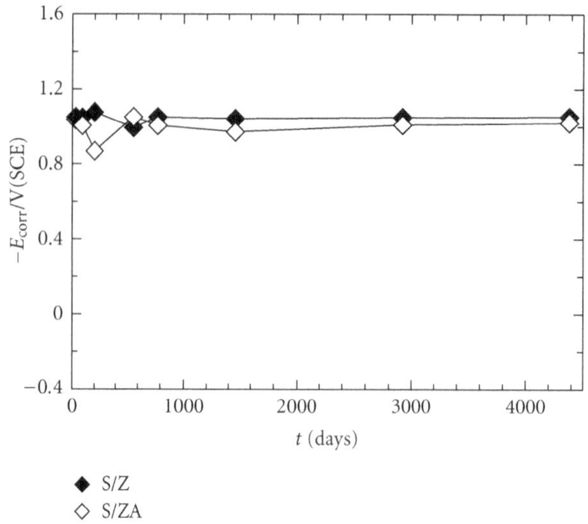

FIGURE 8: Plots showing the time dependence of the E_{corr} values of bare S/Z and S/ZA sheets as a function of the exposure time.

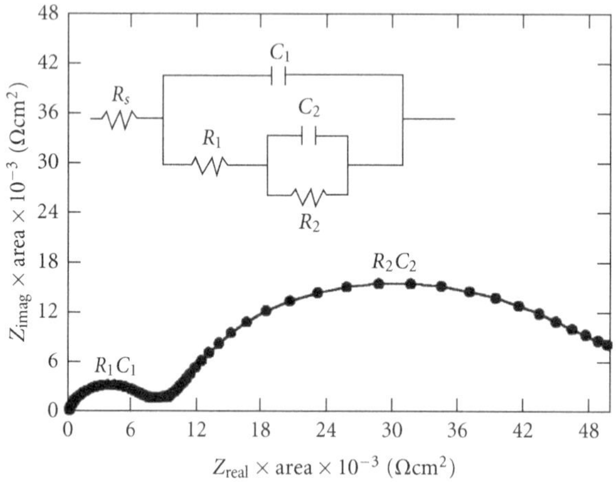

FIGURE 9: Equivalent circuit model used for fitting the tested bare S/Z and S/ZA sheets.

resistance (R_1) values shown by the 55%Al-Zn alloy was slightly higher than that of the zinc layer but the resistive contribution (R_2) of the external and porous layer to the system total impedance was very similar. These results are in accordance with the electrochemical activation demonstrated by both metallic surfaces.

3.2.2. Corrosion Potential and Impedance Results for Painted S/Z and S/ZA Sheets. Rest or corrosion potential (E_{corr}) measurements for painted metals and their time dependence have been questioned with regard to their use as a technique for evaluating the anticorrosive resistance of organic coatings [58]. However, its changes as a function of the exposure time to aqueous media have been successfully used as a simple tool to study the corrosion protection afforded by organic coatings [59–62]. Depending upon the microstructure of the paint coating, especially its polymerization degree, a certain period elapses until electrolyte penetration channels are established through which the underlying metal comes into contact with the medium. So, it is not surprising that, when a compact structure and high crosslinking level are accompanied by an also high film thickness, a few days of testing is not enough time for the electrolyte to enter in contact with the base metal of coated specimens, form the electrochemical double layer, and enable the measurement of a corrosion potential.

Figure 11 shows the corrosion potential (E_{corr}) values measured for each coated steel sheet exposed to the natural atmosphere of La Plata station. As can be seen, the E_{corr} values measured almost from the beginning and up to the end of the test for S/Z/AS, S/ZA/AS, S/Z/ES, and S/ZA/ES were quite similar to those obtained for the bare S/Z and S/ZA sheets (between −0.9 and −1.1 V/SCE). This means that, at least from the thermodynamic point of view, the protective properties offered by the alkyd- and epoxy-based painting systems were not sufficiently effective as to avoid the onset of the underlying zinc or 55%Al-Zn corrosion. On the other hand, the polyurethane-based painting system offered

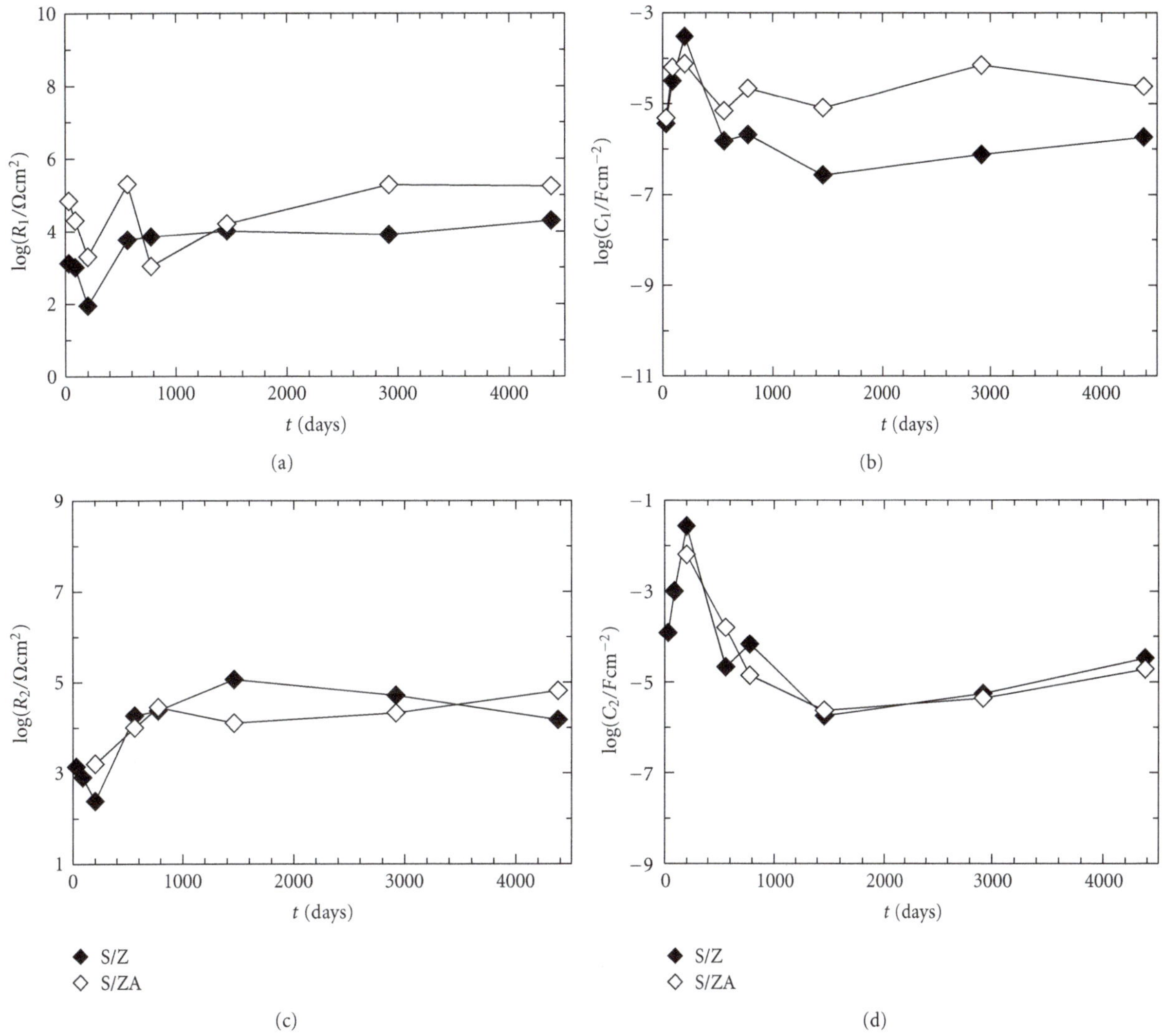

FIGURE 10: Evolution of log R_1, log C_1, log R_2, and log C_2 parameters of the tested bare S/Z and S/ZA sheets.

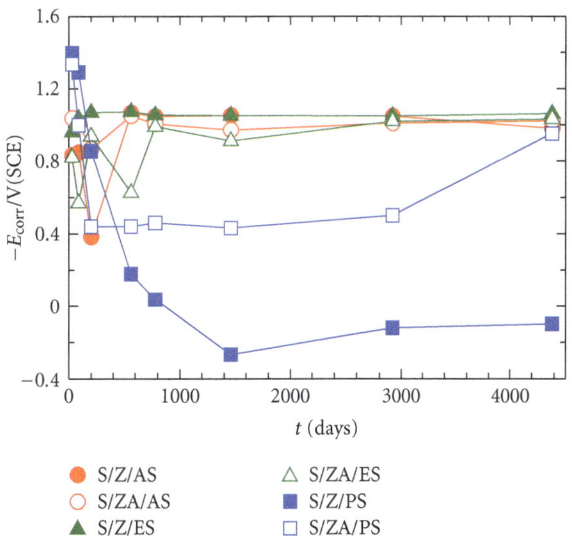

FIGURE 11: Plots showing the time dependence of the E_{corr} values of painted S/Z and S/ZA sheets as a function of the exposure time.

much more promising protective properties, particularly when applied on S/Z sheets, since the S/Z/PS system potential values remained in an electrochemically passive zone. A similar performance was supplied by the S/ZA/PS up to almost reaching the 12 years of exposure where together with the measurement of an $E_{corr} \approx -1.00$ V/SCE, the first sign of a localized corrosion was detected by EIS.

Since the main difference among the S/Z/painting systems and S/ZA/painting systems was the applied paint formulation used in each case, it is assumed that the magnitude of the E_{corr} displacements may be particularly associated with both the relative easiness with which the climatic variables affect the paint film structure and, hence, its protective properties. However, and it will be discussed in the next paragraph, except in the case of the epoxy-based painting system, the other two were able to protect relatively (AS) and effectively (PS) the metallic substrate from the corrosive atmosphere. This conclusion arises from the fact that the corrosion potential values measured for the S/Z/PS and S/ZA/PS panels were mostly nobler than the corresponding to bare S/Z or S/ZA sheets subjected to the same experimental conditions. This effective protection was mainly attributed to the PS barrier painting system, which could resist the strong aggressive action coming from the atmospheric conditions.

The impedance modulus ($|Z|$) of replicated samples as a function of their exposure time to the natural atmosphere of La Plata station illustrates Figure 12. A fast and simple qualitative analysis of this figure allows to infer that both the shape of all the experimental diagrams was fairly similar and it is possible to presume the presence of at least two time constants, one at low frequencies and another at high frequencies.

As seen in Figure 12, all the tested systems showed changes more or less significant of their ($|Z|$) and phase angle (Theta) during the weathering period. The fluctuating impedance values can be attributed to the dynamic behavior of the painting system structure frequently subjected to wet/dry cycles and/or other climatic changes as well as of the metal/paint interface through which the corrosion products gathered at the bottom and/or within the coating defects enhanced the coating barrier protection and, therefore, contribute to an increase in the impedance of the protective system at medium and low frequencies; however, as the time elapses, new defects appear at the weaker (less protective areas) paint layer allowing the inducing corrosion species permeation, and, consequently, the development of new electrochemically active zones.

The fact that the initial substrate attack is localized could be ascribed to the presence of very small defects in the paint layer, which act as an electrical shunt. As the exposure time goes on, the equilibrium between the development rate of the corrosion products and their diffusion rate towards the outdoor medium may be reached and, consequently, the total impedance fluctuations become small.

(1) *Equivalent Circuit Models.* The painting system as well as the S/Z or S/ZA substrates deterioration takes place from processes having a complex nature. Consequently, to interpret and explain in electrochemical terms the time dependence of the acquired impedance data, it has been necessary to propose appropriate equivalent circuit models.

Impedance spectra provide useful information concerning the evolution of both the protective features of the organic coating and the kinetics of the underlying metallic substrate corrosion process as a function of the exposure time to experimental or real service conditions. Thus, the dynamic character of the painting system barrier properties, the anticorrosive action of specific pigments, the corrosion products formation, and also changes in the disbonded area are accounted for the time dependence of the coated steel/medium impedance spectra. In general, an explanation of why and how such changes take place can be given by associating them to the resistive and capacitive parameters derived from fitting impedance data with nonlinear least squares algorithms involving the transfer function of the equivalent circuit model shown in Figure 13, [63–67]. They represent the parallel and/or series connection of some resistors and capacitors, simulating a heterogeneous arrangement of electrolytically conducting paths, where R represents the electrolyte resistance between the reference and working (coated steel) electrodes, R_c (resistance to the ionic flux) describes paths (pores, low crosslinking) of lower resistance to the electrolyte diffusion short-circuiting the paint film, and C_c is the dielectric capacitance representing the intact part of the same paint film [68]. Once the permeating and corrosion-inducing chemicals (water, oxygen and ionic species) reach electrochemically active areas of the substrate, particularly at the bottom of the paint film defects, the metallic corrosion become to be measurable so that its associated parameters, the electrochemical double-layer capacitance, C_{dl}, and the charge transfer resistance, R_t, can be estimated. It is important to remark that the values of these parameters vary direct (C_{dl}) and inversely (R_t) with the size of the corroding area.

On the other hand, distortions observed in these resistive-capacitive contributions indicate a deviation from the theoretical models in terms of a time constants distribution due to either lateral penetration of the electrolyte at the steel/paint interface (usually started at the base of intrinsic or artificial coating defects), underlying metallic substrate surface heterogeneity (topological, chemical composition, surface energy), and/or diffusion processes that could take place along the test [69, 70]. Since all these factors make the impedance/frequency relationship nonlinear, they are taken into consideration by replacing one or more capacitive components (C_i) of the equivalent circuit transfer function by the corresponding constant phase element Q_i (CPE), whose impedance dispersion relation is given by $Z = (j\omega)^{-n}/Y_0$ and n = CPE power $= \alpha/(\pi/2)$ [32, 71].

Difficulties in providing an accurate physical description of the occurred processes are sometimes found. In such cases, a standard deviation value ($\chi^2 < 5 \times 10^{-4}$) between experimental and fitted impedance data may be used as final criterion to define the most probable circuit.

According to the impedance data dispersion, the fitting process was performed using either the dielectric capacitance C_i or the phase constant element Q_i; however, the C_i

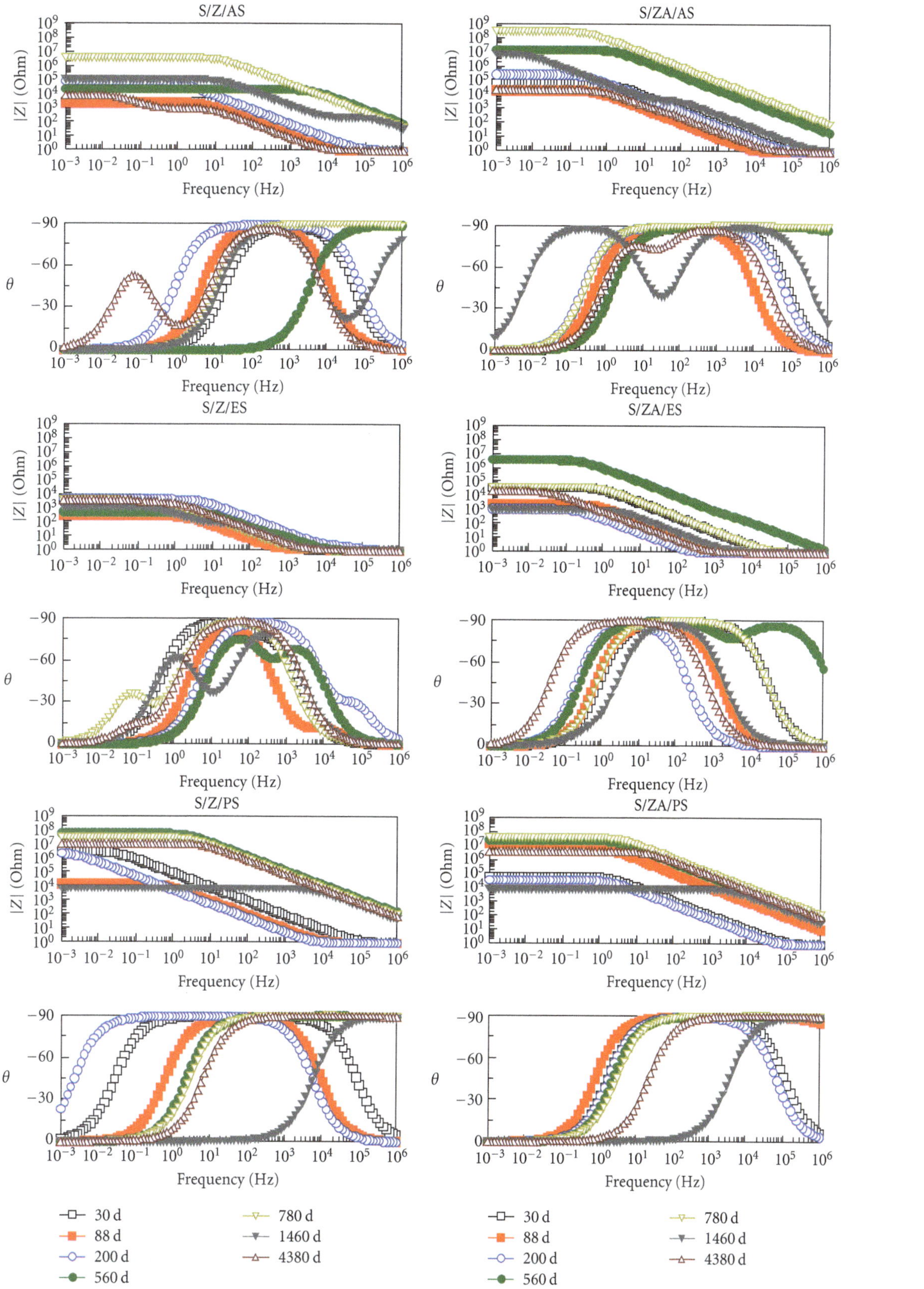

FIGURE 12: Bode plots showing the time dependence of the duplex systems impedance during their exposure to the natural atmosphere of La Plata station for 12 years.

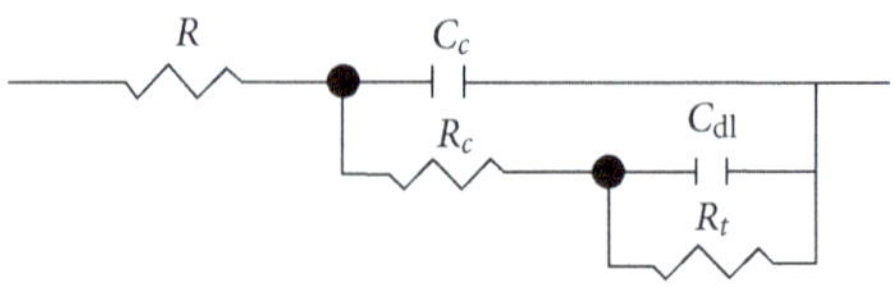

FIGURE 13: Equivalent circuit model used for fitting the tested duplex systems.

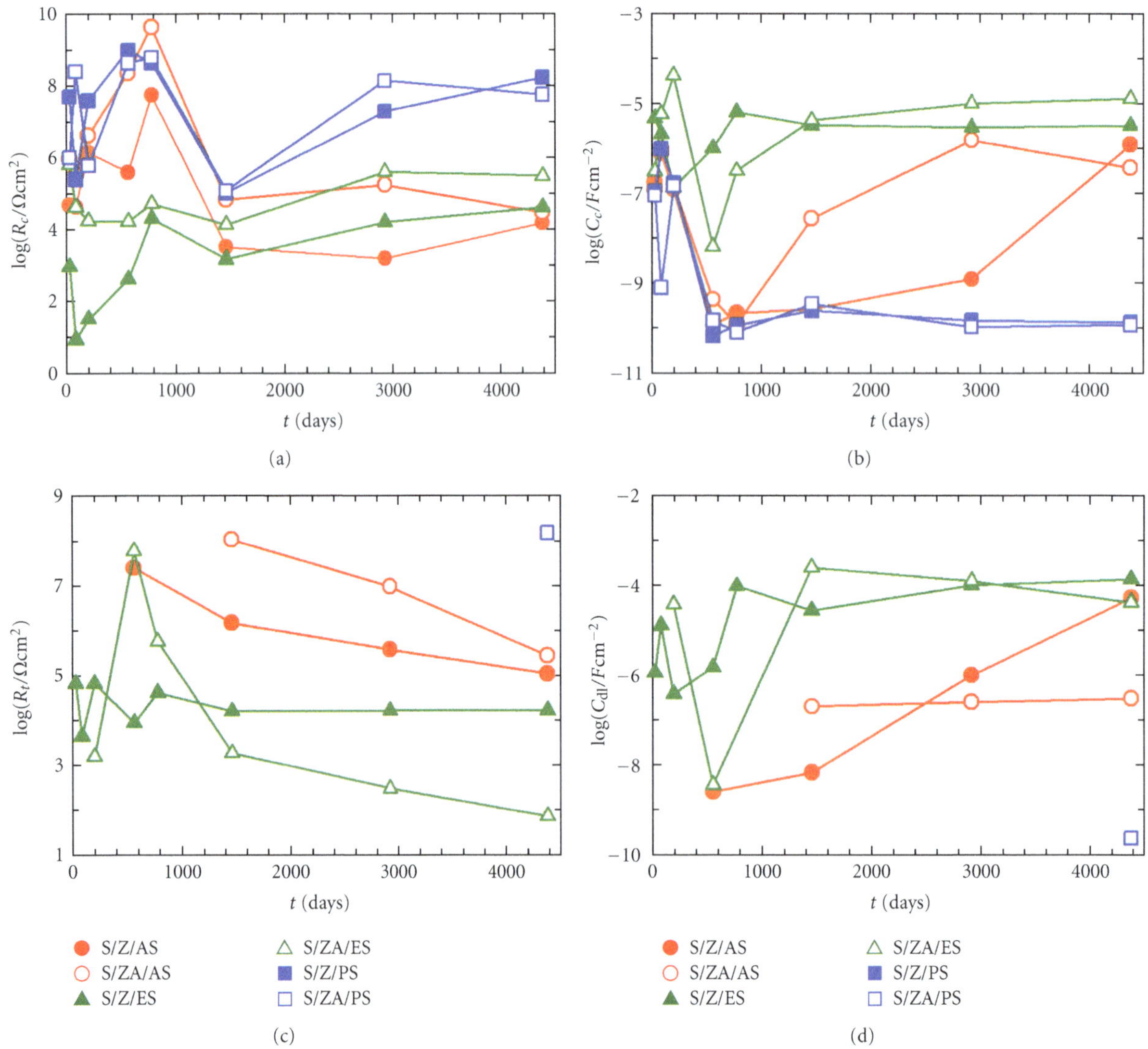

FIGURE 14: Evolution of R_c, C_c, R_t, and C_{dl} parameters of the duplex systems impedance during their exposure to the natural atmosphere of La Plata station for 12 years.

parameter was used in the following plots to facilitate the results visualization and interpretation.

(2) *Time Dependence of the Impedance Resistive and Capacitive Components.* The values of the resistive and capacitive components of the impedance corresponding to all the painted samples exposed to the natural atmosphere of La Plata station for 12 years are shown in Figure 14. As seen, the S/Z/PS and S/ZA/PS samples offered an excellent anticorrosive performance up to the end of the exposure. This behavior could be attributed to its excellent barrier effect due to the structurally homogeneous and strong paint film ($R_c \approx 10^7$–$10^8\ \Omega\text{cm}^2$, $C_c \approx 10^{-10}$–$10^{-9}\ \text{Fcm}^{-2}$), which was able to counteract the significant adhesion loss suffered by this painting system during weathering and slowed down the development of the alloy coating corrosion process up to the end of the exposure.

For the other two sample types, a rather highly fluctuating R_c values (two or more orders of magnitude) were found within the first 1400 days of exposure but then, and up to the end of the test, they remained changing between 10^6–10^4 Ωcm^2; on the other hand, its coupled dielectric capacitance (C_c) followed the same unstable trend at the beginning of the test but then, due to the deterioration degree reached by the AS and ES painting systems as a consequence of the adverse climatic conditions, led to C_c values ($\approx 10^{-6}$–10^{-5} Fcm^{-2}), that is, close to the bare S/Z and S/ZA sheets.

On the other hand, the same Figure shows great differences in the electrochemical response ($R_t C_{dl}$) of the different systems. In the case of the polyurethane systems, the corrosion process was either absent (S/Z/PS) or at least its development was delayed up to the end of exposure (S/ZA/PS). On the contrary, for the rest of the considered duplex systems after variable induction periods, the metallic coating degradation was detected. The worst corrosion protection was afforded by the Epoxy-based painting system since the corrosion process was detected by EIS at 30 (S/Z/ES) and 200 (S/ZA/ES) days. This behavior would be ascribed to the different electrochemical reactivity of the Z and ZA coatings. In the case of the Alkyd system, the induction period was of 600 and 1400 days for S/Z/As and S/ZA/AS, respectively. Again, the better performance of the last system could be accredited to a relatively good, although too short for practice purposes, barrier effect offered by the paint film added to the lower electrochemical reactivity of the S/ZA sheet.

4. Conclusions

At the end of this work, it is possible to summarize some conclusions valid for the studied materials.

All the laboratory and field tests involved in this work were useful to understand the behavior of the studied duplex systems subjected to natural weathering at La Plata Station. The good correlation between visual inspection and electrochemical tests allowed explaining some troubles observed in practice and, on this base, contribute to solve them to maintain its useful life as long as possible.

An almost constant corrosion rate of bare zinc and zinc-aluminum layers acting as galvanic coating of steel sheets was found during the long-term exposure to the natural atmosphere of the La Plata station. Both materials were able to cathodically protect the steel substrate for 12 years.

Regarding the comparative study among the three painting systems applied on S/Z or S/ZA sheets, different $R_c C_c$ and $R_t C_{dl}$ evolutions were obtained depending mainly on the paint. The best protective performance offered by the Polyurethane-based painting system was explained in principle taking in account its better barrier properties. The experimental results coming from the alkyd- and epoxy-based painting systems were not satisfactory due to their low resistance to the atmospheric conditions existing at La Plata station.

Despite the interface degradation (loss of adhesion) shown by all the painting systems, the corrosion process did not progress from the cross cut towards the underlying metallic substrate.

Acknowledgments

The authors thank the Comisión de Investigaciones Científicas de la Provincia de Buenos Aires (CICPBA), the Consejo Nacional de Investigaciones Científicas y Técnicas (CONICET), and the Universidad Nacional de La Plata for the financial support to carry out the present paper.

References

[1] M. Stratmann, K. Bohnenkamp, and W. J. Engell, "An electrochemical study of phase-transitions in rust layers," *Corrosion Science*, vol. 23, no. 9, pp. 969–985, 1983.

[2] L. Ferretti, E. Traverso, and G. Ventura, "Marine corrosion of mild steel in a thermically altered natural environment," *Anti-Corrosion Methods and Materials*, vol. 23, no. 5, pp. 3–5, 1976.

[3] C. E. Bird and F. J. Strauss, "Effect of wet storage staining on subsequent atmospheric corrosion of galvanized iron sheets," *Materials Performance*, vol. 15, no. 11, p. 27, 1976.

[4] F. E. Goodwin, "Mechanism of corrosion of Zinc and Zinc -5% Aluminum steel sheet coatings," in *Zinc-Based Steel Coatings Systems: Metallurgy and Performance*, G. Krauss, Ed., pp. 183–193, The Minerals, Metals & Materials Society, Warrendale, Pa, USA, 1990.

[5] M. Pourbaix, "Une méthode éléctrochimique rapide de predetermination de la corrosion atmospherique," Tech. Rep. 1, CEBELCOR, 1969.

[6] Y. Suzuki, Y. Hisamatsu, and N. Masuko, "Nature of atmospheric rust on iron," *Journal of the Electrochemical Society*, vol. 127, no. 10, pp. 2210–2214, 1980.

[7] *ASM Handbook*, vol. 13 of *Corrosion: Materials*, ASM International, 1992.

[8] J. C. Zoccola, H. E. Townsend, A. R. Borzillo, and J. B. Horton, "Atmospheric factors affecting the corrosion of engineering metals," in *Proceedings of the American Society for Testing and Materials (STP '78)*, vol. 646, pp. 165–184, 1978.

[9] F. C. Porter, *Zinc Handbook: Properties, Processing and Use in Design*, Marcel Dekker, New York, NY, USA, 1991.

[10] J. F. H. van Eijnsbergen, *Duplex Systems*, vol. 7, Elsevier, Amsterdam, The Netherlands, 1994.

[11] K. L. Mittal, "Stresses and adhesion of thin metallic coatings on oxide substrates," in *Adhesion Measurements of Film and Coatings*, K. L. Mittal, Ed., pp. 1–13, VSP, Utrecht, The Netherlands, 1995.

[12] *Steel Structure Painting Manual. Systems and Specifications*, vol. 2, Steel Structures Painting Council, Pittsburgh, Pa, USA, 7th edition, 1995.

[13] J. F. Malone, "Painting hot dip galvanized steel," *Materials Performance*, vol. 31, no. 5, pp. 39–42, 1992.

[14] C. H. Hare, "Corrosion and the preparation of metals for painting," in *Unit 26 Federation Series on Coatings Technology*, Federation of Societies for Coatings Technology, Philadelphia, Pa, USA, 1978.

[15] K. W. Allen, *Strength and Structures. Aspect of Adhesion*, vol. 1, University Press of London, London, UK, 1965.

[16] T. R. Bullet and A. T. S. Rudram, "The coating and the substrate," *Journal of the Oil and Colour Chemists' Association*, vol. 44, pp. 787–807, 1961.

[17] C. H. Hare, *Good Painting Practice. Steel Structure Painting Manual*, Steel Structures Painting Council, Pittsburgh, Pa, USA, 3rd edition, 1995.

[18] R. Barnhart, D. Mericle, Ch. Mobley, T. Hocking, J. H. Bogran, and E. McDaniel, "Why surface preparation is important?"

Journal of Protective Coatings and Linings, vol. 14, no. 9, pp. 61–64, 1997.

[19] H. Leidheiser Jr., "Corrosion of painted metals—a review," *Corrosion*, vol. 38, no. 7, pp. 374–383, 1982.

[20] I. Sekine, M. Yuasa, N. Hirose, and T. Tanaki, "Degradation evaluation of corrosion protective coatings by electrochemical, physicochemical and physical measurements," *Progress in Organic Coatings*, vol. 45, no. 1, pp. 1–13, 2002.

[21] G. Rocchini, "The importance of choosing the correct electrochemical technique for evaluating corrosion rates," *Corrosion Prevention and Control*, vol. 48, no. 4, pp. 125–134, 2001.

[22] S. Maeda, "Surface chemistry of galvanized steel sheets relevant to adhesion performance," *Progress in Organic Coatings*, vol. 28, no. 4, pp. 227–238, 1996.

[23] R. L. Howard, I. M. Zin, J. D. Scantlebury, and S. B. Lyon, "Inhibition of cut edge corrosion of coil-coated architectural cladding," *Progress in Organic Coatings*, vol. 37, no. 1, pp. 83–90, 1999.

[24] R. L. Howard, S. B. Lyon, and J. D. Scantlebury, "Accelerated tests for the prediction of cut-edge corrosion of coil-coated architectural cladding. Part I: cyclic cabinet salt spray," *Progress in Organic Coatings*, vol. 37, no. 1, pp. 91–98, 1999.

[25] R. L. Howard, S. B. Lyon, and J. D. Scantlebury, "Accelerated tests for prediction of cut edge corrosion of coil-coated architectural cladding. Part II: cyclic immersion," *Progress in Organic Coatings*, vol. 37, no. 1, pp. 99–105, 1999.

[26] S. Feliu, V. Barranco, and S. Feliu, "Contradictory results of the UVCON and saline immersion tests regarding the evaluation of some inhibitor/lacquer combinations on galvanised coatings," *Progress in Organic Coatings*, vol. 50, no. 3, pp. 199–206, 2004.

[27] A. R. di Sarli and R. A. Armas, "An assessment of the anticorrosive properties of epoxy paints. Correlation between impedance measurements and the salt-spray cabinet test," *Corrosion Prevention & Control*, vol. 36, no. 5, pp. 127–131, 1989.

[28] P. R. Seré, J. D. Culcasi, C. I. Elsner, and A. R. Di Sarli, "Study of the corrosion process at the galvanized steel/organic coating interface," in *Proceedings of the SCANNING '98*, vol. 20, no. 3, pp. 274–275, Baltimore, Md, USA, 1998.

[29] B. M. Rosales, A. R. di Sarli, F. Fragata et al., "PATINA network—performance of coil coating in natural atmospheres of Ibero-America," *Revista de Metalurgia*, vol. Extra, pp. 201–205, 2003.

[30] B. M. Rosales, A. R. di Sarli, O. de Rincón, A. Rincón, C. I. Elsner, and B. Marchisio, "An evaluation of coil coating formulations in marine environments," *Progress in Organic Coatings*, vol. 50, no. 2, pp. 105–114, 2004.

[31] B. Del Amo, L. Véleva, C. I. Elsner, and A. R. di Sarli, "Performance of coated steel systems exposed to different media: part I. Painted galvanized steel," *Progress in Organic Coatings*, vol. 50, no. 3, pp. 179–192, 2004.

[32] B. A. Boukamp, *Report CT88/265/128, CT89/214/128*, University of Twente, Amsterdam, The Netherlands, 1989.

[33] E. V. Schmid, *Exterior Durability of Organic Coatings*, FMJ International Publications, Redhill, UK, 1988.

[34] S. C. Chung, A. S. Lin, J. R. Chang, and H. C. Shih, "EXAFS study of atmospheric corrosion products on zinc at the initial stage," *Corrosion Science*, vol. 42, no. 9, pp. 1599–1610, 2000.

[35] Q. Qu, C. Yan, Y. Wan, and C. Cao, "Effects of NaCl and SO_2 on the initial atmospheric corrosion of zinc," *Corrosion Science*, vol. 44, no. 12, pp. 2789–2803, 2002.

[36] J. Morales, S. Martín-Krijer, F. Díaz, J. Hernández-Borges, and S. González, "Atmospheric corrosion in subtropical areas: influences of time of wetness and deficiency of the ISO 9223 norm," *Corrosion Science*, vol. 47, no. 8, pp. 2005–2019, 2005.

[37] V. Kucera and E. Mattsson, *Corrosion Mechanisms*, Marcel Dekker, New York, NY, USA, 1987.

[38] T. E. Graedel, "Corrosion mechanisms for zinc exposed to the atmosphere," *Journal of the Electrochemical Society*, vol. 136, no. 4, pp. 193C–203C, 1989.

[39] X. G. Zhang, *Corrosion and Electrochemistry of Zinc*, Plenum Press, New York, NY, USA, 1996.

[40] F. Mansfeld, *Corrosion Mechanisms*, Marcel Dekker, New York, NY, USA, 1987.

[41] H. E. Townsend, L. Allegra, R. J. Dutton, and S. A. Kriner, "Hot-dip coated sheet steels—a review," *Materials Performance*, vol. 25, no. 8, pp. 36–46, 1986.

[42] H. E. Townsend and A. R. Borzillo, "Twenty-year atmospheric corrosion tests of hot-dip coated sheet steel," *Materials Performance*, vol. 26, no. 7, pp. 37–41, 1987.

[43] J. W. Spence, F. H. Haynie, F. W. Lipfert, S. D. Cramer, and L. G. McDonald, "Atmospheric corrosion model for galvanized steel structures," *Corrosion*, vol. 48, no. 12, pp. 1009–1019, 1992.

[44] I. W. Odnevall Wallinder, P. Verbiest, W. He, and C. Leygraf, "Effects of exposure direction and inclination of the runoff rates of zinc and copper roofs," *Corrosion Science*, vol. 42, no. 8, pp. 1471–1487, 2000.

[45] S. Bertling, I. Odnevall, C. Leygraf, and D. Berggren, "Environmental effects of zinc runoff from roofing materials- A new multidisciplinary approach," in *Outdoor Atmospheric Corrosion, ASTM STP 1421*, H. E. Townsend, Ed., pp. 200–215, American Society for Testing and Materials International, West Conshohocken, Pa, USA, 2002.

[46] W. He, I. Odnevall, and C. Leygraf, "Runoff rates of zinc—a four-year field and laboratory study," in *Outdoor Atmospheric Corrosion, ASTM STP 1421*, H. E. Townsend, Ed., pp. 216–229, American Society for Testing and Materials International, West Conshohocken, Pa, USA, 2002.

[47] J. H. Sullivan and D. A. Worsley, "Zinc runoff from galvanised steel materials exposed in industrial/marine environment," *British Corrosion Journal*, vol. 39, no. 4, pp. 282–288, 2002.

[48] S. Matthes, S. Cramer, S. Bullard, B. Covino, and G. Holcomb, "Atmospheric corrosion and precipitation runoff from zinc and zinc alloy surfaces," in *Proceedings of the 58th Annual Conference Corrosion*, NACE, 2003, Paper 03598.

[49] S. Jouen, B. Hannoyer, A. Barbier, J. Kasperek, and M. Jean, "A comparison of runoff rates between Cu, Ni, Sn and Zn in the first steps of exposition in a French industrial atmosphere," *Materials Chemistry and Physics*, vol. 85, no. 1, pp. 73–78, 2004.

[50] D. Reiss, B. Rihm, C. Thöni, and M. Faller, "Mapping stock at risk and release of zinc and copper in Switzerland-dose response functions for runoff rates derived from corrosion rate data," *Water, Air, and Soil Pollution*, vol. 159, no. 1, pp. 101–113, 2004.

[51] R. M. Kain and E. A. Baker, *ASTM STP 947*, ASTM, Philadelphia, Pa, USA, 1986.

[52] S. H. Zhang and S. B. Lyon, "The electrochemistry of iron, zinc and copper in thin layer electrolytes," *Corrosion Science*, vol. 35, no. 1–4, pp. 713–718, 1993.

[53] A. Meneses Garcia, J. Rodarte Corro, O. Vidal Gutierrez, and E. Gonzalez Guzman, "The Mexican National Cancer Institute: case report," *Revista del Instituto Nacional de Cancerologia*, vol. 36, no. 4, pp. 1179–1182, 1990.

[54] M. Pourbaix, *Atlas of Electrochemical Equilibria in Aqueous Solutions*, NACE, Houston, Tex, USA, 1974.

[55] E. A. Sacco, J. D. Culcasi, C. I. Elsner, and A. R. di Sarli, "Evaluation of the protective performance of several duplex systems exposed to industrial atmosphere," *Latin American Applied Research*, vol. 32, no. 4, pp. 307–311, 2002.
[56] J. M. Costa and M. Vilarrasa, "Effect of air pollution on atmospheric corrosion of zinc," *British Corrosion Journal*, vol. 28, no. 2, pp. 117–120, 1993.
[57] M. W. Kendig, A. T. Allen, and F. Mansfeld, "Optimized collection of ac impedance data," *Journal of the Electrochemical Society*, vol. 131, no. 4, pp. 935–936, 1984.
[58] J. Wolstenholme, "Electrochemical methods of assessing the corrosion of painted metals-a review," *Corrosion Science*, vol. 13, no. 7, pp. 521–530, 1973.
[59] R. A. Armas, C. Gervasi, A. R. di Sarli, S. G. Real, and J. R. Vilche, "Zinc-rich paints on steels in artificial seawater by electrochemical impedance spectroscopy," *Corrosion*, vol. 48, no. 5, pp. 379–383, 1992.
[60] J. E. O. Mayne, "The mechanism of the inhibition of the corrosion of iron and steel by means of paint," *Official Digest*, vol. 24, pp. 127–136, 1952.
[61] L. Meszáros and S. A. Lindquist, "Study of the performance of zinc-rich paints coatings," in *Proceedings of the 115th Event of the European Federation of Corrosion (EUROCORR '82)*, pp. 147–156, Budapest, Hungary, October 1982, Section 11.
[62] M. Morcillo, R. Barajas, S. Feliu, and J. M. Bastidas, "A SEM study on the galvanic protection of zinc-rich paints," *Journal of Materials Science*, vol. 25, no. 5, pp. 2441–2446, 1990.
[63] B. del Amo, L. Véleva, A. R. di Sarli, and C. I. Elsner, "Performance of coated steel systems exposed to different media: part I. Painted galvanized steel," *Progress in Organic Coatings*, vol. 50, no. 3, pp. 179–192, 2004.
[64] O. Ferraz, E. Cavalcanti, and A. R. di Sarli, "The characterization of protective properties for some naval steel/polimeric coating/3% NaCl solution systems by EIS and visual assessment," *Corrosion Science*, vol. 37, no. 8, pp. 1267–1280, 1995.
[65] P. R. Seré, D. M. Santágata, C. I. Elsner, and A. R. di Sarli, "The influence of the method of application of the paint on the corrosion of the substrate as assessed by ASTM and electrochemical method," *Surface Coatings International*, vol. 81, no. 3, pp. 128–134, 1998.
[66] D. M. Santágata, P. R. Seré, C. I. Elsner, and A. R. di Sarli, "Evaluation of the surface treatment effect on the corrosion performance of paint coated carbon steel," *Progress in Organic Coatings*, vol. 33, no. 1, pp. 44–54, 1998.
[67] P. R. Seré, A. R. Armas, C. I. Elsner, and A. R. di Sarli, "The surface condition effect on adhesion and corrosion resistance of carbon steel/chlorinated rubber/artificial sea water systems," *Corrosion Science*, vol. 38, no. 6, pp. 853–866, 1996.
[68] H. Leidheiser Jr. and M. W. Kendig, "Mechanism of corrosion of polybutadiene-coated steel in aerated sodium chloride," *Corrosion*, vol. 32, no. 2, pp. 69–75, 1976.
[69] T. Szauer and A. Brandt, "Impedance measurements on zinc-rich paints," *Journal of the Oil and Colour Chemists' Association*, vol. 67, pp. 13–17, 1984.
[70] D. J. Frydrych, G. C. Farrington, and H. E. Townsend, "The barrier properties of thin carbonaceous films formed by ion beam assisted deposition," in *Corrosion Protection by Organic Coatings*, M. W. Kendig and H. Leidheiser Jr., Eds., vol. 87-2, p. 240, The Electrochemical Society, Pennington, NJ, USA, 1987.
[71] E. P. M. van Westing, G. M. Ferrari, F. M. Geenen, and J. H. W. van de Wit, "In situ determination of the loss of adhesion of barrier epoxy coatings using electrochemical impedance spectroscopy," *Progress in Organic Coatings*, vol. 23, no. 1, pp. 89–103, 1993.

Effect of Equal-Channel Angular Pressing on Pitting Corrosion of Pure Aluminum

Injoon Son,[1] Hiroaki Nakano,[2] Satoshi Oue,[2] Shigeo Kobayashi,[3] Hisaaki Fukushima,[2] and Zenji Horita[2]

[1] *Department of Materials Science and Metallurgy, Kyungpook National University, 1370 Sangyeok-dong, Buk-gu, Daegu 702-701, Republic of Korea*
[2] *Department of Materials Science and Engineering, Kyushu University, 744 Motooka, Nishi-ku, Fukuoka 819-0395, Japan*
[3] *Department of Applied Chemistry and Biochemistry, Kyushu Sangyo University, 2-3-1 Matsukadai, Higashi-ku, Fukuoka 813-8503, Japan*

Correspondence should be addressed to Hiroaki Nakano, nakano@zaiko.kyushu-u.ac.jp

Academic Editor: Flavio Deflorian

The effect of equal-channel angular pressing (ECAP) on the pitting corrosion of pure Al was investigated using electrochemical techniques in solutions containing 0.1 m mol$\cdot$dm^{-3} of Na_2SO_4 and 8.46 mol$\cdot$dm^{-3} of NaCl (300 ppm Cl^-) and followed by surface analysis. The potential for pitting corrosion of pure Al was clearly shifted in the noble direction by the ECAP process indicating that this process improves resistance to pitting corrosion. The time dependence of corrosion potential and the anodic potential at 1 A$\cdot$m^{-2} revealed that the rate of formation of Al oxide films increased due to a decrease in the grain size of the Al after ECAP. Since there exists a negligible amount of impurity precipitates in pure Al, the improvement in pitting corrosion resistance of pure Al by ECAP appears to be attributable to an increase in the rate of formation of Al oxide films.

1. Introduction

Although aluminum is inherently an active metal, it shows excellent corrosion resistance over a neutral pH range of 4–8 due to its superficial oxide film. In solutions containing Cl^-, however, pitting corrosion occurs locally where the oxide film is attacked by Cl^- [1–7]. On the other hand, reducing the grain size of metallic materials to the submicron range or even the nanometer range using equal-channel angular pressing (ECAP), high pressure torsion (HPT), or severe torsion straining processing (STSP) is increasingly being studied with the aim of improving mechanical properties such as strength and ductility [8–13]. Although the literature is reasonably scarce with investigations reporting the effects of severe plastic deformation upon the pitting and general corrosion behaviour [14–20] of Al-based alloys, mainly AA1100 and AA5052 (Al-Mg), those reported their majority suggests that corrosion resistance decreases with decreasing the grain size [14, 15]. It should also be remarked that chloride content, pH and other characteristics and nature of solution significantly affect the corrosion behaviour. Considering the ECAP, it has been reported both increase and decrease on the corrosion resistance [15, 17, 18]. Ralston et al. [15] recently reported that finer grains of commercially pure (c.p.) grade Al samples provide significant gains into the corrosion resistance in acidic and alkaline sodium chloride media whilst having lesser, but noticeable, impact in near neutral electrolytes [15]. It was also reported [15] that ECAP has provided low corrosion rates by using a single cast ingot of ultra-high purity Al. Precipitate compounds of Fe-Al and Si were found around the pitting area. Because the corrosion potential of the precipitates is nobler than the Al matrix, the pitting corrosion of Al seems to occur forming local cells where precipitates act as cathodes. The improvement in pitting corrosion resistance from ECAP seems to be attributable to both a decrease in the size of precipitates, which act as cathodes during pitting corrosion, and an increase in the rate of formation of Al oxide films due to the reduction of the grain size of Al. However, the reason for improvement of the pitting corrosion resistance of Al alloys with ECAP was not made clear.

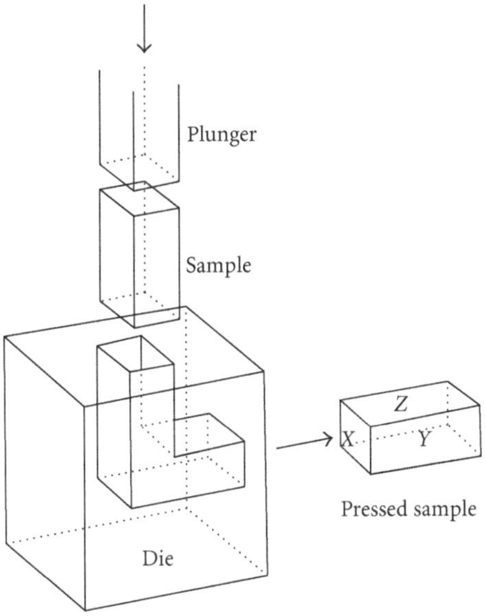

FIGURE 1: Schematic diagram of ECAP.

In this study, pure Al of 4 N (>99.99 mass%) without precipitates was used to determine whether the increase in the formation rate of Al oxide films contributes to the improvement of the pitting corrosion resistance in Al or not. The effect of ECAP on the pitting corrosion of pure Al and anodized pure Al was investigated. The pitting behaviour of Al was investigated using electrochemical techniques. In addition, the reason for the improvement in the pitting corrosion resistance of Al with ECAP was further analyzed using surface analysis techniques.

2. Experimental

Pure Al (>99.99 mass%) was used as experimental specimens. Figure 1 shows a schematic diagram of the ECAP process. ECAP was conducted for eight passes at room temperature (20–25°C) using a die with a channel angle of 90° (error range $< \pm 0.01°$) which creates an equivalent strain of about 1 during one passage through the die. The samples were rotated 90° about the longitudinal axis in the same sense between consecutive passes, the generally designated processing route Bc [11]. Each press was performed at a rate of $\sim 19\,mm\cdot s^{-1}$ using MoS_2 as a lubricant. The initial average grain size of Al prior to ECAP was 30–50 μm. The average grain size of Al after the ECAP process was confirmed by TEM observation to be 0.5–1.0 μm [21].

Specimens of 1.0 cm in diameter were prepared for corrosion testing. Their corrosion resistance was investigated over an area 28.26 mm^2; the remaining area was sealed with waterproof tape to prevent any corrosion due to edge effects. After pure Al was carefully polished down with No. 600 emery paper, it was immersed in a solution containing $0.1\,mol\cdot dm^{-3}$ of Na_2SO_4 and $8.46\,mmol\cdot dm^{-3}$ of NaCl (300 ppm Cl^-) at 25°C for one hour in an air atmosphere. Potentiodynamic polarization curves were measured by polarizing from the less noble potential to the anodic potential direction using the potential sweep method at $0.5\,mV\cdot s^{-1}$. In addition, the time dependence of anodic current density was measured while being maintained at 0.3 V versus NHE after immersion for 60 minutes. The electrode potentials were measured using a saturated KCl, Ag/AgCl reference electrode (0.197 V versus NHE, 25°C). Since mercury is contained in calomel electrode, Ag/AgCl electrode was used as reference electrode in this study. Polarization curve potentials were plotted with reference to NHE. Platinum was used as the counterelectrode in all electrochemical measurements.

Since the naturally formed Al oxide films were of the barrier type, the time dependence of anodic potentials was examined during galvanostatic electrolysis at $1\,A\cdot m^{-2}$ in a solution containing $0.5\,mol\cdot dm^{-3}$ of H_3BO_3 and $0.05\,mol\cdot dm^{-3}$ of $Na_2B_4O_7\cdot 10H_2O$ used to form a barrier-type oxide film [22, 23]. The effect of ECAP on the rate of formation of the Al oxide film was assumed to be revealed by the time dependence of the anode potentials.

Prior to anodizing, pure Al was carefully polished down with No. 1500 emery paper and immersed in a $0.75\,mol\cdot dm^{-3}$ NaOH solution at 25°C for 30 s, after which they were neutralized in $0.48\,mol\cdot dm^{-3}$ of HNO_3 solution for 30 s and electropolished in a 20°C solution containing methanol and perchloric acid ($MeOH:HClO_4 = 4:1$) at 10 V for 5 min. Anodizing was conducted in a solution containing $1.53\,mol\cdot dm^{-3}$ of H_2SO_4 and $0.0185\,mol\cdot dm^{-3}$ of $Al_2(SO_4)_3\cdot 16H_2O$ at 20°C under galvanostatic conditions of three different current densities, that is, 100, 200, and $400\,A\cdot m^{-2}$ for 20 minutes while agitated at 100 rpm using a magnetic stirrer. The time dependence of anodic current density was measured while the system was maintained at 1.2 V after immersion for 30 min in a $0.2\,mol\cdot dm^{-3}$ $AlCl_3$ solution with a pH of 2.8 at 25°C.

The morphology of Al with and without ECAP was observed using SEM after electrolytic polishing. Pitting corrosion occurred while the samples were held at 0.3 V for 5 min in a solution containing $0.1\,mol\cdot dm^{-3}$ of Na_2SO_4 and $8.46\,mmol\cdot dm^{-3}$ of NaCl; SEM micrographs were also obtained in order to microstructural observations.

3. Results and Discussion

3.1. Effect of ECAP on the Pitting Corrosion of Pure Al. Figure 2 shows the effect of ECAP on the potentiodynamic polarization curve of pure Al. In all figures below, the reported results [20] of AA1100 (mass%, >99.0% Al, 0.15% Cu, 0.09% Si, 0.60% Fe, 0.005% Mn, 0.006% Zn, 0.002% Mg, <0.001% Cr) were also shown to compare with pure Al. The anode current densities of pure Al and Al (AA1100) rapidly increased at certain potentials when the anode potentials were shifted from their corrosion potentials in the noble direction. This rapid increase in current density is caused by initiation of pitting corrosion. The pitting corrosion potential, the point at which pitting corrosion is initiated, of pure Al as well as Al (AA1100) was shifted in the noble direction after application of ECAP, indicating improvement in the pitting corrosion resistance. On the other hand, the corrosion potential of pure Al was also slightly shifted in the

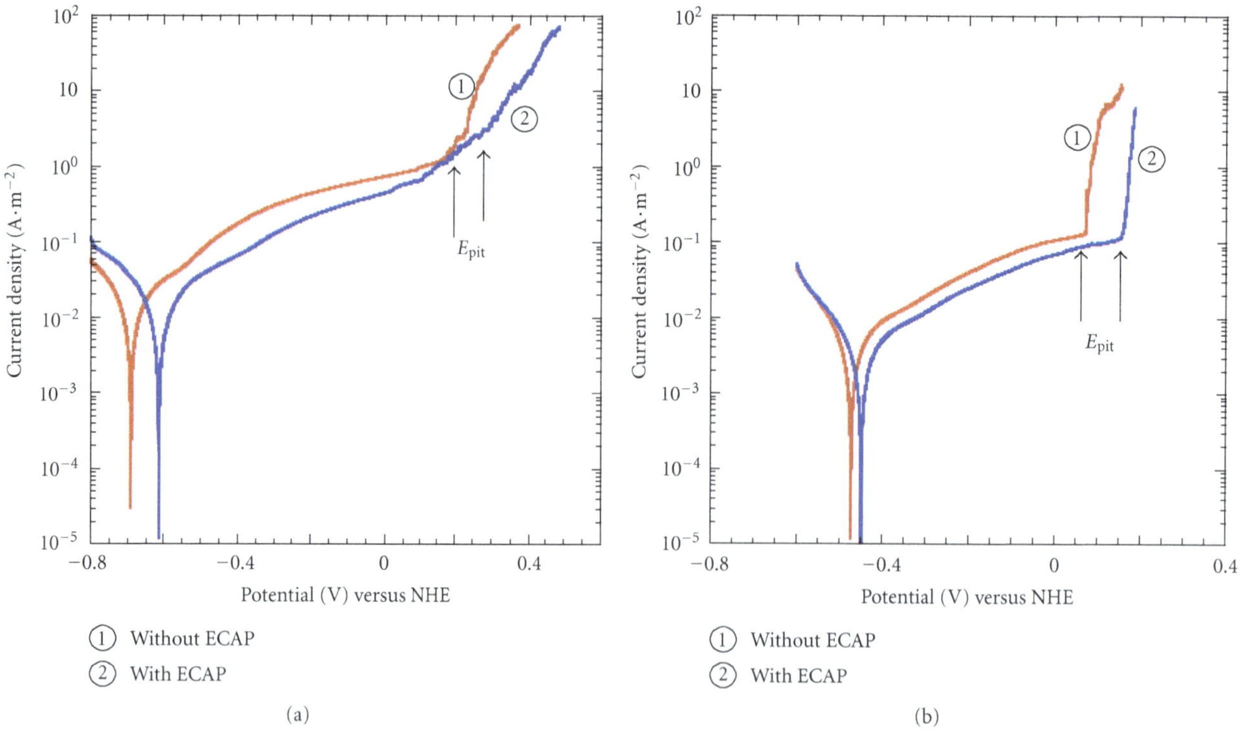

FIGURE 2: Effect of ECAP on the potentiodynamic polarization curves of pure Al (a) and Al (AA1100) (b) in a solution containing 0.1 mol·dm^{-3} of Na_2SO_4 and 300 ppm of Cl^-.

noble direction by ECAP. The pitting corrosion potential of pure Al was nobler than that of Al (AA1100). The current density needed to initiate pitting corrosion was larger for pure Al than for Al (AA1100), and the degree of increase in current density at the pitting corrosion potential was smaller for pure Al than for Al (AA1100). These results indicate that the pitting corrosion resistance of pure Al is better than that of Al (AA1100). It is known that the pitting potential can be coincident with the corrosion potential when the anodic Tafel slopes are practically zero (null), as previously reported [24, 25]. It was demonstrated into the literature that, in a dilute sodium chloride solution (0.01 M), the pitting corrosion potential of a commercially pure Al sample attains about −545 mV versus SCE [26]. It is clearly observed that both corrosion potential and pitting corrosion potential in this study are displaced to nobler-side potential than the literature, while passive current densities have considerably decreased since about 1 decade.

Figure 3 shows the time-dependence of the anodic current density of pure Al and Al (AA1100) while kept at constant potentials of, respectively, 0.3 V and 0.2 V nobler than their pitting corrosion potentials. The anode current density of pure Al increased sharply after a certain period of time due to the initiation of pitting corrosion. The time required before the initiation of pitting corrosion for pure Al was longer with ECAP than without; however, the degree of increase in time of pure Al with ECAP was smaller than that of Al (AA1100). Furthermore, the anodic current density after initiating pitting corrosion, or the growth rate of pitting corrosion, was smaller with ECAP than without for both pure Al as well as Al (AA1100).

Figure 4 shows the morphology of pure Al after a 5 min anodic reaction at 0.3 V nobler than the pitting corrosion potential in a solution containing 0.1 mol·dm^{-3} of Na_2SO_4 and 8.46 mmol·dm^{-3} of NaCl. Crystallographic pits 100–200 μm in size were observed in both pure Al samples made with and without ECAP, as shown in Figures 4(a) and 4(c). This phenomenon can be seen from the enlarged pitting areas in Figures 4(b) and 4(d). Since the grain size of Al greatly decreased after ECAP, the crystals in the pitting area were more compact. Considering the typical faceted-type crystallographic morphology of the pits, which are shown in Figure 4, these evidenced pit morphology are slightly connected by filaments provoked by hydrogen evolution [24, 27, 28] and it seems that they increase with ECAP.

Figure 5 shows the morphology of electropolished pure Al and Al (AA1100) with and without ECAP. As shown in Figures 5(a) and 5(b), precipitates were observed in minute amounts in pure Al, while many from Fe-Al intermetallic compounds were observed in Al (AA1100) (Figures 5(c) and 5(d)). These precipitates were present in Al (AA1100) regardless of having undergone the ECAP process or not; however, the number of large precipitates in Al (AA1100) was greatly decreased due to an increase in small precipitates caused by ECAP. This phenomenon is shown by the arrow in Figure 5(d). Since the precipitates of impurities act as cathodes in the Al matrix, the precipitates nucleate causing the pitting corrosion of Al. Pitting corrosion is less likely to occur

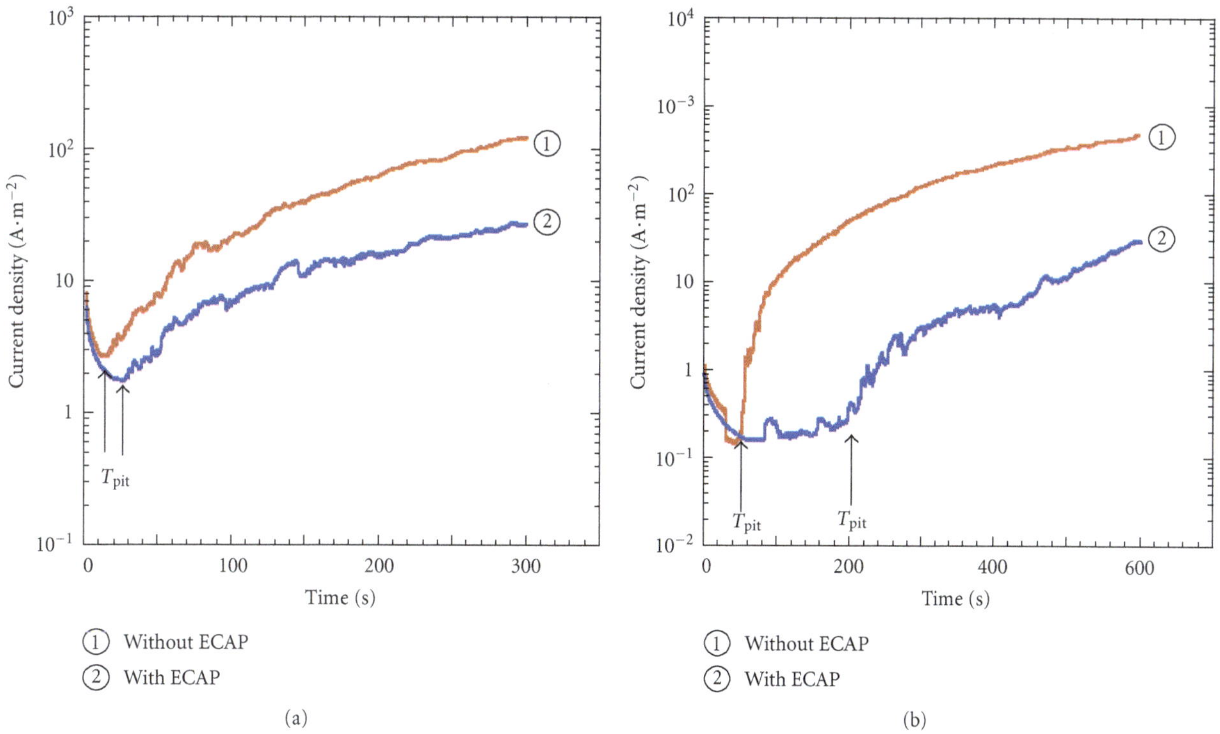

FIGURE 3: Time-dependence of the anodic current density of pure Al (a) and Al (AA1100) (b) while kept at constant potentials of, respectively, 0.3 V and 0.2 V nobler than their pitting corrosion potential in a solution containing 0.1 mol·dm^{-3} of Na_2SO_4 and 300 ppm of Cl^-.

FIGURE 4: Morphology of Al after anodic oxidation for 5 min at 0.3 V nobler than the pitting corrosion potential in a solution containing 0.1 mol·dm^{-3} of Na_2SO_4 and 300 ppm of Cl^-. (a, b) Pure Al without ECAP. (c, d) Pure Al with ECAP.

with a decrease in the size of the precipitate, that is, with a decrease in cathode area. Only pits that grow readily around large precipitates during the initial stage seem to ultimately develop into macropits, this is attributed to the fact that most pits are unstable and are immediately repassivated [2]. The improvement of pitting corrosion resistance of Al (AA1100) with ECAP can thus be partially ascribed to a decrease in the size of the precipitates. However, the improvement of pitting corrosion resistance of pure Al with ECAP cannot be explained by a decrease in the size of the precipitates since pure Al contains only trace amounts of precipitates.

Figure 6 shows the time-dependence of the corrosion potential of pure Al and Al (AA1100) in a solution containing 0.1 mol·dm^{-3} of Na_2SO_4 and 300 ppm of Cl^- at 25°C. The

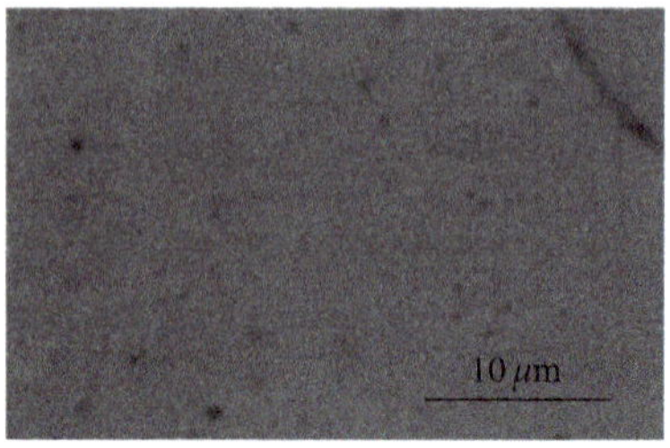

(a) Pure Al without ECAP

(b) Pure Al with ECAP

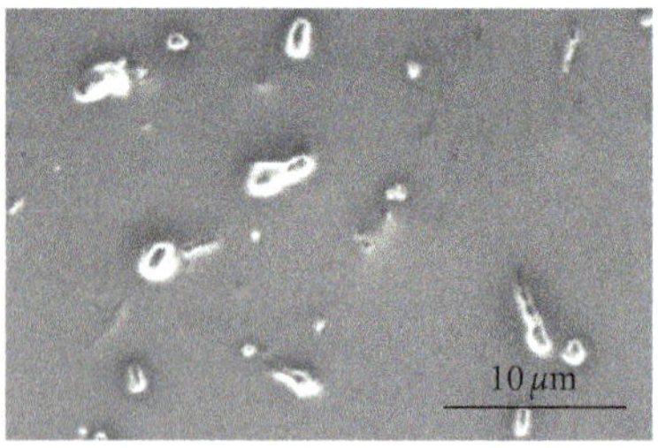

(c) Al (AA1100) without ECAP

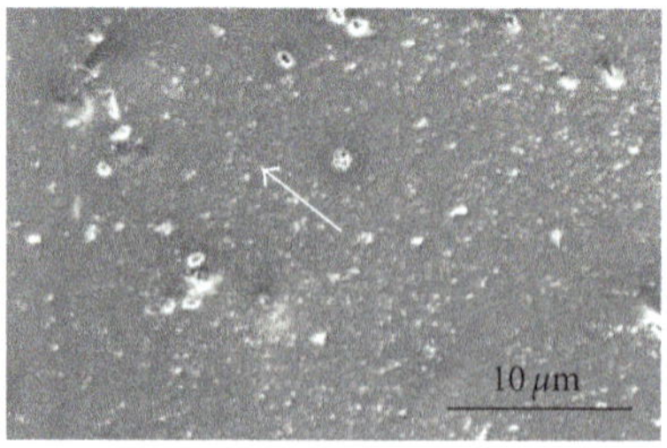

(d) Al (AA1100) with ECAP

FIGURE 5: Effect of ECAP on the morphology of pure Al and Al (AA1100).

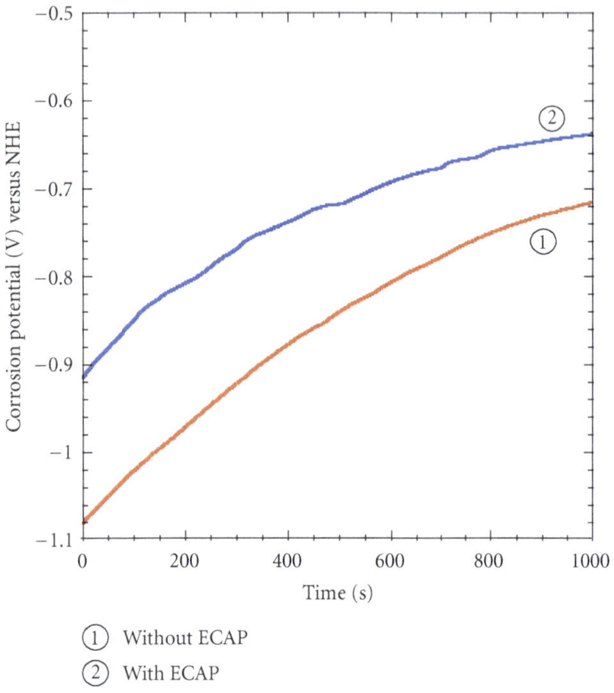

(a)

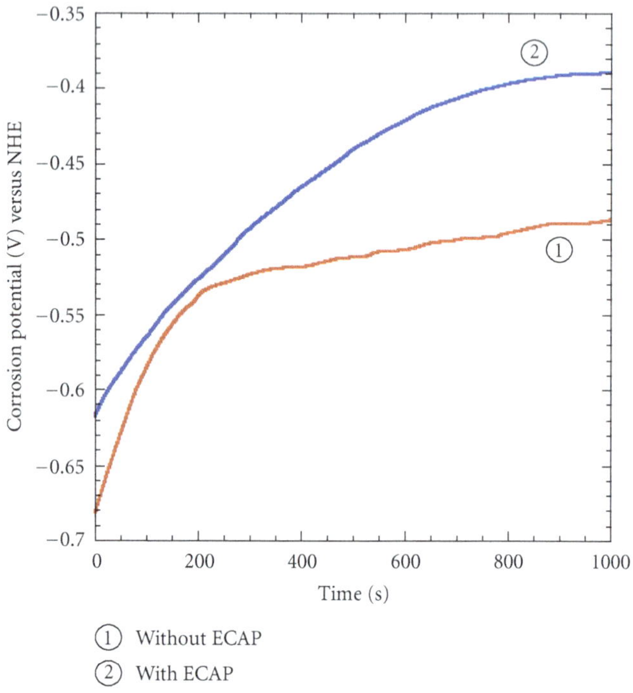

(b)

FIGURE 6: Time-dependence of corrosion potential of pure Al (a) and Al (AA1100) (b) in a solution containing 0.1 mol·dm^{-3} of Na_2SO_4 and 300 ppm of Cl^- at 25°C.

corrosion potential of pure Al shifted in the noble direction with increased time. This can be ascribed to the Al oxide film that naturally forms in the solution. The corrosion potential of pure Al shifted in the noble direction more rapidly with ECAP than without, suggesting that the rate of formation of Al oxide film increased with ECAP. It is well known that the oxidation of metal occurs more rapidly at crystalline lattice defects such as grain boundaries and dislocations [1, 2]. The ECAP process markedly increases the number of grain boundaries and the dislocation density in Al. The oxidation rate of Al appears to increase as a result of the increase in grain boundaries and dislocation density caused by ECAP. The oxidation rate of Al (AA1100) is faster than that of pure Al because the impurities contained in Al (AA1100) segregate

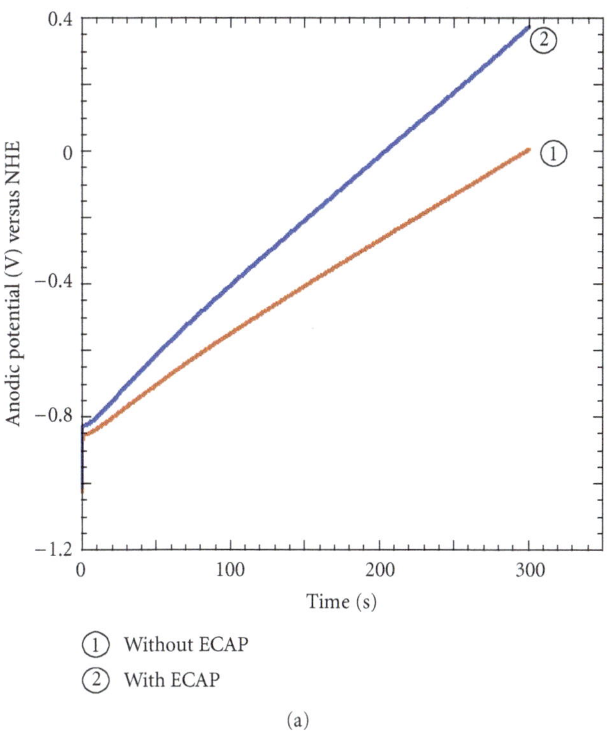

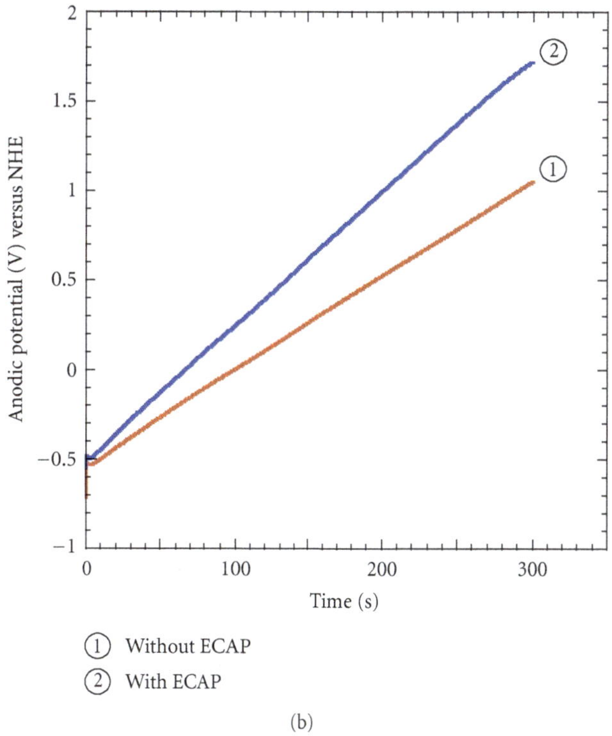

FIGURE 7: Time-dependence of anodic potential of pure Al (a) and Al (AA1100) (b) at 1 A·m^{-2} in a solution containing 0.5 mol·dm^{-3} of H_3BO_3 and 0.05 mol·dm^{-3} of $Na_2B_4O_7$.

at lattice defects such as grain boundaries and dislocations. As a result, the corrosion potential of Al (AA1100) shifted in the noble direction more rapidly than pure Al.

In the phenomenon described above, the average oxidation rate magnitude of Al over the entirety of the Al surface is taken into consideration since it was estimated from the corrosion potential. Since pitting corrosion occurred around precipitates, it is necessary to investigate the rate of formation of Al oxide films immediately around the precipitates. In this study; however, it was assumed that the oxidation rate of Al around the precipitates was nearly identical to the average of the entire Al surface.

Since the naturally formed Al oxide films were of the barrier type, the anode potentials were measured using galvanostatic electrolysis in a neutral solution containing a boric salt used to form a barrier-type oxide film. Figure 7 shows the time-dependence of anode potentials of pure Al and Al (AA1100) at 1 A·m^{-2} in a solution containing 0.5 mol·dm^{-3} of H_3BO_3 and 0.05 mol·dm^{-3} of $Na_2B_4O_7$. The anode potential of pure Al shifted in the noble direction with increased electrolysis duration; the shift was faster with ECAP than without. The shift of the anodic potential corresponds to the rate of formation of Al oxide films. Therefore, it was apparent that the barrier type of Al oxide film was formed more quickly with ECAP than without in a neutral solution. These results wholly correspond to the time-dependence of the corrosion potentials in a Na_2SO_4 solution containing Cl^- as shown in Figure 6. The pitting corrosion resistance of pure Al was improved by ECAP despite the fact that pure Al contains a negligible amount of precipitates. As discussed above, this cannot be explained by a decrease in the size of the precipitates with ECAP. It can be assumed that there is a decrease in the average grain size in the Al matrix, and therefore the increase in rate of formation of the Al oxide film contributes to the improvement of pitting corrosion resistance of pure Al after ECAP. When the formation rate of Al oxide films is rapid, pits from the initial stage appear to be immediately repassivated. Authors have previously reported that the pitting corrosion resistance of Al (AA1100) and Al-Mg alloy (AA5052) was improved by ECAP [20]. The improvement of pitting corrosion resistance of pure Al with ECAP suggests that the decrease in the grain size of the Al matrix significantly contributes to the improvement of pitting corrosion resistance of Al (AA1100) and Al-Mg alloy (AA5052) with ECAP.

3.2. Effect of ECAP on the Pitting Corrosion of Anodized Pure Al. The time-dependence of the anodic current density of pure Al and Al (AA1100) anodized at various current densities is shown in Figure 8. Samples were held at 1.2 V in a solution containing 0.2 mol·dm^{-3} of $AlCl_3$. The anodic current density of anodized pure Al and Al (AA1100) increased rapidly after a certain period of time due to the initiation of pitting corrosion. ECAP produced no observable effect on the pitting corrosion of anodized pure Al at any anodizing current density. Since the anodization duration of pure Al was fixed at 20 min in this study, the thickness of the anodic oxide films increased with increased anodizing

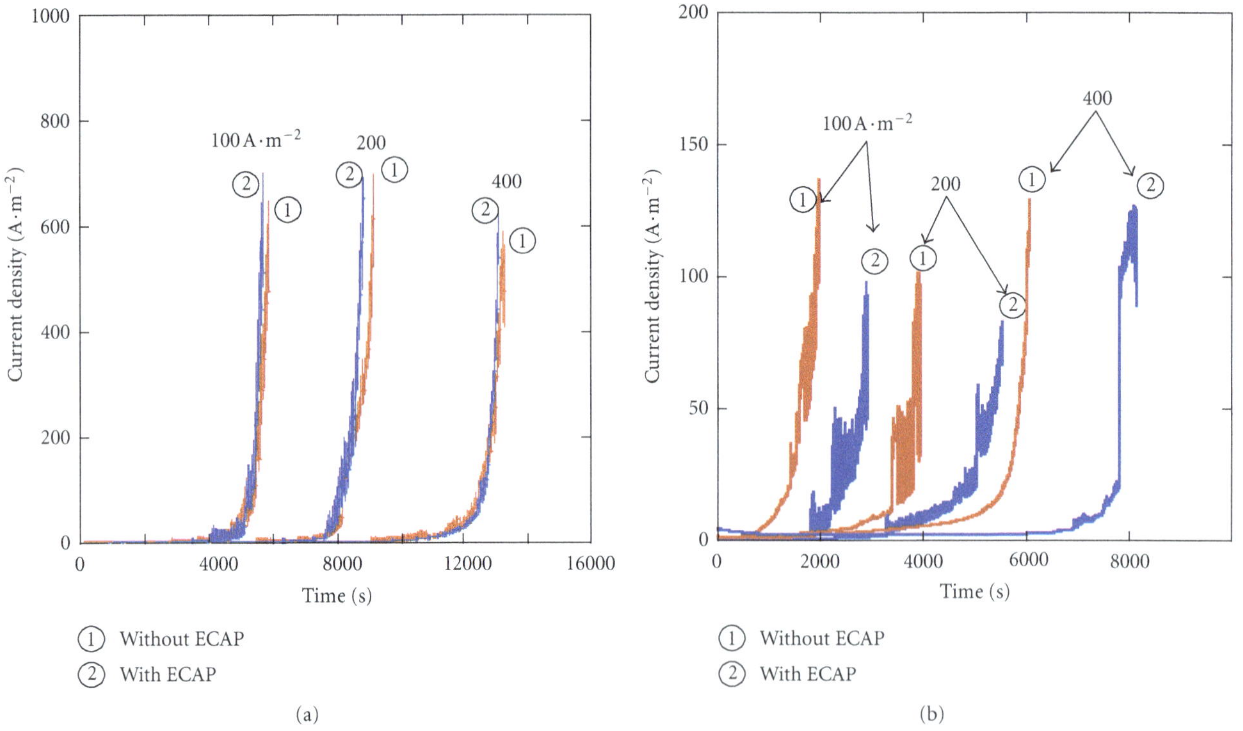

FIGURE 8: Time-dependence of the anodic current density of pure Al (a) and Al (AA1100) (b) anodized at 100, 200, and 400 $A\cdot m^{-2}$ while kept at a constant potential of 1.2 V in 0.2 $mol\cdot dm^{-3}$ of $AlCl_3$ solution.

current density. Therefore, the time required for initiating pitting corrosion increased with anodizing current density. On the other hand, the time to initiate pitting corrosion of Al (AA1100) anodized for 40 min was longer with ECAP than without regardless of anodizing current density. This suggests that ECAP enhanced the pitting corrosion resistance of anodized Al.

Figure 9 shows SEM images of anodized pure Al and Al (AA1100). Precipitates were rarely present in the anodized pure Al shown in Figures 9(a) and 9(b) regardless of having undergone ECAP or not, while many small precipitates of Fe-Al intermetallic compounds were observed in the anodized Al (AA1100), as shown in Figures 9(c) and 9(d). In Al (AA1100) without anodization, many precipitates were also present, as shown in Figure 5. In anodized Al (AA1100), the precipitates remained unoxidized and their size was smaller with ECAP than without, as shown in Figures 9(c) and 9(d). The size of precipitates in Al hardly changed after anodizing.

The pitting corrosion resistance of anodized Al (AA1100) was improved using ECAP, as shown in Figure 8. Since the size of the Fe-Al intermetallic compound precipitates remaining in the anodic oxide films decreased after ECAP (Figures 9(c) and 9(d)) and pitting corrosion occurred preferentially around these precipitates [20], the improvement of pitting corrosion resistance from ECAP can predominantly be ascribed to a decrease in the size of the precipitates in anodic oxide films. On the other hand, there is no effect from a decrease in grain size of Al by ECAP on the pitting corrosion of anodized pure Al because anodized pure Al rarely contains precipitates. Although the natural oxidation rate of Al is assumed to increase due to a decrease in average grain size after ECAP, the effect of the natural oxidation rate on the formation rate of Al oxide films seems to be neglected in anodization with power supply. As a result, no effect of ECAP on the pitting corrosion of anodized pure Al was observed.

Within the context of this work, grain refinement of pure Al appears to offer significant gains in corrosion resistance in acidic and alkaline NaCl environments whilst having lesser, but noticeable, impact in near neutral electrolytes [15]. According to Ralston et al. [15], this illustrates that within specific material and environment combinations, grain refinement or grain coarsening may be an effective way to maximize corrosion resistance.

4. Conclusion

The effect of reducing the grain size of pure Al by ECAP on the pitting corrosion was investigated using electrochemical techniques. The potential for pitting corrosion of pure Al was clearly shifted in the noble direction using ECAP, producing an overall improvement in pitting corrosion resistance. It is evident from the time-dependence of corrosion and anode potentials during galvanostatic electrolysis in a neutral solution that the rate of formation of Al oxide films increases due to the decrease in Al grain size from the ECAP process.

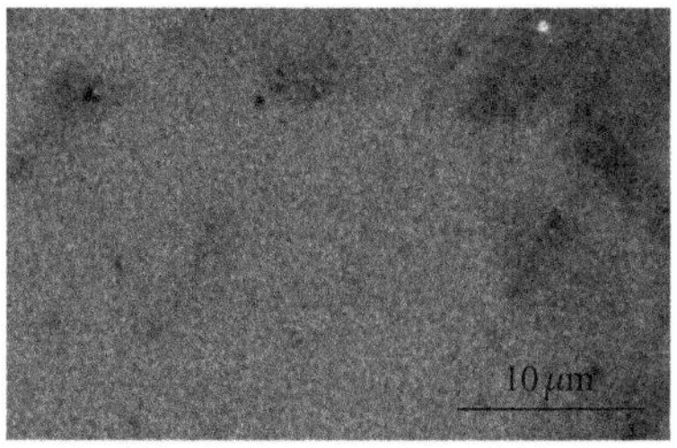

(a) Anodized pure Al without ECAP

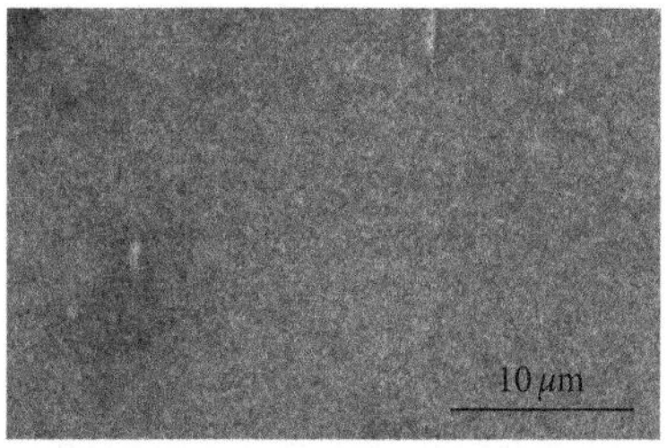

(b) Anodized pure Al with ECAP

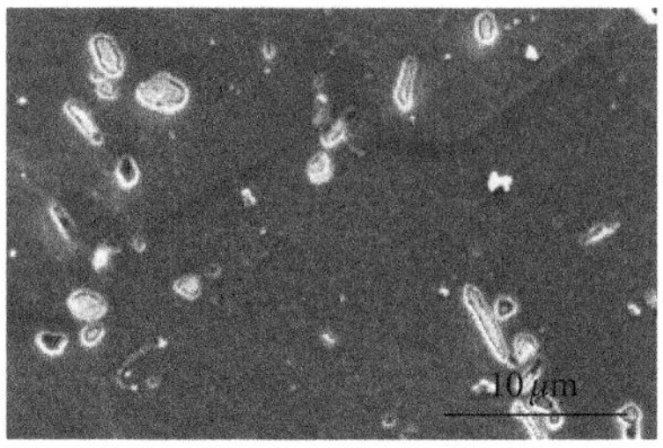

(c) Anodized Al (AA1100) without ECAP

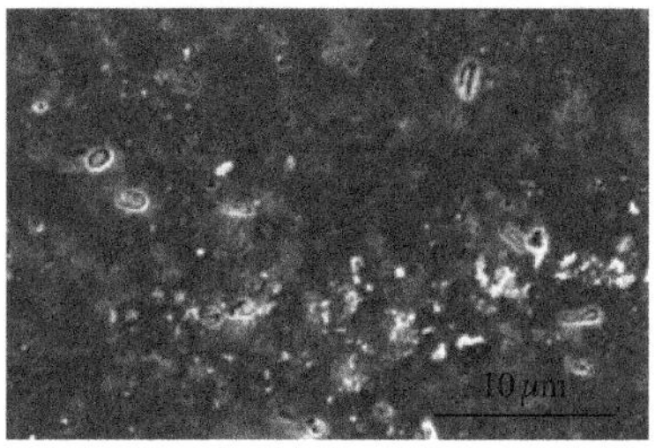

(d) Anodized Al (AA1100) with ECAP

Figure 9: Effect of ECAP on the morphology of pure Al and Al (AA1100) anodized at $400\,A\cdot m^{-2}$.

It is concluded that the improvement of pitting corrosion resistance of pure Al with ECAP is caused by the increased oxidation rate of Al since pure Al contains few impurity precipitates of Fe-Al intermetallic compounds.

Acknowledgments

This study was supported by the Grant-in-Aid for Scientific Research on Innovation Areas (Research in a Proposed Research Area no. 23102505) of the Ministry of Education, Culture, Sports, Science and Technology of Japan in 2011.

References

[1] H. H. Uhlig, *Corrosion and Corrosion Control*, Sangyo Tosho, Tokyo, Japan, 1968.

[2] G. Ito, *Corrosion Science and Engineering*, Corona, Tokyo, Japan, 1973.

[3] Z. Szklarska-Smialowska, "Pitting corrosion of aluminum," *Corrosion Science*, vol. 41, no. 9, pp. 1743–1767, 1999.

[4] L. Fedrizzi, F. Deflorian, S. Rossi, and P. L. Bonora, "Study of aluminium filiform corrosion by using electrochemical techniques," *Materials Science Forum*, vol. 289–292, no. 1, pp. 485–498, 1998.

[5] S. Y. Yu, W. E. O'Grady, D. E. Ramaker, and P. M. Natishan, "Chloride ingress into aluminum prior to pitting corrosion. An investigation by XANES and XPS," *Journal of the Electrochemical Society*, vol. 147, no. 8, pp. 2952–2958, 2000.

[6] M. Shao, Y. Fu, R. Hu, and C. Lin, "A study on pitting corrosion of aluminum alloy 2024—T3 by scanning microreference electrode technique," *Materials Science and Engineering A*, vol. 344, no. 1-2, pp. 323–327, 2003.

[7] G. Bereket and A. Yurt, "The inhibition effect of amino acids and hydroxy carboxylic acids on pitting corrosion of aluminum alloy 7075," *Corrosion Science*, vol. 43, no. 6, pp. 1179–1195, 2001.

[8] R. Z. Valiev, N. A. Krasilnikov, and N. K. Tsenev, "Plastic deformation of alloys with submicron-grained structure," *Materials Science and Engineering A*, vol. 137, no. C, pp. 35–40, 1991.

[9] R. Z. Valiev, R. K. Islamgaliev, and I. V. Alexandrov, "Bulk nanostructured materials from severe plastic deformation," *Progress in Materials Science*, vol. 45, no. 2, pp. 103–189, 2000.

[10] J. J. Park and D. H. Shin, in *Ultrafine Grained Materials II*, The Minerals, Metals & Materials Society, Warrendale, Pa, USA, 2008.

[11] Z. Horita, M. Furukawa, M. Nemoto, and T. G. Langdon, "Development of fine grained structures using severe plastic deformation," *Materials Science and Technology*, vol. 16, no. 11-12, pp. 1239–1245, 2000.

[12] J. Horita, "Production of ultrafine-grianed structures using equal-channel angular pressing," *Journal of the Japan Welding Society*, vol. 74, p. 88, 2005.

[13] Z. Horita, "Nano—structural control using severe plastic deformation," *Materia Japan*, vol. 41, p. 422, 2002.

[14] W. R. Osório, C. M. Freire, and A. Garcia, "The role of macrostructural morphology and grain size on the corrosion resistance of Zn and Al castings," *Materials Science and Engineering A*, vol. 402, no. 1-2, pp. 22–32, 2005.

[15] K. D. Ralston, D. Fabijanic, and N. Birbilis, "Effect of grain size on corrosion of high purity aluminium," *Electrochimica Acta*, vol. 56, no. 4, pp. 1729–1736, 2011.

[16] W. R. Osório, J. E. Spinelli, I. L. Ferreira, and A. Garcia, "The roles of macrosegregation and of dendritic array spacings on the electrochemical behavior of an Al-4.5 wt.% Cu alloy," *Electrochimica Acta*, vol. 52, no. 9, pp. 3265–3273, 2007.

[17] D. Song, A. Ma, J. Jiang, P. Lin, D. Yang, and J. Fan, "Corrosion behavior of equal-channel-angular-pressed pure magnesium in NaCl aqueous solution," *Corrosion Science*, vol. 52, p. 481, 2010.

[18] N. Birbilis, K. D. Ralston, S. Virtanen, H. L. Fraser, and C. H. J. Davies, "Grain character influences on corrosion of ECAPed pure magnesium," *Corrosion Engineering Science and Technology*, vol. 45, no. 3, pp. 224–230, 2010.

[19] W. R. Osório, J. E. Spinelli, C. R. M. Afonso, L. C. Peixoto, and A. Garcia, "Microstructure, corrosion behaviour and microhardness of a directionally solidified Sn-Cu solder alloy," *Electrochimica Acta*, vol. 56, no. 24, pp. 8891–8899, 2011.

[20] I. J. Son, H. Nakano, S. Oue, S. Kobayashi, H. Fukushima, and Z. Horita, "Pitting corrosion resistance of ultrafine-grained aluminum processed by severe plastic deformation," *Materials Transactions*, vol. 47, no. 4, pp. 1163–1169, 2006.

[21] Z. Horita, T. Fujinami, M. Nemoto, and T. G. Langdon, "Equal-channel angular pressing of commercial aluminum alloys: grain refinement, thermal stability and tensile properties," *Metallurgical and Materials Transactions A*, vol. 31, no. 3, pp. 691–701, 2000.

[22] N. Baba, *Denkaiho Niyoru Sankahimaku*, Maki Shoten, Tokyo, Japan, 1996.

[23] H. Takahashi, K. Kasahara, K. Fujiwara, and M. Seo, "The cathodic polarization of aluminum covered with anodic oxide films in a neutral borate solution-I. The mechanism of rectification," *Corrosion Science*, vol. 36, no. 4, pp. 677–683, 1994.

[24] B. Zaid, D. Saidi, A. Benzaid, and S. Hadji, "Effects of pH and chloride concentration on pitting corrosion of AA6061 aluminum alloy," *Corrosion Science*, vol. 50, no. 7, pp. 1841–1847, 2008.

[25] W. R. Osório, L. C. Peixoto, M. V. Canté, and A. Garcia, "Electrochemical corrosion characterization of Al-Ni alloys in a dilute sodium chloride solution," *Electrochimica Acta*, vol. 55, no. 13, pp. 4078–4085, 2010.

[26] N. Birbilis and R. G. Buchheit, "Electrochemical characteristics of intermetallic phases in aluminum alloys : an experimental survey and discussion," *Journal of the Electrochemical Society*, vol. 152, no. 4, pp. B140–B151, 2005.

[27] W. R. Osório, L. C. Peixoto, P. R. Goulart, and A. Garcia, "Electrochemical corrosion parameters of as-cast Al-Fe alloys in a NaCl solution," *Corrosion Science*, vol. 52, no. 9, pp. 2979–2993, 2010.

[28] R. Ambat, A. J. Davenport, G. M. Scamans, and A. Afseth, "Effect of iron-containing intermetallic particles on the corrosion behaviour of aluminium," *Corrosion Science*, vol. 48, no. 11, pp. 3455–3471, 2006.

19

Corrosion of Modified Concrete with Sugar Cane Bagasse Ash

R. E. Núñez-Jaquez,[1] J. E. Buelna-Rodríguez,[1] C. P. Barrios-Durstewitz,[1] C. Gaona-Tiburcio,[2] and F. Almeraya-Calderón[2,3]

[1] *Facultad de Ingeniería Los Mochis, Universidad Autónoma de Sinaloa, Fuente de Poseidón y Prol. Ángel Flores S/N, C.P. 87223 Los Mochis, SIN, Mexico*
[2] *Centro de Investigación e Innovación en Ingeniería Aeronáutica CIIIA, FIME-Universidad Autónoma de Nuevo León, Avenida Universidad S/N, Ciudad Universitaria, C.P. 87223 San Nicolás de los Garza, NL, Mexico*
[3] *Grupo Corrosión, Centro de Investigación en Materiales Avanzados, Miguel de Cervantes No. 120, C.P. 31109 Chihuahua, CHIH, Mexico*

Correspondence should be addressed to C. P. Barrios-Durstewitz, durstewitz.uas@gmail.com

Academic Editor: Andres A. Torres Acosta

Concrete is a porous material and the ingress of water, oxygen, and aggressive ions, such as chlorides, can cause the passive layer on reinforced steel to break down. Additives, such as fly ash, microsilica, rice husk ash, and cane sugar bagasse ash, have a size breakdown that allows the reduction of concrete pore size and, consequently, may reduce the corrosion process. The objective of this work is to determine the corrosion rate of steel in reinforced concrete by the addition of 20% sugar cane bagasse ash by weight of cement. Six prismatic specimens ($7 \times 7 \times 10$ cm) with an embedded steel rod were prepared. Three contained 20% sugar cane bagasse ash by weight of cement and the other three did not. All specimens were placed in a 3.5% NaCl solution and the corrosion rate was determined using polarization resistance. The results showed that reinforced concrete containing sugar cane bagasse ash has the lowest corrosion rates in comparison to reinforced concrete without the additive.

1. Introduction

Corrosion is the loss of material and occurs when metals are exposed to the environment. This is the result of the interaction between chemical and biological agents [1]. Corrosion of reinforced steel has two principal problems that affect the service life of concrete reinforced structures [2].

(a) Metal lost in the reinforced steel rods reduces the mechanical properties of the structure.

(b) Corrosion products occupy a higher volume than the base metal that produces internal stresses, which may cause the concrete to crack.

Degradation of concrete structures by corrosion processes is a serious problem and has major economic implications [3].

In service conditions, the concrete covering the steel bars gives physical and chemical protection to the reinforcement. The concrete cover provides an alkaline environment surrounding the steel, resulting in the formation of an oxide layer, called a passive film, which protects the steel from corrosion. However, the passive layer does not provide a perfect and permanent barrier in aggressive environments containing initiators of corrosion (primarily aggressive ions like sulphates and chlorides, humidity, carbon dioxide, and oxygen) [4]. These factors, together with inadequate construction practices in difficult environmental conditions, lead to corrosion of the reinforcing steel. Carbon dioxide can cause corrosion primarily by reducing the pH of the concrete. Once the alkaline environment is destroyed, the protective layer of oxide on the steel surface is destroyed and corrosion may start. Reinforced concrete structures are exposed to severe environments but often are expected to continue with little or no repair or maintenance for long periods of time. Therefore, reinforced concrete structures need to be durable. One of the main forms of attack from the environment is the ingress of chloride ions, which leads to corrosion of steel

Figure 1: Experimental setup.

reinforcement and subsequent reduction in the resistance, utility, and esthetics of the structure [5]. Chlorides can come from several sources. They can be introduced into concrete during batching or can be introduced from the environment during service. Chlorides can be introduced by [6]

(a) the deliberate addition of chlorides in additives and use of seawater or contaminated aggregates in the mixture,

(b) saline mist and seawater, which moistens directly,

(c) deicing salts.

Mineral additives are commonly used in many applications ranging from high performance concrete bridges, buildings and on-shore and off-shore structures. Well-known mineral additives are silica fume, rice husk ash, and fly ash. The addition of these products in the production of concrete has positive environmental effects, while minimizing problems associated with its disposal [7].

Sugar cane bagasse is the by-product of sugar production. Fresh bagasse contains 40% water and often is used as fuel in the sugar factories. Sugar cane fiber has a particular significance because it is a combustible material for the sugar industry and is a valuable raw material for paper, artificial wood, and others [8]. When this waste is burned in controlled conditions, the generated ash contains amorphous silica, which has pozzolanic properties [9]. Ash from bagasse sugar cane is one of the main waste products generated worldwide and can be used as a mineral additive, mainly due to its high content of silica (SiO_2). Adding sugar cane bagasse ash (SCBA) as a replacement for cement may provide additional enhancements in resistance to chloride ion penetration and waterproofing properties. Thus, tests were conducted to determine whether SCBA can be used as an effective mineral additive for concrete reinforced structures [10].

2. Experimental Procedure

2.1. Materials. The materials used were portland cement (CPC-30R), river sand with a fineness modulus of 2.59, triturate river coarse aggregate with a maximum size of 3/4″, and water. The sugar cane bagasse ash (SCBA) used in the present study was obtained from a local sugar factory in Los Mochis, Sinaloa, México. The ash was obtained from burning the bagasse at 650°C for one hour.

2.2. Thermogravimetric Analysis (TGA). The burn temperature of cane sugar bagasse required to obtain ash with amorphous SiO_2 was determined using TGA. The equipment employed was TA Instruments SIMULTANEOUS DTA-TGA STD 2.960 with a temperature scan rate of 10°C/min in a static atmosphere.

2.3. X-Ray Diffraction. After burning the bagasse, the ash was powdered and examined in an X-ray PANalytical Mod. X'Pert Pro.

2.4. Concrete Specimens. Four prismatic concrete specimens with dimensions of 100 × 70 × 70 mm were constructed with a single 12 cm long 1018 steel rebar (0.9525 mm diameter) embedded in the concrete. In order to confine the exposed area of the steel, the rebar was partially coated with a noncorrosive paint, leaving an exposed length of 5 cm. The concrete was made with water to cement ratio of 0.60. Two specimens were made by replacing 20% by weight of the cement with sugar cane bagasse ash, and the remaining did not contain sugar cane bagasse ash. The specimens were cured by immersing in water for 28 days. After the curing period, the specimens were exposed to a solution of 3.5% by weight of sodium chloride.

Three 100 mm diameter by 200 mm long concrete cylinders with and without the addition of 20% bagasse cane ash were cast for the compressive strength tests.

In order to accelerate chloride ingress, the specimens were cyclically exposed to the sodium chloride solution by a 3-day partial immersion in a 3.5% weight NaCl solution followed by 4 days of drying.

2.5. Linear Polarization Resistance. The linear polarization resistance was determined with an ACM Instruments one-channel galvanostat/potentiostat GILL A.C. A standard calomel electrode (SCE) was used as the reference electrode, the rebar was the working electrode, and a stainless steel plate placed around the specimen was the counter electrode (Figure 1).

Table 1: Corrosion probability.

E_{corr} versus SCE (mV)	Corrosion probability
>−125	10% probability
−125 to −275	Uncertain zone
<−275	90% probability

Table 2: I_{corr} in terms of corrosion level.

I_{corr} (μA/cm^2)	Corrosion level
<0.1	Negligible
0.1–0.5	Low
0.5–1	Moderate
>1	High

The sweep potential was ±20 mV with respect to the corrosion potential and the sweep rate was 10 mV/minute. The IR drop potential was considered. This monitoring was conducted weekly with specimens immersed in the sodium chloride solution. Figure 2 shows the experimental setup. The results were analyzed using "Analysis" of ACM version 4 [11].

The corrosion current density (I_{corr}) and corrosion rate (CR) were estimated from resistance to charge transference (R_{ct}) using:

$$\text{Corrosion Current Density } (I_{corr}) = \frac{B}{R_{ct}} \ (\mu\text{A/cm}^2), \quad (1)$$

$$\text{Corrosion rate (CR)} = I_{corr}\ (\mu\text{A/cm}^2) * 0.011\ [=]\ \text{mmpy}, \quad (2)$$

where B is Stern-Geary constant (B = 26 mV for uniform corrosion) [12].

There are several techniques to determine the degree of deterioration of concrete structures. One is to determine the corrosion potential (E_{corr}) in accordance with ASTM-C-876, which establishes criteria that relate the potential with corrosion probability, as shown in Table 1 [13].

From the corrosion current density values (I_{corr}), obtained using (1), it is possible to determine the corrosion severity using the values shown in Table 2 [14].

3. Results and Discussion

Figure 2 shows the results of the thermogravimetric analysis (TGA). The thermograph shows that around 550°C there is no further decrease in weight. At this temperature, all of the sugar cane bagasse completely burned, but this is with a small sample of bagasse cane. With a greater amount of bagasse ash, it was necessary to burn at a higher temperature. For this reason, the sugar cane bagasse was burned at 650°C for one hour to obtain the ash for this study.

Figure 3 shows the X-ray powder diffraction pattern of SCBA burned at 650°C. The characteristic wide hump (between 2θ = 20 and 2θ = 35) and the peak of quartz show the presence of an amorphous phase, which is important to confirm because amorphous SiO_2 contributes to the pozzolanic properties. The ash pozzolanic reaction

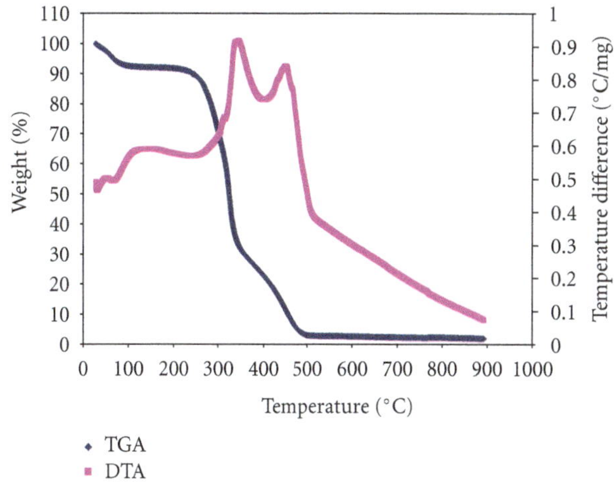

Figure 2: Thermograph of sugar cane bagasse.

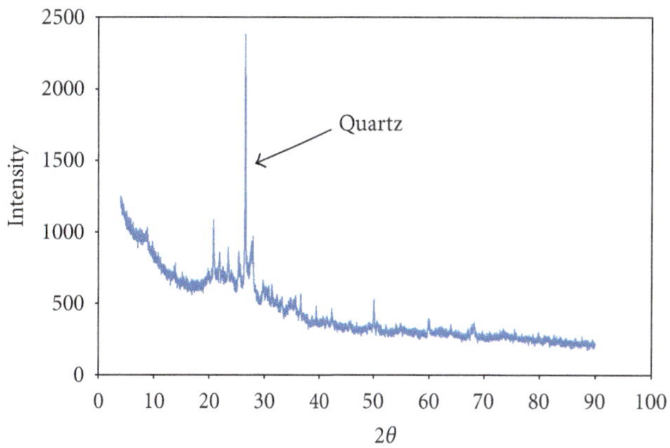

Figure 3: X-ray diffractogram of SCBA burned at 650°C.

decreases the pore size in the concrete, which restricts entry of aggressive ions.

Figure 4 shows the corrosion potential (E_{corr}) results obtained from linear polarization resistance scans of the six specimens. The corrosion potentials ranged between −100 and −550 mV. The specimens with SCBA (Bagasse 1 and 2) present smaller probability of corrosion compared with specimens without the additive (Normal 1 and 2). The horizontal dotted line represents the limit between regions of probability of corrosion as given in Table 1. During the first 15 days of exposure, the values are in the uncertain zone. Gradually, the normal concretes (without SCBA additive) became more negative to values in the 90% of corrosion probability. However, the concretes with SCBA continued in the uncertain zone, in the uncertain zone, after 63 days of cyclical exposure a great difference in the corrosion potential is observed between the specimens with and without SCBA. The normal concrete specimens have started to corrode.0020The specimens with SCBA show the smaller corrosion potential, which is probably due to SCBA reacting with calcium hydroxide (produced during the hydration of portland cement), and thenhydrated compounds formed

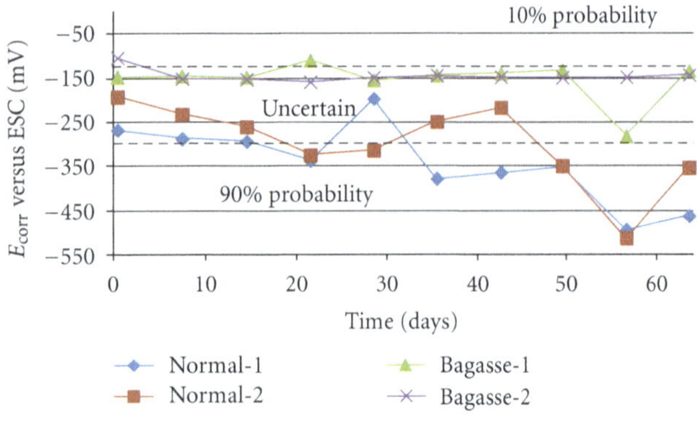

FIGURE 4: Corrosion potential of specimens.

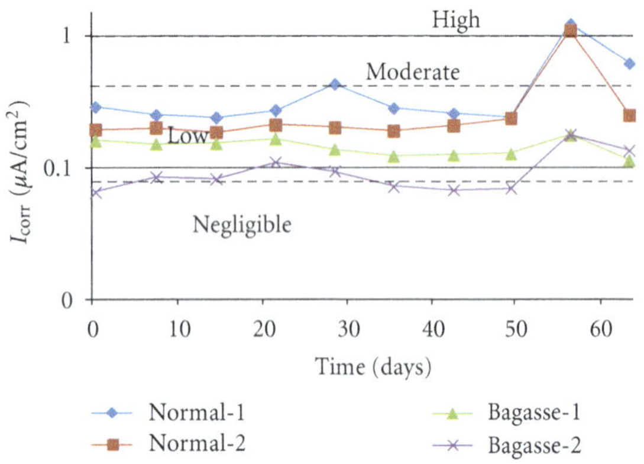

FIGURE 5: Corrosion current density of specimens.

during the pozzolanic reactions fills the pores of the concrete [15], restricting aggressive ions from reaching the steel rebar.

Figure 5 shows the corrosion current density (I_{corr}) results. I_{corr} ranged between 0.1 and 1.0 μA/cm^2. During the first 50 days, I_{corr} were from less than 0.1 μA/cm^2 for the concrete with SCBA. According to Table 2, the corrosion level is negligible. For normal concretes (without SCBA), I_{corr} values ranged between 0.1 μA/cm^2 and 0.5 μA/cm^2 for the first 50 days indicating that corrosion was low. However, after 50 days, I_{corr} gradually increased for the specimens without SCBA and is in the high corrosion level but the concrete with SCBA is still in the low corrosion level.

Figure 6 shows the corrosion rate results. At the beginning of the exposure, both types of specimens had similar results, but after day 50, the corrosion rate of the normal concrete without SCBA gradually increased. The specimens with SCBA showed lower corrosion rates. This behavior was due to the pozzolanic properties of SCBA that decreased the size of the concrete pores, which restricts the ingress of chloride ions.

4. Findings and Conclusions

(1) A temperature of 650°C was sufficient to obtain sugar cane bagasse ash with amorphous SiO_2.

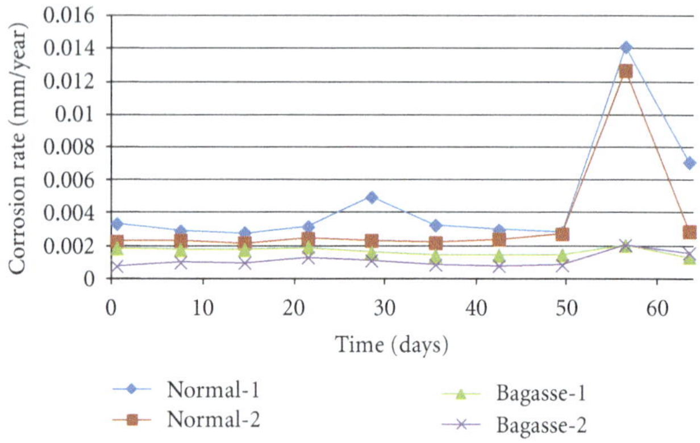

FIGURE 6: Corrosion rate of specimens.

(2) The specimens with SCBA indicated lower corrosion potential values, which corresponded to a low corrosion probability.

(3) The specimens with negligible to moderate corrosion level were the specimens with sugar cane bagasse ash added to the concrete mix.

(4) The use of sugar cane bagasse ash as a partial replacement of cement has a beneficial effect to protect the steel rebar from corrosion because it reduced the pore size in the cement paste, which minimized the ingress of aggressive ions into concrete. Concrete containing SCBA had a lower corrosion rate when compared to concrete without the addition of SCBA.

Acknowledgments

The authors express their gratitude to Ingenio Azucarero de Los Mochis for providing the sugar cane bagasse used in this investigation and also to Universidad Autónoma de Sinaloa, through Programa de Mejoramiento al Profesorado for the financial support of project PROMEP/103.5/08/3289, Folio UASPTC-022.

References

[1] A. Zaki, "Concrete corrosion: introduction to corrosion," in *Principles of Corrosion Engineering and Corrosion Control*, pp. 616–617, Butterworth-Heinemann, 2006.

[2] A. Del Valle, T. Pérez, and M. Martínez, "El Fenómeno de la Corrosión en Estructuras de Concreto Reforzado," Publicación técnica, Secretaría de Comunicaciones. y Transportes Instituto Mexicano Del Transporte, 2001.

[3] O. Covarrubias and P. Rodríguez, "Propiedades de Resistencia a la Corrosión de Aceros Aleados al Si Embebidos en Mortero," Ciencia UANL, 2001.

[4] R. Solís, E. Moreno, and P. Castro, "Durabilidad en la Estructura de Concreto de Vivienda de Zona Costera," Ingeniería de Revista Académica, 2005.

[5] K. Chandramouli, P. Srinivasa Rao, N. Pannirselvam, T. S. Seshadri Sekhar, and P. Sravana, "Chloride penetration resistance studies on concretes modified with alkali resistant glass fibers," *American Journal of Applied Sciences*, vol. 7, no. 3, pp. 371–375, 2010.

[6] P. Broomfield, “Causes and mechanisms of corrosion in concrete,” in *Corrosion of Steel in Concrete*, Taylor and Francis, London, UK, 2007.
[7] G. C. Cordeiro, R. D. Toledo Filho, L. M. Tavares, and E. M. R. Fairbairn, “Pozzolanic activity and filler effect of sugar cane bagasse ash in Portland cement and lime mortars,” *Cement and Concrete Composites*, vol. 30, no. 5, pp. 410–418, 2008.
[8] J. Osorio, F. Varon, and J. Herrera, “Comportamiento Mecánico del Concreto Reforzado con Fibras de Bagazo de Caña de Azúcar,” *Dyna*, no. 153, 2007.
[9] N. B. Singh, V. D. Singh, and S. Rai, “Hydration of bagasse ash-blended portland cement,” *Cement and Concrete Research*, vol. 30, no. 9, pp. 1485–1488, 2000.
[10] K. Ganesan, K. Rajagopal, and K. Thangavel, “Evaluation of bagasse ash as supplementary cementitious material,” *Cement and Concrete Composites*, vol. 29, no. 6, pp. 515–524, 2007.
[11] ASTM G 59–97, “Standard Test Method for Conducting Potentiodynamic Polarization Resistance Measurements,” American Society for Testing and Materials, 1997.
[12] S. Feliú, J. A. González, and C. Andrade, “*Electrochemical Methods for On site Determinations of Corrosion Rates Rebars*,” Techniques to Assess the Corrosion Activity of Steel Reinforced Concrete Structures, ASTM STP 1276. ASTM, 1996.
[13] ASTM-C-876-91, “Standard Test Method for Half Cell Potentials of Reinforcing Steel in Concrete,” ASTM , USA, 1992.
[14] O. Trocónis de Rincón, A. Romero de Carruyo, C. Andrade, P. Helene, and I. Díaz, “Descripción de métodos de ensayos,” in *Manual de Inspección, Evaluación y Diagnostico de Corrosión en Estructuras de Hormigón Armado*, 1985.
[15] M. Frías, E. Villar-Cociña, and E. Valencia-Morales, “Characterisation of sugar cane straw waste as pozzolanic material for construction: calcining temperature and kinetic parameters,” *Waste Management*, vol. 27, no. 4, pp. 533–538, 2007.

20

Corrosion Behaviour of AZ and ZA Magnesium Alloys in Alkaline Chloride Media

Mustafa Ö. Öteyaka,[1] Edward Ghali,[2] and Réal Tremblay[2]

[1] *Department of Mechanical Engineering, Dumlupinar University, 43100 Kütahya, Turkey*
[2] *Department of Mining, Metallurgical and Materials Engineering, Laval University, Quebec City, QC, Canada G1V 0A6*

Correspondence should be addressed to Edward Ghali, edward.ghali@gmn.ulaval.ca

Academic Editor: Chang-Jian Lin

Corrosion behaviour of AZ and ZA magnesium alloys in untreated, aerated, and deaerated chloride media was investigated. The influence of impurity, alloying element, and oxygen on corrosion potential, rate, and pitting was examined in alkaline media saturated with $Mg(OH)_2$ at pH 9 and 25°C. After 8 h immersion, slightly less active corrosion potentials were generally observed in deaerated than oxygen containing solutions. AZ91D and ZA104 alloys recorded much lower corrosion rates in deaerated medium as compared to AZ91E, ZAC 10403 (0.3Ca), and ZACS1040305 (0.3Ca+ 0.5Sr). Generally, ZAC alloy showed the highest corrosion rate, followed by ZACS alloy, and both showed a constant value in the three media with or without oxygen. The effect of oxygen is then dependent on the alloy properties and is not a simple acceleration of the cathodic reaction but could cause also inhibition and/or passivation. Pit depth and distribution were influenced by alloy composition. The pits were found deeper and more localized on AZ and more uniformly dispersed on ZA alloy surfaces.

1. Introduction

Magnesium, as a structural material, has numerous advantages with regard to aluminum (lighter by almost 36%) and plastics (almost the same specific density) [1, 2]. Magnesium alloys, in particular those with aluminum (Al) and Zinc (Zn) as alloying elements, are frequently exposed to degradation caused by corrosion.

The poor corrosion resistance of many magnesium alloys may be the result of internal galvanic corrosion caused by second phases or impurities. It is well documented that the presence of noble metal impurities (Fe, Ni, and Cu) represents the most detrimental factor influencing the corrosion properties of commercial magnesium alloys [3]. Magnesium suffers pitting when exposed to chloride solutions in nonoxidizing media at open circuit potential. In solutions containing 10 and 1000 ppm of NaCl, the Mg corrosion potential is −1060 and −1310 mV versus SHE, respectively [4]. Wilde and Williams [5] measured the corrosion potential changes with time of some corrosion resistant alloys during pitting and found that the corrosion potential decreased in all the alloys when the corrosion potential exceeded the pitting potential. Studies on pitting phenomena on Mg and Mg alloys have compared the pitting behaviour of die-cast alloys and rapidly solidified (thixomolded) Mg alloys [6–9]. For example, the good pitting resistance of die-cast AE alloys was attributed to the presence of Al-rich zones which appeared to act as barrier against pit propagation.

Makar and Kruger [10] showed that rapidly solidified AZ61 alloy exhibited a breakdown potential 200 mV higher than that of cast AZ61 alloy in buffered carbonate solution (pH 10) containing various levels of Cl^-. The surface layer was also composed primarily of ($MgO \cdot H_2O$). The morphology and structure of oxide films on magnesium, formed in air and water, were also shown to be formed by a three-layer system consisting of an inner cellular structure (0.4–0.6 mm), a dense intermediate regime (20–40 nm), and an outer layer with a platelet-like morphology (approximately 2 mm) [11]. Baril and Pébère [12] studied the corrosion behaviour of pure magnesium in aerated and deaerated sodium sulphate solutions (0.01 and 0.1 M) by steady-state current-voltage and electrochemical impedance measurements. It was shown that the anodic current densities were lower and the corrosion resistance values were higher in deaerated media. They concluded that the presence of oxygen did not influence the corrosion of pure magnesium and showed that the corrosion rate was dependent on HCO_3^- concentration in the solution.

AZ91 (Mg-9Al-1Zn) alloy is by far the most commonly used magnesium cast material. The popularity of this particular alloy is a result of its good corrosion resistance as well as favourable castability and mechanical properties [13–15]. AZ91 alloy has a two-phase microstructure typically consisting of a matrix of α-Mg grains with an intermetallic β-phase ($Mg_{17}Al_{12}$) along the grain boundaries [16]. The physical significance of aluminum as the alloying element is largely related to an increased passivity of the oxide film formed on the surface [17]. However, it seems that a threshold of 4% Al in the alloy is required. Precipitation of the more corrosion-resistant $Mg_{17}Al_{12}$ (β-phase), which acts as a corrosion barrier, has also been claimed to improve corrosion [13]. Furthermore, it was reported that the corrosion potential of β-phase in 5% NaCl saturated solution with $Mg(OH)_2$ was approximately 490 and 420 mV more cathodic to pure Mg and AZ91 alloy, respectively [13]. The solubility of rare earths (RE) in the magnesium matrix is very limited in the presence of aluminum, but the intermetallic Al-RE formed is electrochemically passive and does not significantly affect the corrosion rate [18].

Alloying elements such as strontium (Sr) and calcium (Ca) can be chosen to improve magnesium alloy performance in order to be used for structural applications in aggressive environments. Considering the electrochemical potential values, the addition of Ca should result in low microgalvanic effects. Reduction of the anodically active area upon precipitation was reported for Mg-Ca-Cu splats where precipitation of Mg_2Ca neutralises the galvanic effect produced by the Mg_2Cu phase. The addition of 2% Ca decreased the corrosion rate of rapid solidified AZ91 alloy from 0.8 to 0.2 mm/year [19]. Furthermore, AZ91 castings with 0.3% Sr alloy displayed considerably reduced microporosity with no significant loss in strength and ductility at room temperature [20].

AZ alloys are used extensively in many commercial applications and ZA alloys are less expensive and still emerging. Previous studies have examined the corrosion behaviour of these two different categories of magnesium alloys (AZ and ZA) in various solutions [21–24]. It was concluded that pitting corrosion was an important and prevalent form of degradation of these alloys. It was also demonstrated that the pits were more numerous on the surface of ZACS1040305 (Mg-10Zn-4Al-0.3Ca-0.5Sr) alloy than on that of AZ91E alloy. Moreover, uniform corrosion was observed on ZAC10403 (Mg-10Zn-4Al-0.3Ca) and ZACS1040305 alloys, while pitting corrosion was present on AZ91E alloy in 5% NaCl solution saturated with atmospheric oxygen at 25°C, with $Mg(OH)_2$ at pH 9.

As the primary alloying element in AZ magnesium alloys is aluminum, while that in ZA alloys is zinc, this major difference could affect their corrosion behaviour in untreated, aerated, and deaerated chloride medium. To our knowledge the corrosion behaviour of AZ and ZA alloys in such solutions containing different contents in oxygen has never been investigated. The objectives of the present study were to examine and compare, using conventional techniques, the corrosion characteristics and particularly the pitting behaviour of these two groups of magnesium alloys (AZ and ZA) in untreated, aerated, and deaerated alkaline 5% NaCl solution.

2. Experimental Procedure

2.1. Material and Sample Preparation. In accordance with ASTM standard B93/B93M-09 [25], permanent mold castings were produced from five different alloys: AZ91D, AZ91E, ZA104, ZAC10403, and ZACS1040305. Table 1 presents the chemical composition of these alloys. Corrosion test samples of $1 \times 1 \times 2$ cm were taken from the castings and encased in epoxy resin. Each sample was then polished with 600 and 1200 grit SiC papers and 3 micron diamond paste, cleaned with alcohol, dried, and subsequently exposed to the test solution.

2.2. Electrochemical Technique. The free corrosion potential (E_{corr}) and potentiodynamic curves of magnesium alloys AZ91D, AZ91E, ZA104, ZAC10403, and ZACS1040305 were determined in the following aqueous media: untreated, aerated, and deaerated 5% NaCl solutions saturated with $Mg(OH)_2$ at pH 9 and 25°C, using a Model 273 potentiostat/galvanostat (EG&G). ASTM standard G5-94 [26] was used to perform the potentiodynamic tests. Following immersion in the solution, the test began within one minute. The open circuit potential (OCP or E_{corr}) measurement was performed for a period of 8 h. Potentiodynamic curves were traced at a scanning rate of $0.1\,mV{\cdot}s^{-1}$ on the polished surfaces over a potential range of 1000 mV (cathodic polarisation −10 mV versus corrosion potential). All of the potentials are given with respect to a saturated calomel reference electrode $Hg/Hg_2Cl_{2\,(solid)}/KCl_{sat.}$ (0.242 V versus SHE). Experimental corrosive media were obtained as follows: the untreated 5% NaCl solution corresponded to air saturation with oxygen ($\approx 5.8\,mL{\cdot}L^{-1}$ of dissolved oxygen), the aerated 5% NaCl solution was completely saturated with air bubbling ($\approx 12.5\,mL{\cdot}L^{-1}$ of dissolved oxygen [27]), and the deaerated 5% NaCl solution was obtained by nitrogen bubbling to eliminate dissolved oxygen. At least three tests were performed under each condition. The reproducibility of corrosion rate was of the order of ±5–10%.

2.3. Surface Observations. Following 8 h of immersion, the specimens were examined under an optical microscope (OM) equipped with an image analyser to characterise the corroded surface. Surface topography was investigated using an atomic force microscope (AFM) in the tapping mode. Images were taken with 5000×5000 data acquisition at a scan rate of 1 Hz in air at room temperature. Tridimensional surface topography was mapped for an area of $50 \times 50\,\mu m$ using the integrated software program.

3. Results and Discussion

3.1. Corrosion Potential Analysis. Figure 1 shows the time variations of E_{corr} for AZ91D and AZ91E alloys during 8 h of immersion in the untreated, aerated, and deaerated

Table 1: Chemical composition (wt%)* of the magnesium alloys used.

Alloy	Al	Zn	Ca	Sr	Mn	Si	Fe	Cu	Ni	Mg
AZ91D	9.00	0.67	—	—	0.33	0.010	0.005	0.030	0.002	Bal.
AZ91E	8.70	0.70	—	—	0.26	0.200	0.005	0.015	0.001	Bal.
ZA104	3.69	9.73	—	—	0.25	0.015	0.030	<0.001	0.001	Bal.
ZAC10403	3.78	10.5	0.31	—	0.03	0.001	0.003	<0.001	0.002	Bal.
ZACS1040305	4.12	11.7	0.31	0.54	0.35	18 ppm	189 ppm	2 ppm	2 ppm	Bal.

*Throughout the text, the percentage of elements and solution concentration is expressed in weight percent.

5% NaCl solutions saturated with $Mg(OH)_2$. For AZ91D, the E_{corr} value increases somewhat during the first 30 min and remained stable thereafter, showing a slight rise in the untreated and deaerated solutions (Figure 1(a)). This behaviour may reflect several phenomena called constant process of breakdown and repair of passivation: film initiation, attack by the Cl^- ion, and possibly formation of protective corrosion products on the surface. Moreover, the average corrosion potential values in the aerated solution were generally similar to those obtained in the untreated solution. The E_{corr} values of AZ91D in the aerated solution increased suddenly after 3 h immersion, indicating a possible slowing down of the corrosion reaction. A possible explanation for this behaviour may be the presence of oxygen which may have accelerated the production of corrosion products on the surface of AZ91D specimen as compared to the untreated solution.

Figure 1(b) presents the variation of E_{corr} over time for the AZ91E alloy in the untreated, aerated, and deaerated 5% NaCl solutions saturated with $Mg(OH)_2$. In the first 30 min of immersion, a rapid drop of potential was observed toward more negative values in all three corrosive media. During the rest of the immersion period, the E_{corr} of AZ91E in the untreated and aerated solutions remained relatively stable, with a significant difference of ≈20 mV observed in the deaerated solution, which can suggest an important corrosion rate.

Figure 2 illustrates the time variation of E_{corr} for the ZA alloys in the untreated, aerated, and deaerated 5% NaCl solutions saturated with $Mg(OH)_2$. For alloy ZA104 (Figure 2(a)), the E_{corr} gradually increased during the immersion period in all three corrosive media. However, in the deaerated medium, the corrosion potential remained less negative than that of the other two media throughout the entire testing process. This behaviour may be attributed to a stable growth of the protective surface film. Figure 2(b) shows the effect of adding 0.3% Ca to ZA104 alloy. The E_{corr} value for alloy ZAC10403 increased slightly up to 6 h of immersion and was higher in the aerated and deaerated solutions. For alloy ZACS1040305, containing an additional 0.5% Sr, the free corrosion potential (E_{corr}) remained relatively constant or stationary in the three corrosive media after the first 30 min of immersion (Figure 2(c)).

Table 2 summarises the average E_{corr} values obtained for the AZ and ZA magnesium alloys after 8 h of immersion in the untreated, aerated, and deaerated 5% NaCl solutions saturated with $Mg(OH)_2$. As determined from the free corrosion measurements, the corrosion potential values were shown to be relatively close to one another or in the same zone of potential values independent of oxygen content in the solutions. Baril and Pébère [12] have already stated that oxygen did not play a major role in the corrosion potential of pure magnesium. The present results support the same statement for the AZ and ZA groups of alloys. However, in comparing the corrosion potential of AZ91E alloy with that of the other alloys, and based on several measurements, it can be statistically confirmed that the purity of the alloy showed a trend toward more active potentials (Table 2).

More specifically, it can be stated that the active E_{corr} of AZ91E was ≈10–30 mV more negative than that of AZ91D in the examined three corrosive media. This may be explained by the low impurity content of the AZ91E alloy. The corrosion potential observed for ZA104 was more positive (≈20 mV) in the deaerated solution than in the oxygen containing solutions, likely due in part to the presence of an active aluminum phase. For ZAC10403 alloy, the increase towards less negative values of corrosion potential observed after 6 h of immersion in the aerated solution (Figure 2(b)) was most likely resulting from two events, namely, the initiation or formation of oxide layer (including CaO) and the probable obstruction of corrosion product on the surface. The recorded corrosion potentials of alloy ZACS1040305 appear to be similar to ZAC alloy regardless of the media used (Figure 2(c)). After 8 h of immersion in deaerated solution, the E_{corr} values for ZAC and ZACS alloys (Table 2) indicate the same results despite probable different corrosion rates related to various chemical compositions and microstructures [22].

3.2. Potentiodynamic Study. Figure 3 shows the potentiodynamic curves of the AZ and ZA magnesium alloys in the untreated, aerated, and deaerated 5% NaCl solutions saturated with $Mg(OH)_2$ at pH 9. With one exception, it can be generally stated from the potentiodynamic curves that the presence of dissolved oxygen in different quantities had minor influence on the form of the cathodic polarisation curve and consequently no significant effect on the corrosion rates of the studied alloys. There was no one standard or conventional cathodic polarisation curve that obeys Tafel law and could lead us to determine corrosion rate with great precision. It is nevertheless clear that the corrosion behaviour of these alloys is particularly complex in these three passive chloride alkaline media; the term passive referring to the theoretical approach put forth by the Pourbaix E-pH diagram.

In previous studies [21–23], a passive zone was observed in alloys ZA104 and ZAC10403 in untreated 5% NaCl

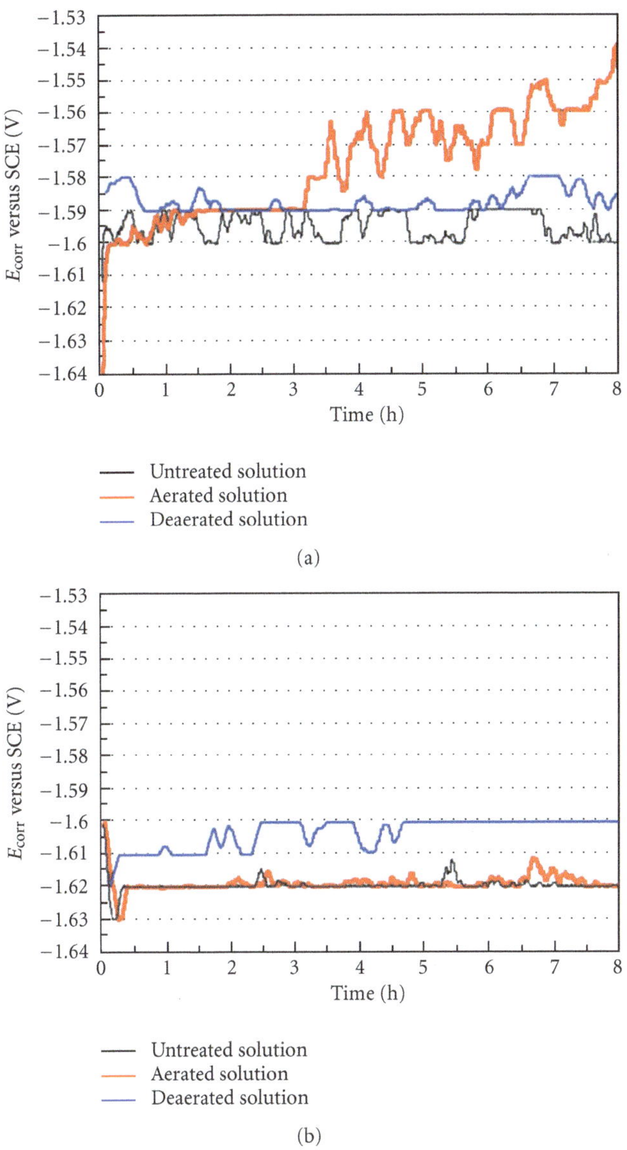

Figure 1: Time variation of E_{corr} for (a) AZ91D and (b) AZ91E in untreated, aerated, and deaerated 5% NaCl solution saturated with $Mg(OH)_2$ at pH 9 and 25°C.

solution saturated with $Mg(OH)_2$ at pH 9. In this study, a passive zone was monitored for alloy ZAC10403 in aerated and deaerated solutions. This passive zone can be associated with the sudden observed increase of Ecorr after 6 h of immersion (Figure 2(b)).

In yet another study [24], using different electrochemical techniques with ZA alloy compositions for long immersion periods in aerated and deaerated 5% NaCl solutions with the same hydroxide saturation and pH, it was concluded that the corrosion resistance or passivation was better with the addition of Sr associated with Ca than with the addition of Ca alone.

3.3. Corrosion Rates. Results reported in Table 3 show that in the untreated solution alloy AZ91E recorded the lowest

Table 2: Average E_{corr} values (V) for AZ and ZA magnesium alloys after 8 h of immersion in deaerated, untreated and aerated 5% NaCl solutions saturated with $Mg(OH)_2$ at pH 9 and 25°C. (Standard deviation <10 mV).

Alloys	Deaerated solution	Untreated solution	Aerated solution
AZ91D	−1.587	−1.593	−1.550
AZ91E	−1.600	−1.620	−1.618
ZA104	−1.546	−1.562	−1.565
ZAC10403	−1.560	−1.562	−1.562
ZACS1040305	−1.581	−1.600	−1.586

corrosion current rate as earlier reported in [28], with higher but equal values for alloys AZ91D and ZA104, followed by ZACS1040305, and finally ZAC10403, with the highest corrosion rate. Concerning AZ91E, this result confirms the fact that it had better corrosion resistance over AZ91D. It is also well documented that the presence of more noble metal impurities such Cu, Ni, and Fe in AZ91D is the most detrimental factor influencing the corrosion properties of alloys in this solution [3, 4]. The corrosion rates of ZAC10403 and ZACS1040305 were twice or more than those of AZ91E in the untreated solution. It can thus be stated that the superior performance of alloy AZ91E was due to the formation of magnesium hydroxide at the surface, which halted the progression of dissolution. The relative poor performance of ZA to AZ91E was most likely due to the galvanic effect of intermetallic $Mg_xZn_yAl_z$ with the magnesium matrix.

For a finely polished surface of ZAC alloy specimens tested in the three examined media, the corrosion rate increased by 40% to a maximum of 5 $\mu A \cdot cm^{-2}$ for ZAC10403 due to the addition of calcium and by 20% to a value of 4 $\mu A \cdot cm^{-2}$ for ZACS1040305 due to the presence of (Ca + Sr) comparatively to ZA104 (Table 3). Calcium appeared to have a strong accelerating effect due to more active galvanic cells at the corrosion potential, while the presence of Sr partially inhibited this effect. It is known that strontium acts as a beneficial element by decreasing the corrosion rates and improving the corrosion resistance of AZ magnesium alloy group in 5% NaCl solution saturated with $Mg(OH)_2$ at pH 9 on atmospheric oxygen [22, 24].

As can be seen in Table 3, all the alloys were not really influenced by the quantitative difference in oxygen concentration from atmospheric saturation (untreated solution) to complete saturation by air bubbling (aerated solution). The only exception was alloy AZ91E, where the relatively low quantity of atmospheric oxygen in the solution appeared to have had a favourable passive effect at pH 9, resulting in the lower corrosion rate. This result can be logically attributed to the composition and microstructure of the alloy.

The deaerated medium appeared to be very beneficial for alloys AZ91D and ZA104 which displayed the lowest corrosion rates (Table 3). These two alloys, containing Zn or impure elements such as Fe and Ni, were largely influenced by oxygen in serving as cathodes to increase corrosion rate. Baril and Pébère [12] reported that the corrosion potential

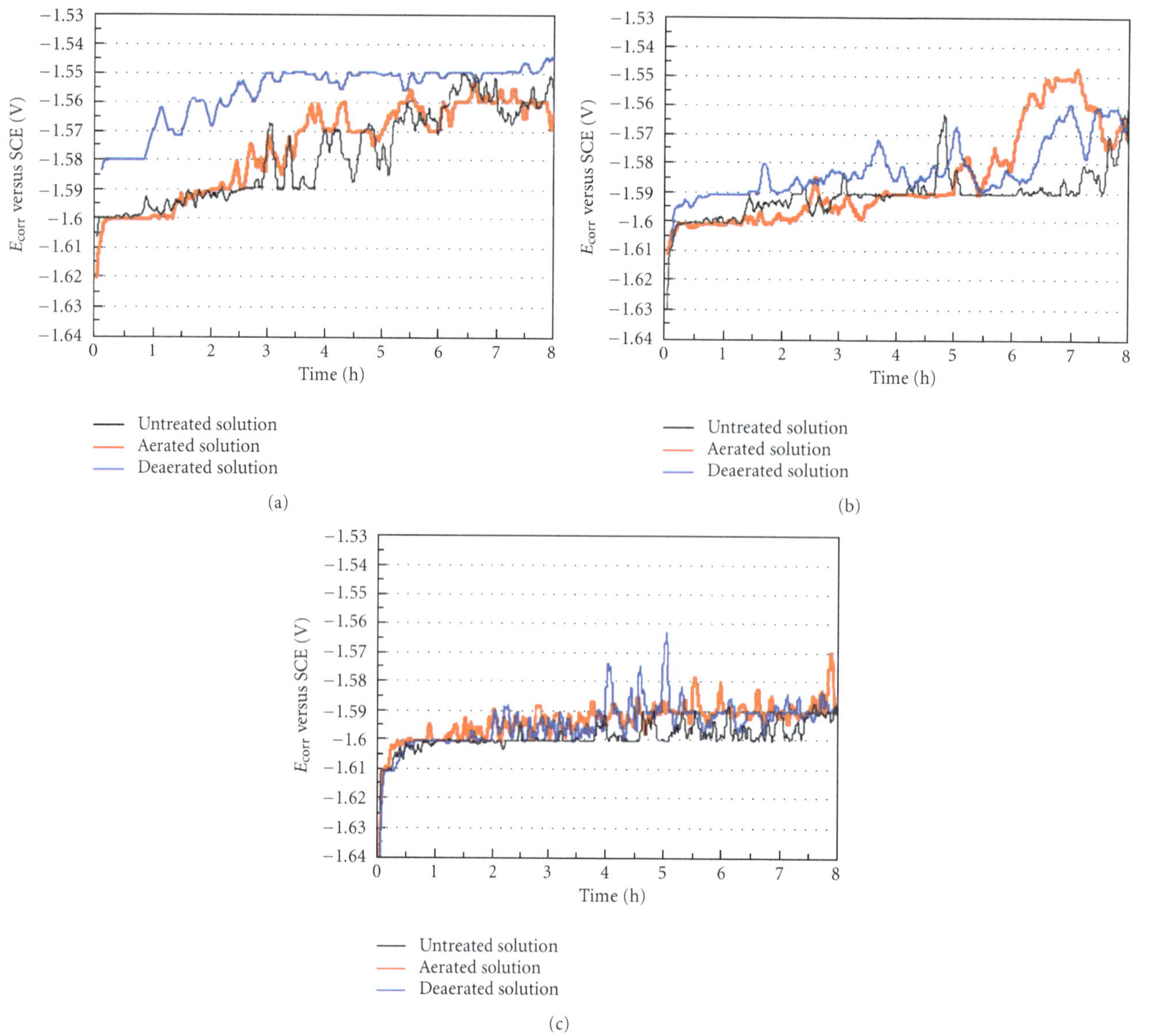

FIGURE 2: Time variation of E_{corr} for (a) ZA104, (b) ZAC10403, and (c) ZACS1040305 in untreated, aerated, and deaerated 5% NaCl solution saturated with $Mg(OH)_2$ at pH 9 and 25°C.

TABLE 3: Approximate i_{corr} values ($\mu A \cdot cm^{-2}$) for the AZ and ZA magnesium alloys calculated from the potentiodynamic curves by the method of linear polarization.

Alloy	Deaerated solution	Untreated solution	Aerated solution	Model[1]
AZ91D	2	3	3	III
AZ91E	4	2	3	IV
ZA104	2	3	3	III
ZAC10403	5	5	5	V
ZACS1040305	4	4	4	V

[1]Mechanistic corrosion models are described in Section 3.4.

of pure magnesium (in absence of alloying element) was more active in deaerated Na_2SO_4 solution and that the anodic current densities were lower than those found under aerated conditions. In this study, content in impurity and alloying elements in AZ91E alloy seems to act as active cathodes increasing corrosion in deaerated alkaline medium, and causing a corrosion rate twice that of AZ91D and ZA104 alloys in the same solution (Table 3). Then, it can be stated that corrosion rate in absence of dissolved oxygen depends mainly on alloy composition and microstructure.

3.4. Interpretations of Oxygen Effect on E_{corr} and i_{corr}. Some phenomena should be considered for potential values of magnesium alloys in active-passive media. The metal is originally very active and sufficient metallic ions are formed giving nonconductive corrosion products (oxides or hydroxides) on the metallic surface in alkaline medium. Some metallic sites relatively poor in metallic ions could give active potentials for the entire surface. The potentials recorded in this study could reflect the average of both active and passive

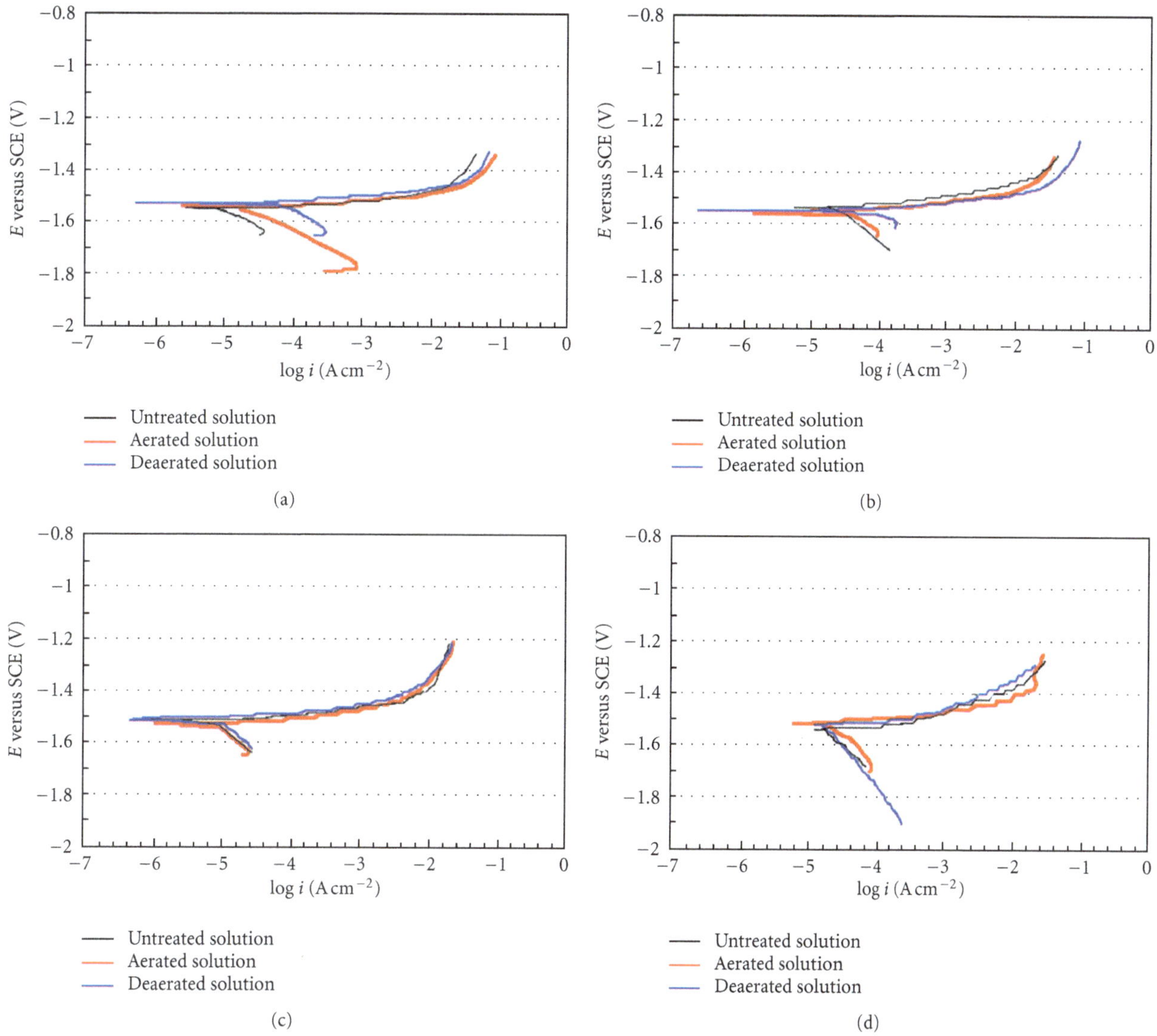

FIGURE 3: Potentiodynamic curves at a scan rate of 0.1 mV·s^{-1} for (a) AZ91D, (b) AZ91E, (c) ZA104, and (d) ZAC10403 in untreated, aerated, and deaerated 5% NaCl solution saturated with $Mg(OH)_2$ at pH 9 and 25°C.

regions but very frequently the potential of the active sites dominates. Also, a quantity of other metallic impurities such as Cu, Fe, and Ni could cause very efficient galvanic corrosion and polarize the open circuit potential. For example, in case of iron and mild steels, the accelerating effect of oxygen is commonly accompanied by a certain shift to more positive or noble potentials because of the exothermic reaction accompanying the hydrogen reduction in presence of oxygen. Also, principally different corrosion rates could exist for the same potential and generally no relation could be suggested between potential values and corrosion rates in this study [29].

However, it is interesting to compare the variation of corrosion rates of magnesium alloys in the alkaline medium with those of mild steels in neutral water medium since these two types of alloys show similar active-passive behaviours in such media. The effect of oxygen content on corrosion rate of mild steels in water at neutral passive solutions has been extensively examined by Uhlig and Revie [30]. In their study, the effect of oxygen concentration on corrosion rates of mild steels in slightly agitated distilled water during tests of 48 h at 25°C was considered. Three mechanistic corrosion models were illustrated based on the amount of dissolved oxygen in the solution. For low concentration of dissolved oxygen (0 to 4 mL/L), acceleration of the cathodic reaction is dominant with increasing oxygen concentration and the corrosion rate increases quasi-linearly (model I). This acceleration is principally due to the depolarization of the cathodic reaction or to replacement of the slow hydrogen evolution by an exothermic cathodic reaction in the presence of oxygen at 25°C and atmospheric pressure:

$$2H^{+}+2e^{-} \longrightarrow H_2O \quad (E = 0.00–0.0592\ \text{pH})\ \text{V} \qquad (1)$$

$$2H^{+}+2e^{-}+\frac{1}{2}O_2 \longrightarrow H_2 \quad (E = 1.23–0.0592\ \text{pH})\ \text{V} \qquad (2)$$

The complete and maximum depolarization by oxygen depends on the presence of sufficient oxygen concentration and enough cathodic sites at the interface Fe-H_2O. For dissolved oxygen concentration between ≈5 to 17 mL/L, a slowing down of the quasi-linear acceleration of the corrosion rate and the appearance of progressive inhibition-passivation phenomenon at the interface are observed (model (I + II)). This inhibition could be caused by a controlled oxygen quantity at the interface, less cathodic or anodic sites on the specimen due to corrosion product (oxides) or adsorbed oxygen, and the presence of passive layer at the interface. For higher dissolved oxygen concentration (≈18 to 25 mL/L), the corrosion rate is dominated by the inhibition-passivation phenomenon and decreases progressively (model II). It is clearly shown that the corrosion rate could return to low values close to that measured in absence of oxygen.

In the present study, the corrosion rates of magnesium alloys reported in Table 3 reveal the presence of three intermediate situations based on the mechanistic corrosion models described previously [31]. Then, for AZ91D and ZA104 alloys, increasing the amount of dissolved oxygen concentration in alkaline chloride media from atmospheric saturation (5.8 mL/L) to complete saturation (12.5 mL/L) has no influence on the corrosion rate (model III). It could be assumed also that the inhibiting effect of the additional dissolved oxygen eliminated the expected accelerating one.

AZ91E alloy shows a fast corrosion rate which could be related to abundant anodic and cathodic sites in absence of dissolved oxygen (model IV). In this case, atmospheric saturated oxygen is able to imply 50% inhibition passivation, while higher quantities of dissolved oxygen implied only 25% inhibition of corrosion rate, very possibly because of a certain accelerating effect of additional oxygen. A last model (model V) is observed for the two other alloys (ZAC10403, ZACS1040305) where a moderate or high amount of dissolved oxygen in the solution does not show a clear evidence of acceleration or inhibition of the corrosion rate. These two alloys have shown active and strong corrosion rates in a way that oxygen concentration has no significant effect on dissolution rate or that possibly inhibition eliminated or neutralized the accelerating effect of dissolved oxygen concentration.

3.5. Pitting Analysis. The morphology of the corrosion products formed on the surface of the AZ and ZA magnesium alloy specimens after 8 h of immersion in untreated 5% NaCl solution saturated with $Mg(OH)_2$ at pH 9 was investigated. Following the corrosion test, the samples were submitted to slight mechanical polishing which enabled us to observe pits underneath the corrosion product with no alteration. Pitting phenomena in the aerated and deaerated solutions produced similar results and followed the same trend as that recorded in the untreated solution. Figure 4 shows the pitting formation at the corroded surface of AZ91D and ZA10403 specimens following 8 h of immersion in the untreated chloride solution.

For AZ91D, it appears evident that pitting corrosion resulted in a localised surface attack. This highly focused corrosion thus created cavities or pits on the surface (Figure 4(a)). The size of the affected zones varied extensively, from very deep and localised to small and hollow over a large region. Pit shape also varied: narrow and deep, shallow and wide, elliptical, as well as vertical or horizontal attack sites, or subsurface cavities. For example, on alloy ZA10403, the type of corrosion pitting was wide and more uniform (Figure 4(b)).

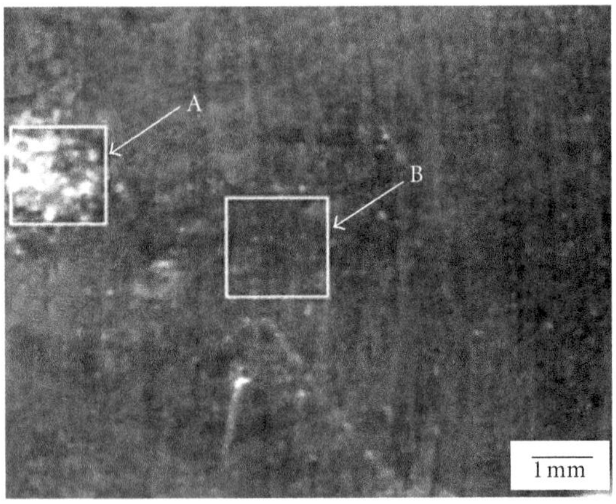

(a)

(b)

FIGURE 4: Optical observation of pitting for (a) AZ91D and (b) ZA10403 after 8 h of immersion in untreated 5% NaCl solution saturated with $Mg(OH)_2$ at pH 9 and 25°C.

In Figure 5, two zones (inserts A and B) of the AZ91D specimen from Figure 4(a) were further investigated by means of an AFM apparatus. Figure 5(a) shows a 3D surface analysis of the corroded zone (insert A) while Figure 5(b) illustrates an unattacked zone (insert B). The corroded region was completely different from the nonattacked area and was free from scratches. Although no specific mechanism for (1) pit nucleation or (2) pit growth has been acknowledged, the pitting process can generally be mapped by examining cell concentration phenomena. These important passive and active regions are a demonstration of the presence of dominated accelerated effect of dissolved oxygen concentration on corrosion rate as well as strong passive sites very possibly due to magnesium hydroxide and/or passive film. This model associates pit nucleation

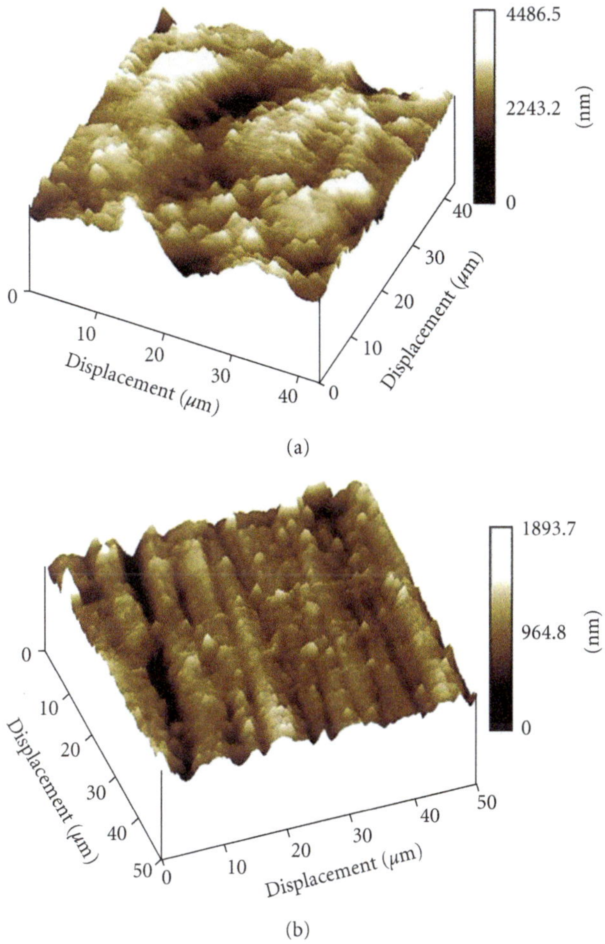

FIGURE 5: 3D AFM image of AZ91D showing (a) a corroded surface (Figure 4(a), insert A) and (b) an nonattacked zone (Figure 4(a), insert B) after 8 h of immersion in untreated 5% NaCl solution saturated with $Mg(OH)_2$ at pH 9 and 25°C.

or initiation with a localised loss of passivity in the alloy. According to one proposed mechanism, breakdown of the passive film is the result of the presence of strong complex agents in the solution (i.e., Cl^- ions) interacting chemically with this film and catalytically dissolving the passive layer until the metal surface is reexposed. The formation of oriented scratches in Figure 5(b) is due simply to mechanical polishing prior to optical observation.

Figure 6 illustrates the pitting on the corroded surface of AZ and ZA alloy specimens following 8 h of immersion in untreated solution. Figure 6(a) presents an optical micrograph of AZ91E surface cross-section specimen showing two pits (number 1 and number 2) measuring ≈105 and 162 μm in depth, respectively. The SEM high magnification micrograph in Figure 6(b) shows a typical pit shape on AZ91E surface which resembles a deep conical crater. AZ91D surface specimen presented comparable results of pitting corrosion (Figure 6(c)). For the ZA alloys, pit depth measurement was difficult because of the uniform pit distribution and the fact that the pits were narrow and well integrated in the surface (Figure 6(d)). It can be stated at this point that pit depth was more significant in the AZ alloys than in the ZA alloys.

4. Conclusions

The main conclusion of this work is that the performance of magnesium alloy in alkaline chloride passive medium is not as predictable as it is in acid medium and depends on several characteristics of the studied alloy. In this context, the corrosion behaviour of two groups of magnesium alloys (AZ and ZA) was investigated in deaerated, untreated, and aerated 5% NaCl solutions saturated with $Mg(OH)_2$ at pH 9 and 25°C. The following conclusions are emphasized.

(i) Open circuit potential in untreated and aerated with respect to the deaerated solutions did not show important potential shifts. This could be explained by the complex effect of dissolved oxygen concentration in the solution (acceleration of the cathodic reaction, inhibition, formation of corrosion products, and/or passivation). Principally, no relation has been observed between the potential shift and corrosion rates as a function of dissolved oxygen in the solution.

(ii) Five mechanistic corrosion models showing the oxygen effect on corrosion rates of magnesium alloys are suggested: (model I) dominant acceleration, (model (I + II)) acceleration, inhibition and/or passivation, (model II) dominant inhibition and/or passivation, (model III) no influence on corrosion rates of oxygen saturation by bubbling as compared to atmospheric oxygen saturation, (model IV) higher corrosion rate without oxygen, and (model V) almost no influence was identified due to oxygen presence.

(iii) Corrosion rates of AZ91D and ZA104 alloys were sensitive to the presence of oxygen in alkaline medium. However, untreated and aerated solutions showed the same acceleration of corrosion rates by 50%.

(iv) Despite the presence of fewer impurities in the AZ91E alloy, it recorded a higher corrosion rate than did AZ91D alloy in the deaerated medium, very possibly caused by the absence of the conditions required for passivation.

(v) ZAC10403 alloy has shown the highest corrosion rates, followed by ZACS1040305 and ZA104 alloys. Minor additions of Ca or (Ca + Sr) in ZAC10403 and ZACS1040305 alloys, respectively, produced a constant corrosion rate in the three corrosive solutions.

(vi) Observations of corroded surfaces in the untreated solution confirm that the pits were more localized and deeper on AZ alloys while more uniformly distributed on ZA alloys.

(vii) Pitting phenomena on AZ and ZA alloy surfaces in aerated and deaerated solutions showed similar results and followed the same trend as that recorded in untreated solution.

Acknowledgments

The authors would like to thank the Natural Sciences and Engineering Research Council of Canada (NSERC) for their

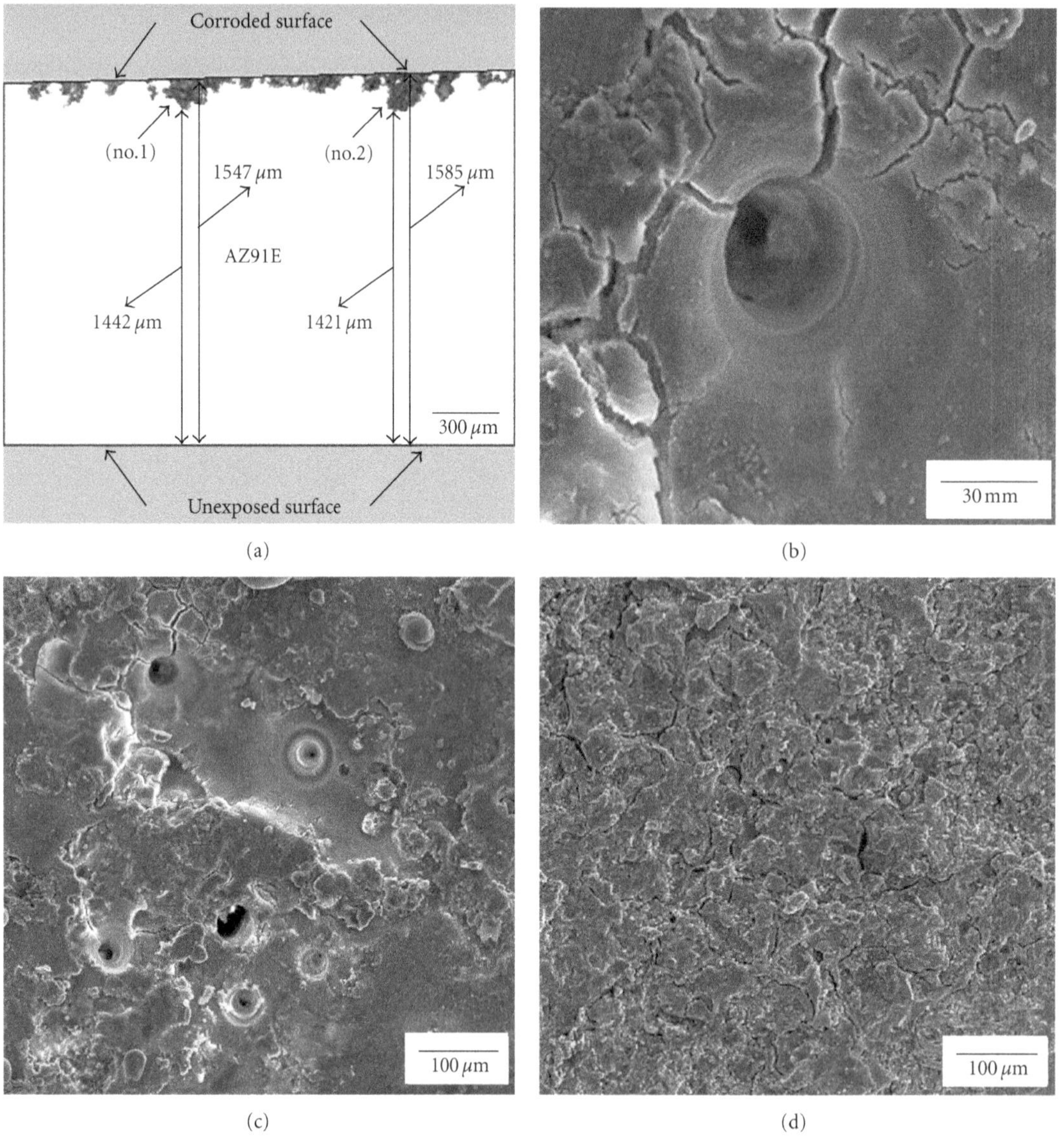

Figure 6: Micrographs of ((a) and (b)) optical pit depth and SEM pit shape on AZ91E surfaces, respectively, and ((c) and (d)) SEM corroded surfaces on AZ91D and ZA104, respectively, after 8 h of immersion in untreated 5% NaCl solution saturated with $Mg(OH)_2$ at pH 9 and 25°C.

financial support. They also express their profound gratitude to Dr. A.-M. Lafront (Université Laval), for her supervision and participation during the realisation of this work. Special thanks to the personnel of the Department of Mining, Metallurgical and Materials Engineering (Université Laval) for their professional and technical assistance.

References

[1] B. Landkof, "Magnesium applications in the electronic industries," in *Proceedings of the 2nd Israeli International Conference on Magnesium Science & Technology*, E. Aghion and D. Eliezer, Eds., pp. 50–56, Magnesium Research Institute, Sdom, Israel, February 2000.

[2] D. Magers and J. Willekens, "Global outlook on the use of magnesium die-castings in automotive applications," in *Proceedings of the Magnesium Alloys and their Application*, B. L. Mordike and K. U. Kainer, Eds., pp. 105–112, Werkstoff-Informationsgesellschaft Gmbh, Frankfurt, Germany, 1998.

[3] K. Nisancioglu, O. Lunder, and T. K. Aune, "Corrosion mechanism of AZ91 magnesium alloy," in *Proceedings of the International Magnesium Association*, pp. 43–50, Detroit, Minn, USA, 1990.

[4] Z. Szklarska-Smialowska, *Pitting Corrosion of Metals*, NACE International, Houston, Tex, USA, 1986.

[5] B. E. Wilde and E. Williams, "On the correspondence between electrochemical and chemical accelerated pitting corrosion tests," *Journal of Electrochemical Society*, vol. 117, no. 6, pp. 775–779, 1970.

[6] S. Lebeau, Y. Yamamoto, and K. Sakamoto, "Thixomolding of magnesium automotive components," SAE Technical Paper 980087, International Congress and Exposition, Detroit MI, USA, 1998.

[7] I. Nakatsugawa, F. Yamada, H. Takayasu, T. Tsukeda, and K. Saito, "Corrosion behavior of thixomolded Mg-Al alloys," in

Proceedings of the International Symposium on Environmental Degradation of Materials and Corrosion Control in Metals, pp. 113–123, Québec, Canada, August 1999.

[8] R. Decker, R. Carnahan, R. Vining, D. Walukas, S. Lebeau, and N. Prewitt, "Thixomolding magnesium based alloys," in *Proceedings of the Magnesium Alloys and Their Applications*, B. L. Mordike and K. U. Kainer, Eds., pp. 545–550, Werkstoff-Informationsgesellschaft Gmbh, Frankfurt, Germany, April1998.

[9] T. Yamaguchi, T. Sukeda, and K. Saito, "Thixomolding—its features and assignments," *Kikai-Gijutsu, Nikkan Kogyo Shinbun*, vol. 47, no. 3, pp. 59–64, 1999.

[10] G. L. Makar and J. Kruger, "Corrosion studies of rapidly solidified magnesium alloys," *Journal of the Electrochemical Society*, vol. 137, no. 2, pp. 414–421, 1990.

[11] J. H. Nordlien, K. Nisancioglu, S. Ono, and N. Masuko, "Morphology and structure of water-formed oxides on ternary MgAl alloys," *Journal of the Electrochemical Society*, vol. 144, no. 2, pp. 461–466, 1997.

[12] G. Baril and N. Pébère, "Corrosion of pure magnesium in aerated and deaerated sodium sulphate solutions," *Corrosion Science*, vol. 43, no. 3, pp. 471–484, 2001.

[13] H. Alves, U. Koster, E. Aghion, and D. Eliezer, "Environmental behavior of magnesium and magnesium alloys," *Materials Technology*, vol. 16, no. 2, pp. 110–126, 2001.

[14] R. Ambat, N. N. Aung, and W. Zhou, "Evaluation of microstructural effects on corrosion behaviour of AZ91D magnesium alloy," *Corrosion Science*, vol. 42, no. 8, pp. 1433–1455, 2000.

[15] G. Song, A. Atrens, and M. Dargusch, "Influence of microstructure on the corrosion of diecast AZ91D," *Corrosion Science*, vol. 41, no. 2, pp. 249–273, 1998.

[16] C. Suman, "Effects of direct aging on mechanical properties and corrosion resistance of diecast magnesium alloys AZ91D and AM60B," *SAE Transactions*, vol. 99, no. 5, pp. 849–859, 1990.

[17] O. Lunder, M. Videm, and K. Nisancioglu, "Corrosion resistant magnesium alloys," SAE Technical Paper 950428, International Congress and Exposition, Detroit, MI, USA, 1995.

[18] G. Neite, K. Kubota, K. Higashi, and F. Hehmann, "Magnesium-based alloys," *Materials Science and Technology A*, vol. 8, pp. 113–213, 1996.

[19] I. A. Anyanwu, T. Honda, S. Kamado, Y. Kojima, S. Takeda, and T. Ishida, "Heat and corrosion resistance of Mg-Zn-Al-Ca alloys," in *Proceedings of the Magnesium Alloys and Their Applications*, K. U. Kainer, Ed., pp. 110–115, Wiley-VCH, Weinheim, Germany, 2006.

[20] E. Øvrelid, G. B. Floøistad, T. Rosenqvist, P. Bakke, and T. A. Engh, "The effect of Sr addition on the hydrogen solubility and hydride formation in pure Mg and the alloy AZ91," *Scandinavian Journal of Metallurgy*, vol. 27, no. 3, pp. 133–140, 1998.

[21] A. M. Lafront, M. Ö. Öteyaka, R. D. Klassen, P. R. Roberge, and E. Ghali, "Study of the corrosion of zinc and aluminium magnesium alloys by electrochemical noise (EN) and scanning reference electrode technique (SRET)," in *Proceedings of the NACE International Seminar*, Montréal, Canada, August 2002.

[22] M. Ö. Öteyaka, A. M. Lafront, R. Tremblay, and E. Ghali, "Pitting corrosion of some magnesium alloy by scanning reference technique (SRET)," in *Proceedings of the NACE International Seminar*, Montréal, Canada, August 2002.

[23] M. Ö. Öteyaka, A. M. Lafront, E. Ghali, and R. Tremblay, "Pitting corrosion potential of magnesium alloys in different corrosive media," in *Proceedings of the NACE International Seminar*, Montréal, Canada, August 2002.

[24] M. Ö. Öteyaka, A. M. Lafront, E. Ghali, and R. Tremblay, "Potentiodynamic study of some AZ and ZA magnesium alloys in different corrosive media," in *Proceedings of the Magnesium Alloys and Their Applications*, K. U. Kainer, Ed., pp. 517–523, International Magnesium Association, Wolfsburg, Germany, 2003.

[25] ASTM B93/B93M-09, "Standard specifications for magnesium alloys in ingot form for sand castings, permanent mold castings, and die castings," ASTM International, West Conshocken, Pa, USA, 2009.

[26] ASTM G5-94, "Standard reference test method for making potentiostatic and potentiodynamic anodic polarization measurements," ASTM International, West Conshocken, Pa, USA, 2004.

[27] H. H. Uhlig, D. N. Triadis, and M. Stern, "Effect of oxygen, chlorides, and calcium ion on corrosion inhibition of iron by polyphosphates," *Journal of Electrochemical Society*, vol. 102, no. 2, pp. 59–66, 1955.

[28] G. L. Song and A. Atrens, "Corrosion mechanisms of magnesium alloys," *Advanced Engineering Materials*, vol. 1, no. 1, pp. 11–33, 1999.

[29] L. L. Shreir, R. A. Jarman, and G. T. Burnstein, "Basic concepts of corrosion, oxygen reduction, corrosion in aqueous solutions," in *Corrosion*, vol. 1, pp. 99–105, Butterworth Heinemann, Oxford, UK, 1995.

[30] H. H. Uhlig and R. W. Revie, "Effect of dissolved oxygen," in *Corrosion and Corrosion Control, Iron and Steel*, pp. 91–96, John Wiley & Sons, NY, NY, USA, 3rd edition, 1985.

[31] E. Ghali, "Magnesium performance," in *Corrosion Resistance of Aluminum and Magnesium Alloys, Understanding, Performance, and Testing*, pp. 338–343, John Wiley & Sons, Hoboken, NJ, USA, 2010.

On the Stress Corrosion Cracking and Hydrogen Embrittlement Behavior of Austenitic Stainless Steels in Boiling Saturated Magnesium Chloride Solutions

Osama M. Alyousif[1] and Rokuro Nishimura[2]

[1] *Department of Mechanical Engineering, Kuwait University, P.O. Box 5969, Safat 13060, Kuwait*
[2] *Department of Applied Materials Science, Osaka Prefecture University, 1-1, Gakuen-cho, Sakai, Osaka 599-8531, Japan*

Correspondence should be addressed to Osama M. Alyousif, osama.alyousif@ku.edu.kw

Academic Editor: W. Ke

The stress corrosion cracking (SCC) and hydrogen embrittlement (HE) behaviors for types 304, 310, and 316 austenitic stainless steels were investigated in boiling saturated magnesium chloride solutions using a constant load method under different conditions including test temperature, applied stress, and sensitization. Both of type 304 and type 316 stainless steels showed quite similar behavior characteristics, whereas type 310 stainless steel showed a different behavior. The time to failure (t_f) parameter was used among other parameters to characterize the materials behavior in the test solution and to develop a mathematical model for predicting the time to failure in the chloride solution. The combination of corrosion curve parameters and fracture surface micrographs gave some explanation for the cracking modes as well as an indication for the cracking mechanisms. On the basis of the results obtained, it was estimated that intergranular cracking was resulted from hydrogen embrittlement due to strain-induced formation of martensite along the grain boundaries, while transgranular cracking took place by propagating cracks nucleated at slip steps by dissolution.

1. Introduction

The stress corrosion cracking behavior of austenitic stainless steels in chloride and other corrosive solutions has been extensively investigated using various methods [1–7]. The constant load method can produce a corrosion elongation curve which can be used to obtain three useful parameters namely, time to failure (t_f), steady-state elongation rate (l_{ss}), and transition time to time to failure ratio (t_{ss}/t_f). These parameters are useful in analyzing the failure behavior. In particular, the steady-state elongation rate (l_{ss}) can be used to predict the time to failure (t_f) [3, 8]. Recent studies found that the characteristics of the steady-state elongation rate can be applied to SCC of type 430 ferritic stainless steels in acidic chloride and sulfate solutions by using the constant load method [9].

In previous papers [10, 11], we have reported that solution-annealed austenitic stainless steels show two different cracking modes depending on the cracking mechanism; transgranular (TG) for SCC and intergranular (IG) for HE depending on the test temperature which was critical in controlling the cracking mode. In other papers [12, 13], we have also reported that sensitized stainless steels have a lower susceptibility for SCC and HE than that for solution-annealed austenitic stainless steels. On the basis of these findings, a new aspect for intergranular hydrogen embrittlement mechanism for solution-annealed types 304, 316 and 310 austenitic stainless steels was proposed [14]. The objectives of this paper are to elucidate the effects of test temperature, applied stress and sensitization on SCC (or HE) susceptibility of austenitic stainless steels, and to discuss a proposed cracking mechanism for austenitic stainless steels in chloride solutions.

2. Experimental

The specimens used were commercial type 304, 316, and 310 austenitic stainless steels whose chemical compositions

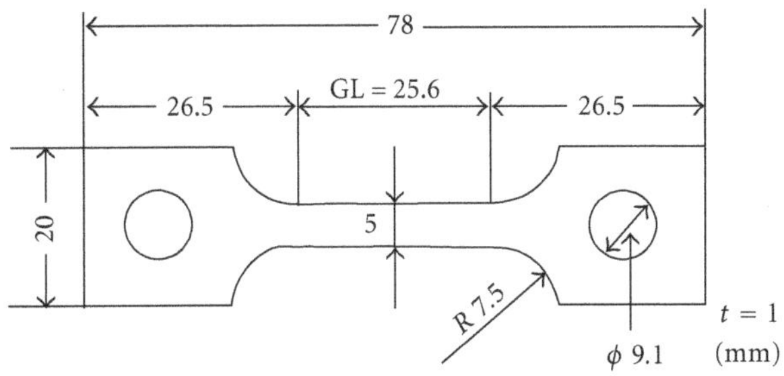

FIGURE 1: Geometry of specimens (dimensions in mm).

(wt%) are shown in Table 1. As shown in Figure 1, the geometry for stress corrosion cracking experiments is as follows: the gauge length is 25.6 mm, the width 5 mm, and the thickness 1 mm. The specimens were solution-annealed at 1373 K for 3.6 ks under an argon atmosphere and then water-quenched. Another batch of specimens of the same materials was subjected to a sensitization heat treatment at 923 K under an argon atmosphere for 24 hours then water quenched. Prior to experiments, the solution-annealed specimens were polished to 1000 grit emery paper, degreased with acetone in an ultrasonic cleaner and washed with distilled water. After the pretreatment, the specimens were immediately set into a stress corrosion cracking cell. Stress corrosion cracking tests were conducted in boiling saturated magnesium chloride solutions, whose boiling temperatures were changed by the change in the concentration of magnesium chloride. The test temperature range was between 403 K–428 K. The applied stress was in the range from 0 to 500 MPa. Some tests were performed at a constant applied stress with varying test temperature to study the effect of test temperature, and some experiments were performed at a constant test temperature with varying applied stress to study the effect of applied stress. All experiments were carried out under an open circuit condition.

A lever-type constant load apparatus (lever ratio 1 : 10) to which three specimens can be separately and simultaneously attached was used with a cooling system on the top to avoid evaporation of the solution during the experiments. The specimens were insulated from rod and grip by surface-oxidized zirconium tube. Elongation of the specimens under a constant load was measured by an inductive linear transducer with an accuracy of ±0.01 mm.

3. Effect of Test Temperature

The effects of test temperature were studied for austenitic stainless steels in boiling saturated magnesium chloride solutions and the details and results of this work are shown in a previous publication [10]. Figure 2 depicts the logarithm of the parameter t_f versus the reciprocal of the test temperature for the three stainless steel types used in the experiments. For type 304, the relationship between t_f and $1/T$ falls in a straight line, shown as region-I, until a test temperature of about 413 K ($1/T = 2.42 \times 10^{-3}$), below which the relationship deviates from the linearity. The region from 413 K ($1/T = 2.42 \times 10^{-3}$) to a threshold test temperature, approximately 403 K ($1/T = 2.48 \times 10^{-3}$), below which little fracture takes place within a laboratory time scale, was called region-II. For type 316, the parameter t_f was also grouped into two regions; region-I at test temperatures above 424 K ($1/T = 2.36 \times 10^{-3}$) and region-II with test temperatures below 424 K ($1/T = 2.36 \times 10^{-3}$) up to a threshold test temperature of about 408 K ($1/T = 2.45 \times 10^{-3}$). The t_f values of type 310 showed slight sensitivity to test temperature. The t_f values for type 310 were grouped in a straight line which constitutes region-I until a threshold test temperature of about 408 K ($1/T = 2.45 \times 10^{-3}$) without region-II. Therefore, in Figure 2, the elongation curve for type 310 at 416 K ($1/T = 2.40 \times 10^{-3}$) falls in region-I, while that for type 316 at 416 K ($1/T = 2.40 \times 10^{-3}$) falls in region-II.

Type 304 is a metastable austenite, and it is very susceptible to martensite formation [15]. This type of austenitic stainless steels was used extensively in investigating hydrogen embrittlement behavior because of its material composition, structure "instability", intergranular cracking mode, and ability to phase transform upon hydrogen charging or induced strain. The material displayed two different behaviors on the basis of the test temperature dependency. Hydrogen embrittlement behavior occurred at test temperatures below the critical temperature and SCC behavior occurred at test temperatures above the critical temperature, where the critical temperature is the temperature between region-I and region-II. It can be concluded based on the results of using the constant load method that type 304 is very susceptible to hydrogen embrittlement as well as stress corrosion cracking in the saturated boiling magnesium chloride solutions. This notion will be further discussed in the following section.

Type 316 has also a metastable austenite, and it is moderately susceptible to martensite formation. The material exhibited a dual cracking behavior similar to type 304 and this may be attributed to the closely identical chemical composition for both types. Type 316 had a higher critical temperature than type 304. This type of austenitic stainless steel can also be branded as susceptibility to both of hydrogen embrittlement and SCC failure depending on the test temperature in magnesium chloride solution as shown in the postfracture micrographs.

But when it comes to type 310 which has a stable austenite structure, the material did not exhibit any hydrogen embrittlement cracking features based on the postfracture analysis of the broken samples. This behavior can be explained by the material's high content of chromium which is a natural inhabitant of martensite transformation, and this may be the reason that this type of stainless steels is susceptible to SCC only in magnesium chloride solutions.

Hydrogen is a product of the electrochemical reactions at the metal surface. For metastable austenitic steels like types 304 and 316, the strain-induced martensite along the grain boundaries will enhance the hydrogen permeation [16–18]. Martensite structure has a very high diffusivity coefficient and very small hydrogen content compared to those of the austenite. This may explain the intergranular cracking as the crack preferential path is through the martensite structure

Table 1: Chemical compositions (wt%) and mechanical properties of the austenitic stainless steels used.

	C	Si	Mn	P	S	Ni	Cr	Mo	σ_{Yield} (MPa)	$\sigma_{Tensile}$ (MPa)
SS304	0.06	0.35	0.96	0.027	0.004	8.13	18.20	—	276	691
SS310	0.05	0.84	1.27	0.016	0.001	19.30	24.76	—	275	575
SS316	0.06	0.70	0.96	0.031	0.004	10.15	16.98	2.22	323	636

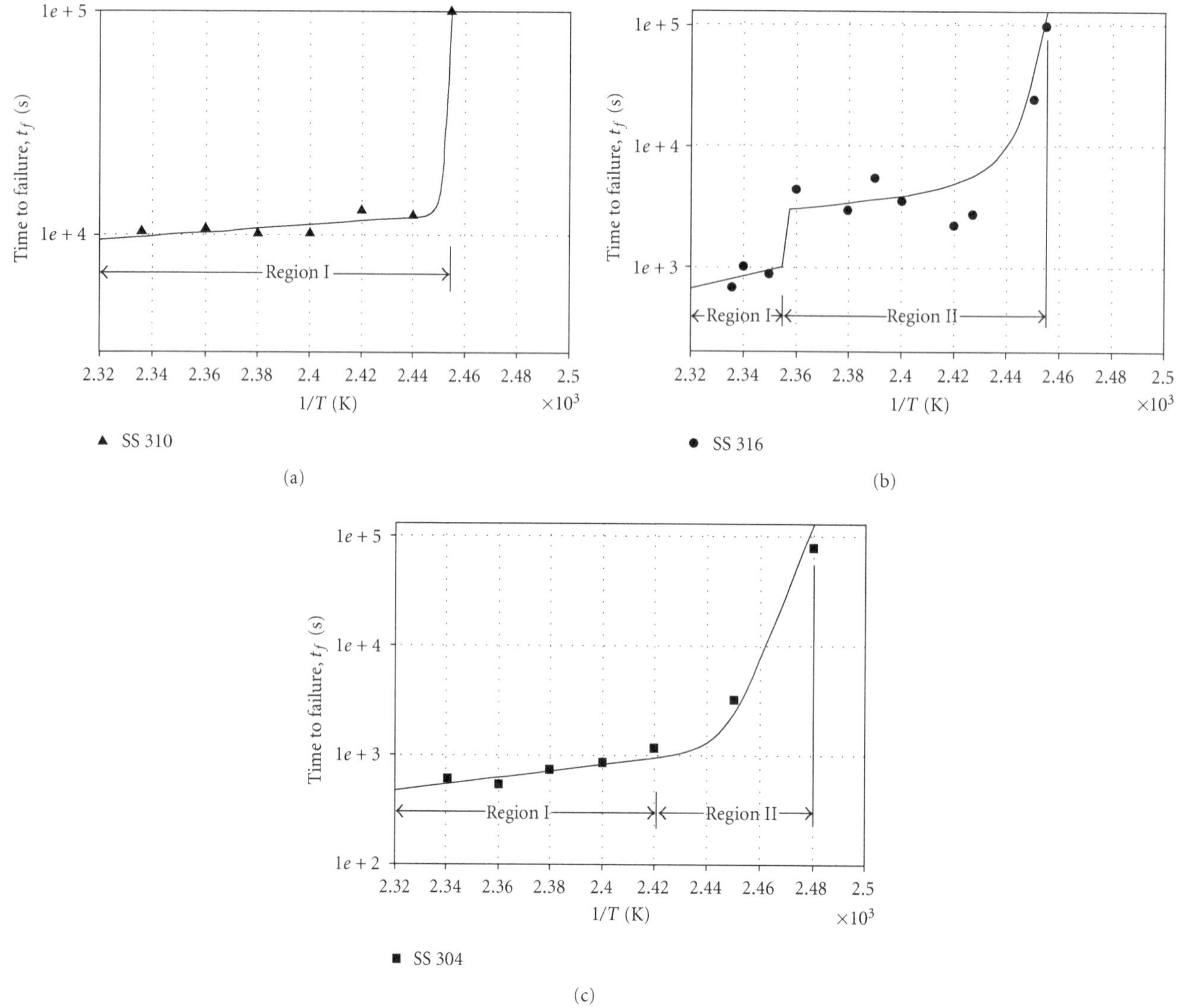

Figure 2: The relationship between reciprocal of test temperature ($1/T$) and logarithm of time to failure (t_f) for type 310, 304, and 316 steels in boiling saturated $MgCl_2$ solutions.

facilitated by the higher diffusion rates of hydrogen. At high temperatures, the corrosion rate will be higher than the corrosion rate at low temperature and this will result in a predominant metal dissolution at the crack tip faster than the hydrogen entry to the structure, and this will enhance the predominance of SCC mechanism. The lower temperature means lower corrosion reaction rates, and this will prohibit the film rupture mechanism hence letting the hydrogen embrittlement to take place and thus controlling the cracking mechanism. It is worth noting that increasing the amount of α'-martensite content in the structure produced by applied stress (or strain) will increase the susceptibility of the material to hydrogen embrittlement [18]. Type 310 is a stable austenitic stainless steel, and no evidence

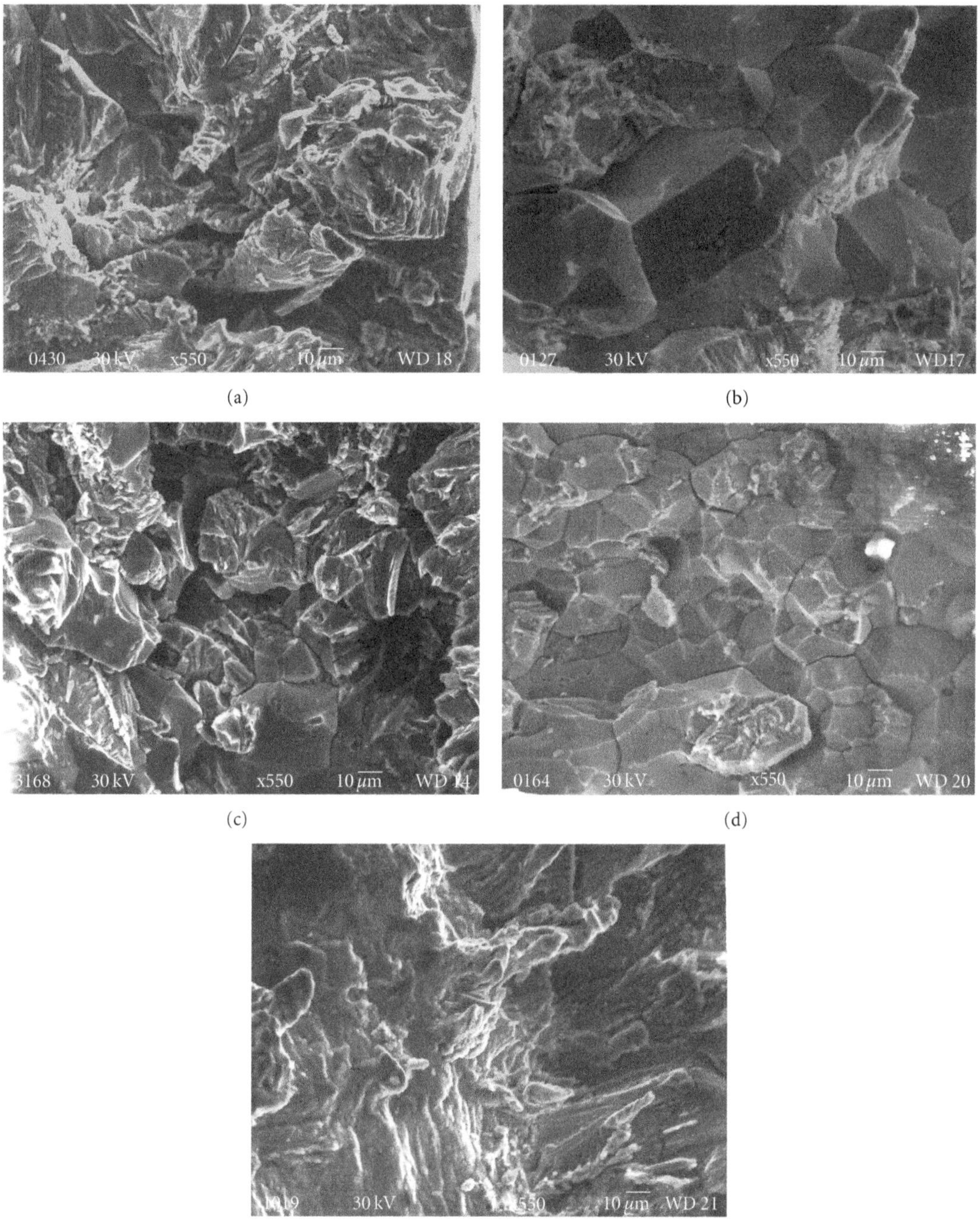

Figure 3: (a) Transgranular cracking for type 304 at $T = 527$ K, (b) intergranular cracking for type 304 austenitic stainless steel at $T = 403$ K (x550), (c) transgranular cracking for type 316 austenitic stainless steel at $T = 527$ K, (d) intergranular cracking for type 316 austenitic stainless steel at $T = 414$ K (x550), and (e) transgranular cracking for type 310 austenitic stainless steel at $T = 414$ K in boiling saturated $MgCl_2$ solutions.

of martensite formation was found so that the only possible cracking mechanism will be the film-rupture mechanism [19].

Figure 3 shows the fracture surface scanning electron microscopic examination (SEM) for the steels used in the experiments. For types 304 and 316, it can be seen that there exist two fracture modes, transgranular cracking mode which occurred at test temperature above the critical test temperature corresponding to the type of steel and intergranular cracking mode which took place at test temperatures below the critical test temperature corresponding to the type of steel. On the other hand, type 310 showed only transgranular cracking mode at all test temperatures.

4. Effect of Applied Stress

Figure 4 depicts the relationships between applied stress (σ) and logarithm of time to failure (t_f) for the three stainless

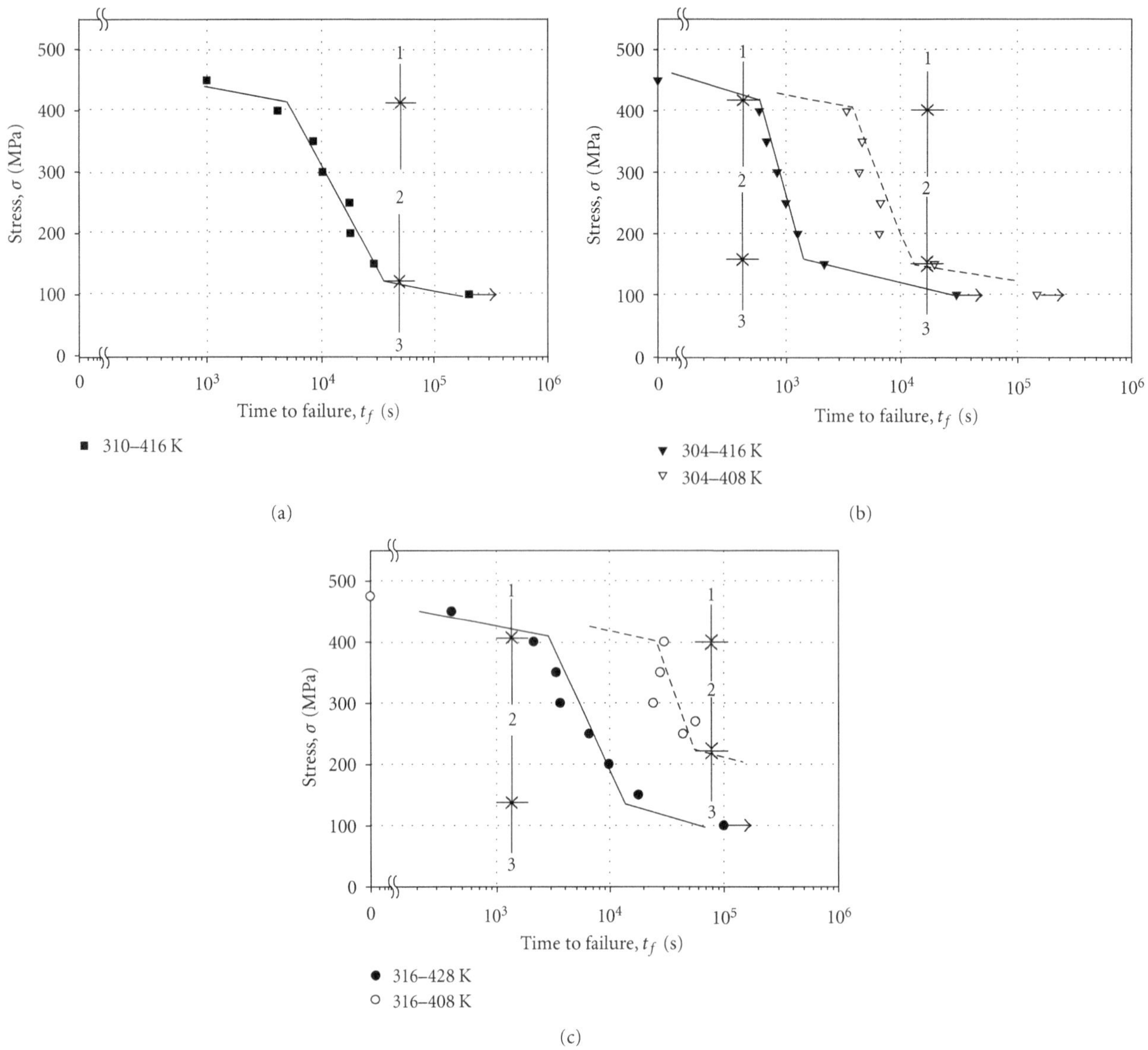

FIGURE 4: The relationship between applied stress (σ) and logarithm of time to failure (t_f) for type 310, 304, and 316 steels in boiling saturated $MgCl_2$ solutions.

steels at three test temperatures used. All relationships can be clearly divided into three regions shown by Arabic numerals 1–3 in Figure 3; region 1 is the stress-dominated region, region 2 the stress corrosion cracking-dominated region, and region 3 the corrosion-dominated region.

In region 2, t_f values for type 304 steel at 416 K and for type 316 steel at 428 K are shorter than those at 408 K, and t_f for type 304 steel is shorter than that for type 316 steel. In addition, the applied stress range in region 2 for type 316 steel at 408 K is narrower than that for type 304 steel at 408 K; the maximum applied stress in region 2 is the same for both steels, but the minimum applied stress for type 316 steel is higher than that for type 304 steel. At higher temperatures, the applied stress range in region 2 is almost the same for three stainless steels, and the time to failure is in the order of type 304 < type 316 < type 310.

In the applied stress region where the stress corrosion cracking-dominated fracture occurred, the fracture mode for type 304 and 316 steels at 416 and 428 K was transgranular, while that at 408 K was mostly intergranular, in particular at higher applied stresses (Figure 5). In other words, for type 304 and 316 steels in boiling magnesium chloride solutions, the fracture mode was transgranular at higher temperatures and intergranular at the lower temperature. The minimum applied stress to induce intergranular cracking for type 316 steel was significantly higher than that for type 304. For type 310 steel, only transgranular fracture was found as far as stress corrosion cracking occurred in any temperatures in boiling magnesium chloride solutions [10].

In general, transgranular cracking was characterized by a higher steady-state elongation rate in both stress corrosion cracking-dominated and stress-dominated regions and

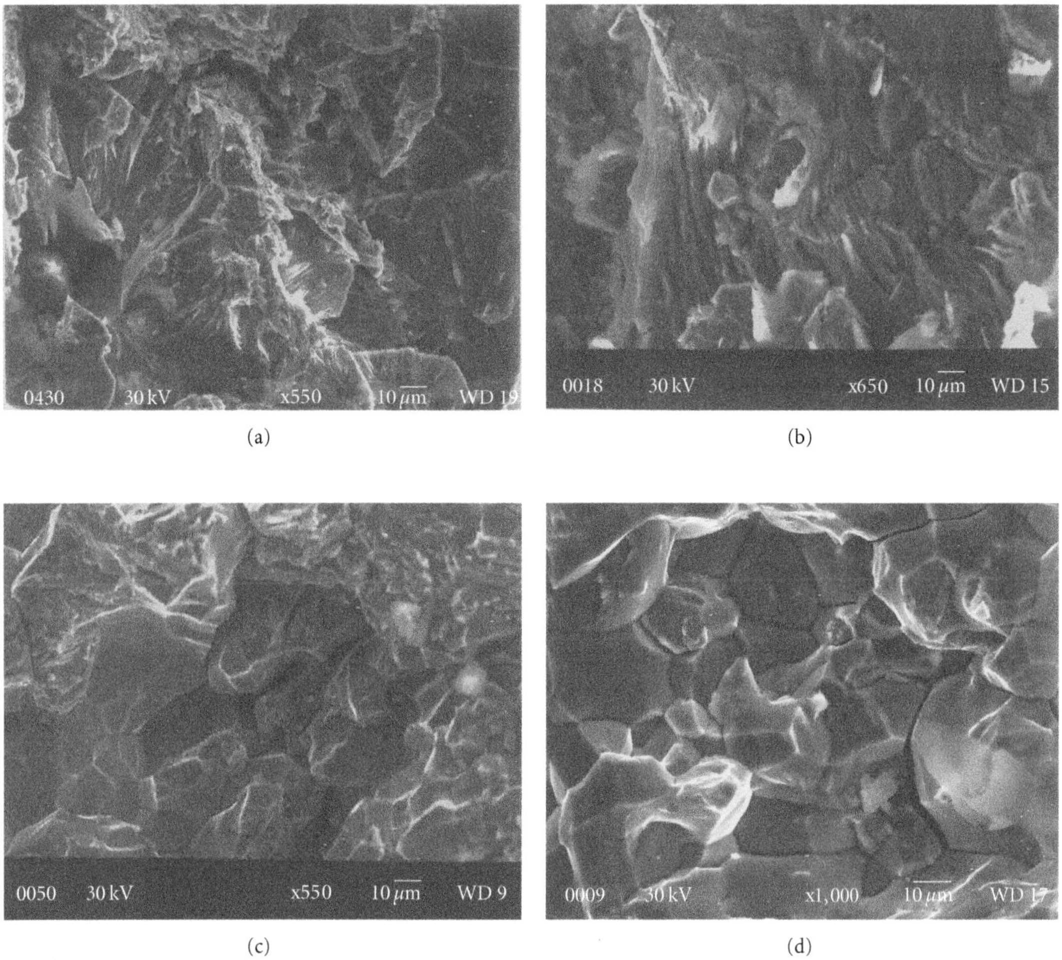

FIGURE 5: Fracture appearances for (a) type 304 at $T = 416$ K and $\sigma = 300$ MPa; x550 (b) for type 310 at $T = 416$ K and $\sigma = 300$ MPa; x650, (c) for type 316 at $T = 408$ K and $\sigma = 300$ MPa; x550 and (d) for type 316 at $T = 408$ K and $\sigma = 400$ MPa; x1000 in boiling saturated $MgCl_2$ solutions.

shorter time to failure. By contrast, the failure in intergranular cracking was slow and most of the time to failure was in the steady-state elongation region indicating the small mechanical elongation until failure. Consequently, intergranular cracking for these austenitic stainless steels in the boiling magnesium chloride solution revealed more brittle nature in comparison with transgranular cracking.

Type 304 steel is a metastable austenite and is highly susceptible to martensite formation [14]. Type 316 steel is also a metastable austenite and is moderately susceptible to martensite formation [18]. On the other hand, type 310 steel is known to be less susceptible to martensite formation [17]. For metastable austenitic stainless steels like type 304 and 316 steels, the strain-induced formation of martensite tends to take place along grain boundaries and particularly is facilitated by hydrogen entry [16, 20]. The presence of martensite is known to induce hydrogen embrittlement (HE) because of very high hydrogen diffusivity coefficient and very small hydrogen content compared to those of the austenite [18]. Thus, martensite formed at grain boundaries is apt to be responsible for HE of intergranular mode, and type 304 and 316 steels have higher susceptibility to martensite-induced HE of intergranular mode, but type 310 has little susceptibility to martensite-induced HE of intergranular mode.

On the other hand, transgranular cracking is caused by propagation of cracks nucleated at slip steps and is not related to martensite formed at grain boundaries. Thus stress corrosion cracking in the narrow sense of the word (SCC) includes such transgranular cracking.

Type 304 and 316 steels suffered intergranular cracking at a lower temperature (408 K) and transgranular cracking at higher temperatures (428 K for type 316 and 416 K for type 304). In determining the cracking mode, a competition between dissolution of material at slip steps inducing SCC and hydrogen entry inducing HE is the decisive factor as well as the formation of martensite. The hydrogen entry is determined by the difference between hydrogen absorption and hydrogen escape. Transgranular SCC for type 304 and 316 steels was caused by propagation of cracks nucleated at slip steps, not at martensite at grain boundaries [20]. Therefore, the cracking mode would be determined by the competition between the dissolution rate at slip steps and the hydrogen entry rate at grain boundaries with martensite. As test temperature increases, the dissolution rate increases. On the other hand, both hydrogen absorption rate and hydrogen escape rate increase with increasing test temperature, whereas the amount of hydrogen entry decreases, because the hydrogen escape rate becomes superior to the hydrogen entry rate with increasing temperature. This means that the amount of hydrogen entry decreases with increasing test temperature [21]. In addition, it is considered that the amount of martensite decreases with increasing test temperature, which suggests that the hydrogen entry rate decreases with increasing test temperature [22]. Thus, the dissolution rate becomes higher than the hydrogen entry rate at higher temperatures. This will result in transgranular SCC at higher temperatures.

On the other hand, it is known that the amount of martensite increases with strain or applied stress [15]. At 408 K for type 304 and 316 steels, the fracture mode changed from a mixed transgranular and intergranular mode to a complete intergranular mode with increasing applied stress. If cracking failure at 408 K is caused by martensite-induced HE, the increase in intergranular mode with increasing applied stress is in agreement with an increase in the amount of martensite with increasing applied stress. Similarly, the minimum applied stress to induce intergranular cracking for type 316 steel was significantly higher than that for type 304 steel in agreement with the fact that type 316 steel is less susceptible to stress-induced martensite formation than type 304 steel. Furthermore, type 304 steel was more susceptible to intergranular cracking in comparison with type 316 steel in agreement with the fact the 304 steel has higher susceptibility to strain-induced martensite formation in comparison with type 316 steel.

Consequently, intergranular cracking observed at the lower temperature, that is, 408 K for type 304 and 316 steels is due to grain boundary martensite-induced hydrogen embrittlement (HE) and transgranular cracking for type 304 and 316 steels at higher temperatures at 416 and 428 K and for 310 steel is ascribed to stress corrosion cracking in the narrow sense of the word (SCC).

5. Effect of Sensitization

Figure 6 depicts the logarithm of t_f versus $1/T$ for the three sensitized stainless steels (solid lines) and for the solution-annealed stainless steels (dashed lines) for comparison. For sensitized type 304, the relationship between t_f and $1/T$ consisted of two straight lines denoted as region-I and region-II below and above a test temperature of about 413 K ($1/T = 2.42 \times 10^{-3}$). The time to failure was found to be larger for the sensitized specimens than for the solution-annealed specimens over the whole test temperature range, and in addition, the critical test temperature between region-I and region-II has shifted from 408 K ($1/T = 2.45 \times 10^{-3}$) for the solution-annealed specimen to 413 K ($1/T = 2.42 \times 10^{-3}$) for the sensitized one.

For type 316, although the relationship between $\log t_f$ and $1/T$ for the solution-annealed specimen was composed of two regions (region-I and region-II) [10], that for the sensitized specimen showed only one region which corresponded to region-II for the solution-annealed specimens and had no region-I over the whole test temperature range as described later. In addition, it can be noted that there was approximately two orders of magnitude difference in the time to failure between the solution-annealed and sensitized specimens. On the other hand, the relationship for the sensitized type 310 showed only one region as well as that for type 316, but the behavior for type 310 was found to correspond to region-I as described later. The t_f for type 310 showed a small difference between the sensitized and solution-annealed specimens. It was, therefore, clearly found that sensitization provided a beneficial effect for types 304 and 316 and a slight effect for type 310 on the fracture susceptibility. The cracking modes for the sensitized steels were similar to the modes observed in the effect of test temperature experiments, for types 304 and 316 the samples exhibited transgranular cracking mode above the critical temperature and intergranular cracking mode below the critical temperature. And the transgranular cracking was dominant for all test temperatures for type 310.

6. A Proposal for Transgranular Hydrogen Embrittlement Mechanism

Sensitized type 316 had higher corrosion resistance than solution-annealed type 316 and sensitized type 304. From Figure 4, it would be also predicted that the sensitized type 316 has a higher transition temperature than that of sensitized type 304. $T_{\rm trans}$ thus predicted would be higher than the maximum test temperature experimentally conducted, since TG-SCC was not detected over the test temperature range conducted. However, the experimental result showed that TG-HE instead of IG-HE was observed for sensitized type 316 at a higher test temperature. Therefore, we need to consider why TG-HE occurred at the higher test temperature, since only IG-HE was predicted to occur over the range of test temperature used.

The difference between TG-HE and IG-HE is where the anodic reaction takes place at grain boundaries or at slip steps. This means that at a transition test temperature ($T^{\rm HE}_{\rm trans}$) from IG-HE to TG-HE or vice versa, the preferential site for anodic reaction changes from slip steps to grain boundaries or vice versa. To explain this phenomenon, the change in the degree of corrosion of sensitized types 304 and 316 with test temperature must be considered. On the basis of the results experimentally obtained, one can make the following assumptions.

(1) The degree of Cr depletion is smaller for sensitized type 316 than for the sensitized type 304 under the identical heat treatment condition; the fracture susceptibility based on Cr depletion for sensitized type 304 is higher than that for sensitized type 316.

(2) The inhibiting effect of chloride ions at Cr depletion becomes larger with increasing the degree of Cr depletion; that is, the inhibiting effect of chloride ions

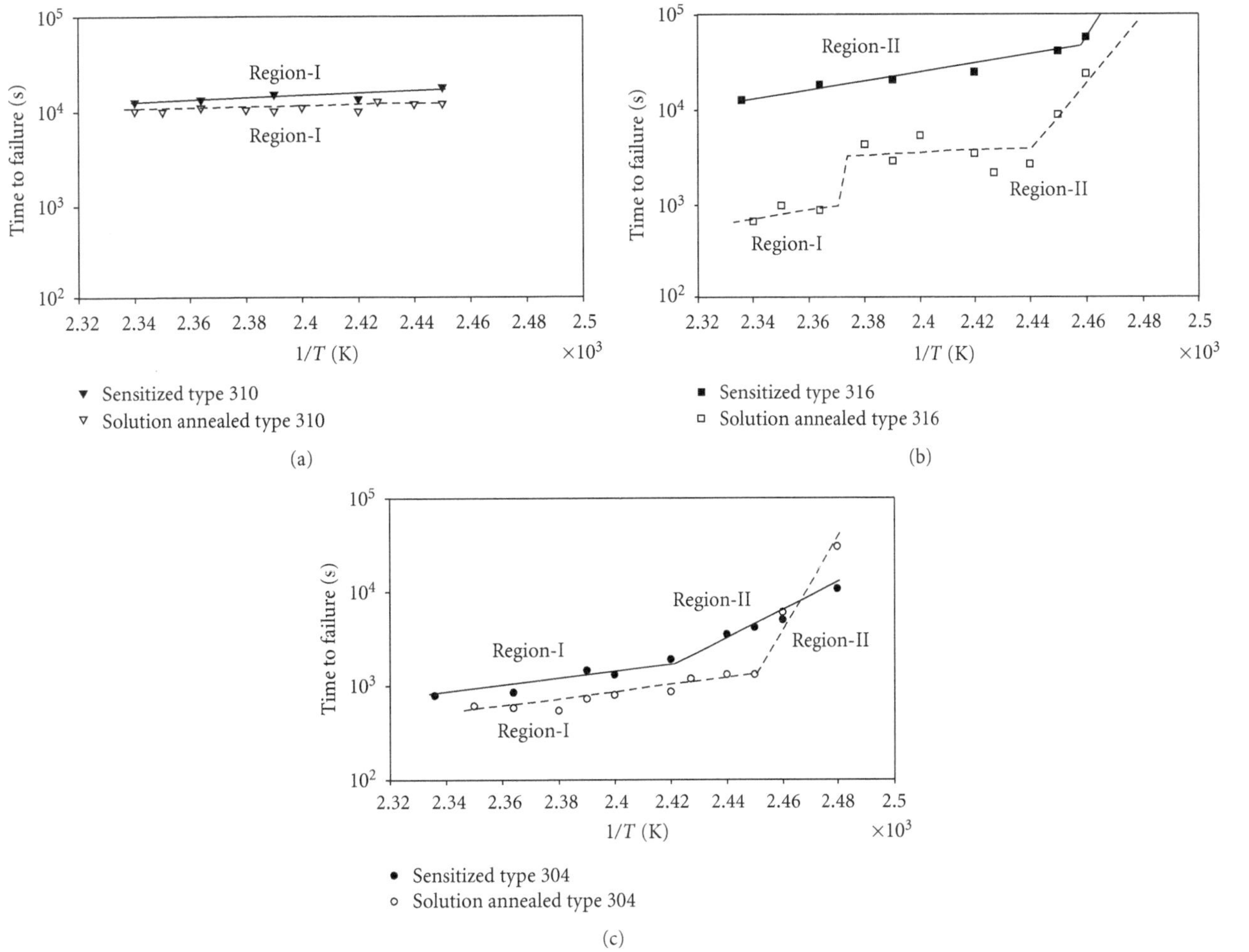

FIGURE 6: The relationship between reciprocal of test temperature ($1/T$) and logarithm of time to failure (t_f) for sensitized type 310, 304 and 316 steels in boiling saturated $MgCl_2$ solutions.

at Cr depletion for sensitized type 304 is higher than that for sensitized type 316.

(3) However, the degree of corrosion for the sensitized type 304 is larger than that for the sensitized 316 even by the inhibiting effect of chloride ions; that is, corrosive dissolution of sensitized type 304 is larger than that of sensitized type 316.

With respect to sensitized type 316 with higher corrosion resistance, the crack nucleation and propagation could not take place at slip steps or grain boundaries with anodic reaction over the range of test temperature used, whereas the fresh surface of slip steps may be able to act as the site for the cathodic hydrogen evolution. Under such a condition, as test temperature increases, anodic reaction would be enhanced more at grain boundaries than at slip steps due to the existence of Cr depleted zones, and the preferential site for anodic reaction would change from slip steps to grain boundaries. This would be the reason why the fracture mode changes from IG-HE to TG-HE.

On the basis of the considerations described in the previous sections, a mechanism for TG-HE can be proposed with a new aspect under the assumptions described below, where the IG-HE mechanism has already been reported in the previous work [14].

(1) The α'-martensite serves as an obstacle for dislocation movement on the slip plane as well as grain boundary sliding (GBS) for IG-HE.

(2) The α'-martensite does not self-destruct, even if it grows with enough thickness.

(3) The α'-martensite and the Cr depleted zone can serve as a preferential site for dissolution (with film formation) or hydrogen permeation. If the phenomenon at grain boundaries is considered, hydrogen permeation at grain boundaries will not induce TG-HE.

(4) The diffusing hydrogen (H_{dif}) leads to the enhancement of dislocation movement at slip plane as well as GBS; that is, HELP.

As soon as a load is applied to the specimens, a crack nucleation would take place at slip steps with the stress- and hydrogen-induced martensite. During the crack nucleation period, α'-martensite grows and simultaneously H_{dif} leads

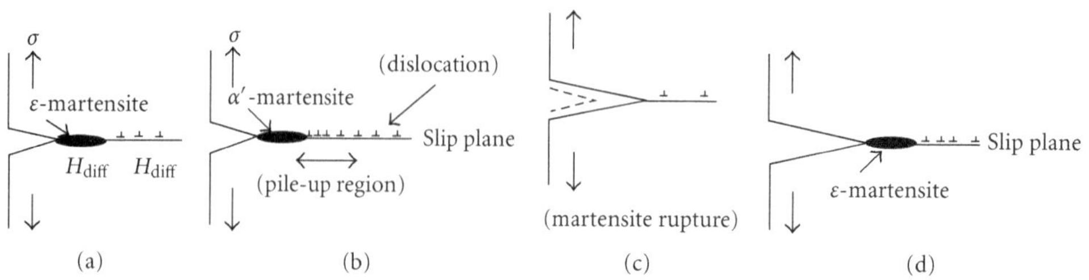

FIGURE 7: Schematic representation of a transgranular hydrogen embrittlement caused by the formation of α'-martensite and dislocation pile-up.

to HELP. An additional local stress (σ_{add}) would be generated by an interaction between α'-martensite and dislocation movement by HELP in addition to a local stress caused by applied stress (σ_{app}). Therefore, a net local stress (σ_{tip}) consists of σ_{app} and σ_{add}

$$\sigma_{tip} = \sigma_{app} + \sigma_{add}. \quad (1)$$

Under a constant applied stress condition, σ_{app} is constant, and σ_{add} increases with time. When σ_{tip} reaches a critical fracture stress (σ_F), α'-martensite rupture takes place:

$$\sigma_F = \sigma_{app} + \sigma_{add}^{Max}, \quad (2)$$

where σ_{add}^{Max} is the maximum additional local stress for σ_{tip} to reach σ_F and σ_F is assumed to be constant independent of applied stress and test temperature but depends upon the materials. Just after α'-martensite rupture, σ_{tip} drops to σ_{app} (= constant under a constant applied stress condition) and then increases with the increase in σ_{add} during α'-martensite growth up to σ_F, that is, $\sigma_{app} \leqq \sigma_{tip} \leqq \sigma_F$ or $0 \leqq \sigma_{add} \leqq \sigma_{add}^{Max}$. After the crack nucleation period, the crack propagation period proceeds with the cyclic formation—rupture event of α'-martensite. Figure 7 shows a schematic representation of the cyclic formation-rupture event of α'-martensite along with the above considerations.

Under a constant applied stress condition, an increase rate in σ_{tip} is provided from(1)

$$\frac{d\sigma_{tip}}{dt} = \frac{d\sigma_{add}}{dt}, \quad (3)$$

where $d\sigma_{app}/dt$ becomes zero as σ = constant. The number of the pile-up dislocation at crack tips (n_{tip}) is composed of n_{app} caused by σ_{app} and n_{add} by σ_{add} during α'-martensite formation

$$n_{tip} = n_{app} + n_{add}. \quad (4)$$

The n_{app} is constant ($dn_{app}/dt = 0$) under the constant applied stress condition and hence,

$$\frac{dn_{tip}}{dt} = \frac{dn_{add}}{dt} \quad \text{(during } \alpha'\text{-martensite formation).} \quad (5)$$

According to Tetelman and McEvily [23], a plastic displacement by dislocation movement is connected with an elastic shear displacement as follows:

$$n_{tip} \cdot b = L_{tip} \frac{(\tau - \tau_i)}{G}, \quad (6)$$

where b is the Burgers vector, L_{tip} is the length of the pile-up dislocation, τ is the shear stress ($\sigma = 2\tau$), τ_i is the friction stress, and G is the shear modulus. In addition, L_{tip} is also composed of L_{app} caused by σ_{app} and L_{add} formed during α'-martensite formation,

$$L_{tip} = L_{app} + L_{add}, \quad (7)$$

where L_{app} is constant, $dL_{app}/dt = 0$ as well as n_{app}, and then

$$\frac{dL_{tip}}{dt} = \frac{dL_{add}}{dt} \quad \text{(during } \alpha'\text{-martensite formation).} \quad (8)$$

Furthermore, a concentrated tensile stress, corresponding to σ_{tip}, is expressed as

$$\sigma_{tip} \propto (\tau - \tau_i)\left(\frac{L_{tip}}{r}\right)^{1/2} F(\varphi) \quad \left(b \ll r \ll L_{tip}\right), \quad (9)$$

where r is the length from the tip of a crack to the tip of the pile up which corresponds to the α'-martensite thickness at crack tips, and $F(\varphi)$ is a function of φ, that is, an angle between planes in adjacent grains.

On the other hand, a hydrogen permeation current (j_{perm}) would be proportional to a hydrogen evolution current density (j_c) as the cathodic reaction. Furthermore, j_{perm} consists of j_{dif} associated with the diffusing hydrogen leading to HELP and $j_{martensite}$ contributing to the formation of α'-martensite,

$$j_{perm} = j_{martensite} + j_{dif}. \quad (10)$$

As only j_{dif} is associated with dn_{dis}/dt, the following relation is obtained

$$\frac{j_{dif}}{zF} \propto \frac{dn_{dis}}{dt}. \quad (11)$$

Equation (11) shows that as j_{dif} increases, dn_{dis}/dt increases proportionally. From (7), (9), and (11), the following relation is given

$$\frac{dn_{dis}}{dt} \propto \frac{dL_{dis}}{dt}. \quad (12)$$

Furthermore, from (9) an increasing rate of σ_{tip} under each constant applied stress condition is expressed as

$$\frac{d\sigma_{tip}}{dt} \propto (\tau - \tau_i)\left(\frac{L_{tip}}{r}\right)^{-1/2} \frac{dL_{tip}}{dt}. \quad (13)$$

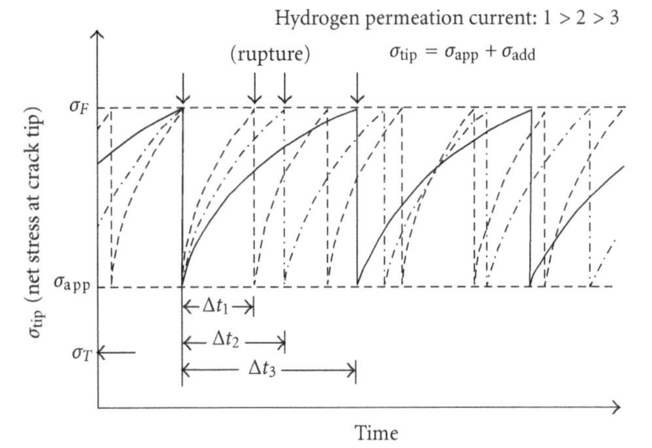

FIGURE 8: Schematic representation of time dependence of the net local stress (σ_{tip}) during the film rupture-formation events at various hydrogen permeation currents under a constant applied stress condition, where σ_T is the threshold stress below which no HE takes place and σ_F is the critical fracture stress at or above which the failure of α'-martensite occurs.

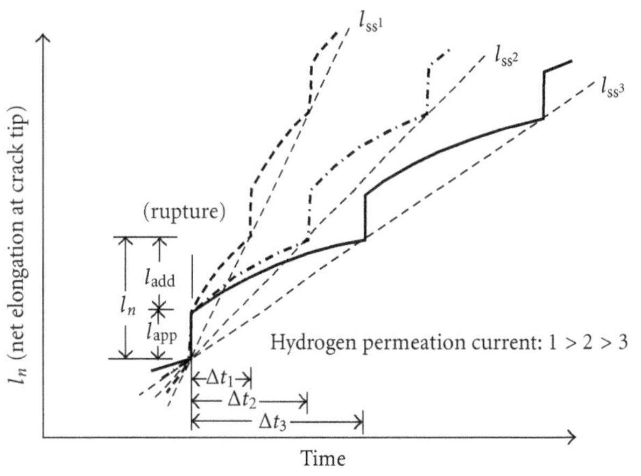

FIGURE 9: Schematic representation of time dependence of the net elongation (l_n) during the film rupture-formation events at various hydrogen permeation currents under a constant applied stress condition, which corresponds to that in Figure 7.

From (11), (12) and (13), (14) is provided as

$$\frac{d\sigma_{tip}}{dt} \propto (\tau - \tau_i)\left(\frac{L_{tip}}{r}\right)^{-1/2}\frac{j_{dif}}{zF}, \quad (14)$$

where L_{tip} and r are the values at σ_F, which are assumed to be constant. Finally, it is recognized that $d\sigma_{tip}/dt$ is dependent on τ and j_{dif}. Thus, the TG-HE mechanism of the sensitized type 316 is considered to take place by the cyclic α'-martensite formation-rupture event with a help of HELP caused by the diffusing hydrogen.

Figure 8 shows the schematic representation of the time variation of σ_{tip} at various permeation current densities under constant applied stress conditions. When σ_{tip} reaches σ_F, the rupture of α'-martensite takes place. Then σ_{tip} returns to σ_{app} and simultaneously begins to increase with increasing σ_{add} and this process is repeated.

As the permeation current density increases, the frequency of the formation—rupture event of α'-martensite increases with a consequent increase in the TG-HE susceptibility. Correspondingly, Figure 9 shows the schematic representation of the time variation of a net elongation (l_n), which corresponds to that of σ_{tip} in Figure 8, where l_n is composed of l_{app} caused by applied stress and l_{add} caused during the process of α'-martensite formation; $l_n = l_{app} + l_{add}$. An average elongation rate becomes the steady-state elongation rate, l_{ss}, although we could not observe the step by step on the elongation increase such as that in Figure 7 for TG-HE of the austenitic stainless steels.

7. Conclusions

(i) Using a constant load method, the effect of test temperature on stress corrosion cracking of austenitic stainless steels can be evaluated. There was two regions obtained, one for SCC-dominated failure and the other for hydrogen embrittlement dominated failure.

(ii) The cracking mechanism for type 304 was transgranular between test temperatures of 413 K and 428 K. The cracking mode for type 304 changed to intergranular at test temperatures between 412 K and 403 K. The cracking mechanism for type 316 was transgranular between test temperatures of 424 K and 428 K. The cracking mechanism for type 316 was a mixture of transgranular and intergranular cracking for test temperatures between 424 K and 408 K. The cracking mechanism for type 310 was transgranular at test temperatures higher than 408 K and no intergranular cracking behavior was observed.

(iii) The reason for the intergranular cracking mechanism for types 304 and 316 might be the formation of martensite at the grain boundary where as for type 310 no martensite transformation existed which contributed to the cracking mechanism uniformity. This behavior for types 304 and 316 requires further investigation.

(iv) The fracture susceptibility (SCC and HE) became lower for the sensitized austenitic stainless steels than for the solution-annealed ones. This was estimated to be caused by the inhibiting effect of chloride ions on anodic reaction (dissolution).

(v) Sensitized type 304 had two different cracking modes (TG-SCC and IG-HE) as well as for the solution-annealed one. However, the critical test temperature from IG-HE to TG-SCC shifted to a higher test temperature for the sensitized type 304 than for the solution-annealed one. On the other hand, the fracture behavior of type 310 had the same with and without sensitizing temperature treatment.

(vi) An HE mechanism with a new aspect was proposed in terms of martensite transformation, hydrogen entry site, sensitization, the role of diffusing hydrogen, dislocation movement (or grain boundary sliding) and so on.

Acknowledgment

The authors would like to acknowledge the funding provided by Kuwait University under Grant no. EM05/02.

References

[1] T. P. Hoar and J. M. West, "Mechano-chemical anodic dissolution of austenitic stainless steel in hot chloride solution," *Proceeding of the Royal Society A*, vol. 268, pp. 304–315, 1962.

[2] T. P. Hoar and J. C. Scully, "Mechanochemical anodic dissolution of austenitic stainless steel in hot chloride solution at controlled electrode potential," *Journal of the Electrochemical Society*, vol. 111, no. 3, pp. 348–352, 1964.

[3] R. Nishimura and Y. Maeda, "Stress corrosion cracking of type 304 austenitic stainless steel in sulphuric acid solution including sodium chloride and chromate," *Corrosion Science*, vol. 46, no. 2, pp. 343–360, 2004.

[4] T. Nakayama and M. Takano, "Application of a slip dissolution-repassivation model for stress corrosion cracking of AISI 304 stainless steel in a boiling 42% $MgCl_2$ solution," *Corrosion*, vol. 42, no. 1, pp. 10–15, 1986.

[5] M. B. Whiteman and A. R. Troiano, "Troiano, hydrogen embrittlement of austenitic stainless steel," *Corrosion*, vol. 21, pp. 53–56, 1965.

[6] M. L. Holzworth, "Hydrogen embrittlement of type 304L stainless steel," *Corrosion*, vol. 25, pp. 107–115, 1969.

[7] P. R. Rhodes, "Mechanism of chloride stress corrosion cracking of austenitic stainless steels," *Corrosion*, vol. 25, no. 11, pp. 462–472, 1969.

[8] R. Nishimura and H. Sulaiman, "Stress corrosion cracking of sensitized type 316 austenitic stainless steel in pure sulfuric acid solution," in *Corrosion Control for Low Cost Reliability*, p. 4325, 1993.

[9] R. Nishimura and Y. Maeda, "Metal dissolution and maximum stress during SCC process of ferritic (type 430) and austenitic (type 304 and type 316) stainless steels in acidic chloride solutions under constant applied stress," *Corrosion Science*, vol. 46, no. 3, pp. 755–768, 2004.

[10] O. M. Alyousif and R. Nishimura, "The effect of test temperature on SCC behavior of austenitic stainless steels in boiling saturated magnesium chloride solution," *Corrosion Science*, vol. 48, no. 12, pp. 4283–4293, 2006.

[11] O. M. Alyousif and R. Nishimura, "The stress corrosion cracking behavior of austenitic stainless steels in boiling magnesium chloride solutions," *Corrosion Science*, vol. 49, no. 7, pp. 3040–3051, 2007.

[12] O. M. Alyousif and R. Nishimura, "Stress corrosion cracking and hydrogen embrittlement of sensitized austenitic stainless steels in boiling saturated magnesium chloride solutions," *Corrosion Science*, vol. 50, no. 8, pp. 2353–2359, 2008.

[13] O. M. Alyousif and R. Nishimura, "On the stress corrosion cracking and hydrogen embrittlement of sensitized austenitic stainless steels in boiling saturated magnesium chloride solutions: effect of applied stress," *Corrosion Science*, vol. 50, no. 10, pp. 2919–2926, 2008.

[14] R. Nishimura and O. M. Alyousif, "A new aspect on intergranular hydrogen embrittlement mechanism of solution annealed types 304, 316 and 310 austenitic stainless steels," *Corrosion Science*, vol. 51, no. 9, pp. 1894–1900, 2009.

[15] D. Eliezer, D. G. Chakrapani, C. J. Altstetter, and E. N. Pugh, "The influence of austenite stability on the hydrogen embrittlement and stress- corrosion cracking of stainless steel," *Metallurgical Transactions A*, vol. 10, no. 7, pp. 935–941, 1979.

[16] H. Hanninen and T. Hakarainen, "On the effects of alpha prime martensite in hydrogen embrittlement of a cathodically charged alsl type 304 austenitic stainless steel," *Corrosion*, vol. 36, no. 1, pp. 47–51, 1980.

[17] S. S. Birley and D. Tromans, "Stress corrosion cracing of 304 austenitic steel and the martensite transformation," *Corrosion*, vol. 27, pp. 63–71, 1971.

[18] X. Sun, J. Xu, and Y. Li, "Hydrogen permeation behavior in metastable austenitic stainless steels 321 and 304," *Acta Metallurgica*, vol. 37, no. 8, pp. 2171–2176, 1989.

[19] T.-P. Perng and C. J. Altstetter, "Effects of deformation on hydrogen permeation in austenitic stainless steels," *Acta Metallurgica*, vol. 34, no. 9, pp. 1771–1781, 1986.

[20] R. Nishimura, "SCC failure prediction of austenitic stainless steels in acid solutions. Effect of pH, anion species, and concentration," *Corrosion*, vol. 46, no. 4, pp. 311–318, 1990.

[21] J. K. Tien, "Diffusion and the dislocation sweeping mechanism," in *Effect of Hydrogen on Behavior of Materials*, A. W. Thompson and I. M. Bernstein, Eds., p. 309, AIME, 1975.

[22] R. E. Reed-Hill, *Physical Metallurgy Principles*, Brooks/Cole Engineering Division, Monterey, Calif, USA, 2nd edition, 1973.

[23] A. S. Tetelman and A. J. McEvily Jr., *Fracture of Structual Materials*, John Wiley & Sons, New York, NY, USA, 1967.

Permissions

The contributors of this book come from diverse backgrounds, making this book a truly international effort. This book will bring forth new frontiers with its revolutionizing research information and detailed analysis of the nascent developments around the world.

We would like to thank all the contributing authors for lending their expertise to make the book truly unique. They have played a crucial role in the development of this book. Without their invaluable contributions this book wouldn't have been possible. They have made vital efforts to compile up to date information on the varied aspects of this subject to make this book a valuable addition to the collection of many professionals and students.

This book was conceptualized with the vision of imparting up-to-date information and advanced data in this field. To ensure the same, a matchless editorial board was set up. Every individual on the board went through rigorous rounds of assessment to prove their worth. After which they invested a large part of their time researching and compiling the most relevant data for our readers. Conferences and sessions were held from time to time between the editorial board and the contributing authors to present the data in the most comprehensible form. The editorial team has worked tirelessly to provide valuable and valid information to help people across the globe.

Every chapter published in this book has been scrutinized by our experts. Their significance has been extensively debated. The topics covered herein carry significant findings which will fuel the growth of the discipline. They may even be implemented as practical applications or may be referred to as a beginning point for another development. Chapters in this book were first published by Hindawi Publishing Corporation; hereby published with permission under the Creative Commons Attribution License or equivalent.

The editorial board has been involved in producing this book since its inception. They have spent rigorous hours researching and exploring the diverse topics which have resulted in the successful publishing of this book. They have passed on their knowledge of decades through this book. To expedite this challenging task, the publisher supported the team at every step. A small team of assistant editors was also appointed to further simplify the editing procedure and attain best results for the readers.

Our editorial team has been hand-picked from every corner of the world. Their multi-ethnicity adds dynamic inputs to the discussions which result in innovative outcomes. These outcomes are then further discussed with the researchers and contributors who give their valuable feedback and opinion regarding the same. The feedback is then collaborated with the researches and they are edited in a comprehensive manner to aid the understanding of the subject.

Apart from the editorial board, the designing team has also invested a significant amount of their time in understanding the subject and creating the most relevant covers. They scrutinized every image to scout for the most suitable representation of the subject and create an appropriate cover for the book.

The publishing team has been involved in this book since its early stages. They were actively engaged in every process, be it collecting the data, connecting with the contributors or procuring relevant information. The team has been an ardent support to the editorial, designing and production team. Their endless efforts to recruit the best for this project, has resulted in the accomplishment of this book. They are a veteran in the field of academics and their pool of knowledge is as vast as their experience in printing. Their expertise and guidance has proved useful at every step. Their uncompromising quality standards have made this book an exceptional effort. Their encouragement from time to time has been an inspiration for everyone.

The publisher and the editorial board hope that this book will prove to be a valuable piece of knowledge for researchers, students, practitioners and scholars across the globe.

List of Contributors

M. Lebrini, F. Robert and C. Roos
Laboratoire Materiaux etMol´ecules enMilieu Amazonien, UAG-UMRECOFOG, Campus Trou Biran, 97337 Cayenne, French Guiana

R. Ademar, J. G. Gonzalez-Rodriguez and J. Uruchurtu
U.A.E.M.-C.I.I.C.A.P. Av. Universidad 1001, Col. Chamilpa, 62500-Cuernavaca, Mor., Mexico

J. Porcayo-Calderon and V. M. Salinas-Bravo
I.I.E., Gerencia de Procesos Termico, Av. Reforma 120, Temixco, Mor., Mexico

G. Dominguez-Patiño
U.A.E.M.-Facultad de Biolog´ıa, Av. Universidad 1001, Col. Chamilpa, 62500-Cuernavaca, Mor., Mexico

A. Bedolla-Jacuinde
Instituto de Investigaciones Metal´urgicas, Universidad Michoacana de San Nicol´as de Hidalgo, Edificio "U", Ciudad Universitaria, 58066—Morelia, Michoac´an, Mexico

Qian-li Fu, Shang-xin Zhang and Yin Xia
Emperor Qin's Terra-Cotta Warriors & Horses Museum, Lin Tong, Shaanxi Province, Xi'an 710600, China

Pu-jun Jin
School of Materials Science and Engineering, Shaanxi Normal University, Chang An South Road 199#, Shaanxi Province, Xi'an 710062, China

Xue Ling
School of Cultural Heritage, Northwest University, Tai Bai North Road 229#, Shaanxi Province, Xi'an 710069, China

Wei-gang Sun
Institute of Archaeology of Shaanxi, Yan Ta Region, Shaanxi Province, Xi'an 710043, China

M. B. Lekala and J.W. van derMerwe
School of Chemical and Metallurgical Engineering, University of the Witwatersrand and DST/NRF Centre of Excellence in Strong Materials, Johannesburg 2001, South Africa

S. L. Pityana
Council for Scientific and Industrial Research, National Laser Centre, Pretoria 0001, South Africa

Juan A. Jaén
Departamento de Qu´ımica F´ısica, CITEN, Lab. No. 105, Edificio de Laboratorios Cient´ıficos-VIP, Universidad de Panam´a, Panama, Panama

Josefina Iglesias and Cecilio Hernández
Laboratorio de An´alisis Industriales y Ciencias Ambientales, Universidad Tecnol´ogica de Panam´a, Panama, Panama

Ying-zi Zhang
Department of Civil Engineering, Harbin Institute of Technology, Weihai, Shandong 264209, China

Ying-fang Fan
Institute of Road and Bridge Engineering, Dalian Maritime University, Dalian, Liaoning 116026, China

Hong-nan Li
State Key Laboratory of Coastal and Offshore Engineering, Dalian University of Technology, Dalian, Liaoning 116023, China

G. Salinas
CIICAP, Universidad Aut´onoma del Estado de Morelos, Avendia Universidad 1001, Colonia Chamilpa, 62209 Cuernavaca, MOR, Mexico

M. A. Espinoza-Medina
Facultad de Ingeniera Mecanica, Universidad Michoacana de San Nicol´as de Hidalgo, 58000 Morelia, MICH, Mexico

Christopher D. Taylor
Materials Technology - Metallurgy (MST-6), Materials Science and Technology, Los Alamos National Laboratory, Los Alamos, NM 87545, USA

Ayssar Nahlé, Ideisan I. Abu-Abdoun and Ibrahim Abdel-Rahman
Department of Chemistry, College of Sciences, University of Sharjah, P.O. Box 27272, Sharjah, UAE

Olfa Hammami
Unit of Research "COPROMET" ENIT, University of Tunis-El Manar, BP 371002 Belvedere, Tunisia
Institut UTINAM, CNRS UMR 6213, ´ Equipe Sonochimie et R´eactivit´e des Surfaces, Universit´e de Franche-Comt´e, 16 Route de Gray, 25030 Besanc¸on Cedex, France

Leila Dhouibi and Ezzeddine Triki
Unit of Research "COPROMET" ENIT, University of Tunis-El Manar, BP 371002 Belvedere, Tunisia

Patrice Bercot and ElMustafa Rezrazi
Institut UTINAM, CNRS UMR 6213, ´ Equipe Sonochimie et R´eactivit´e des Surfaces, Universit´e de Franche-Comt´e, 16 Route de Gray, 25030 Besancon Cedex, France

Inemesit A. Akpan and Nnanake-Abasi O. Offiong
Corrosion and Materials Science Unit, Department of Chemistry, University of Uyo, PMB 1017, Akwa Ibom State, Uyo 520001,Nigeria

Laura A. Hernandez-Alvarado
Faculty of Chemical Sciences, Universidad Autonoma de San Luis Potosi, Avenida Salvador Nava No. 6, 78290 San Luis Potosi, SLP, Mexico

Orlando Salas, Oladis Troconis de Rincon, Daniela Rojas, Adriana Tosaya, Nathalie Romero, Miguel S´anchez and William Campos
Centro de Estudios de Corrosion, Facultad de Ingenier´ıa, Universidad del Zulia, Maracaibo, Zulia, Venezuela

Luis S. Hernandez
Institute of Metallurgy, Universidad Autonoma de San Luis Potosi, Avenida Sierra Leona No. 550, 78210 San Luis Potosi, SLP, Mexico

Sandra L. Rodriguez-Reyna
Faculty of Engineering, Universidad Autonoma de San Luis Potosi, Avenida Salvador Nava No. 8, 78290 San Luis Potosi, SLP, Mexico

B. E. Amitha Rani and Bharathi Bai J. Basu
Surface Engineering Division, CSIR-National Aerospace Laboratories, Bangalore 560037, India

Ayssar Nahle, Ideisan I. Abu-Abdoun and Ibrahim Abdel-Rahman
Department of Chemistry, College of Sciences, University of Sharjah, P.O. Box 27272, Sharjah, United Arab Emirates

J.W. van der Merwe
School of Chemical and Metallurgical Engineering, University of the Witwatersrand, Johannesburg, Private Bag 3, Wits 2050, South Africa

C. I. Elsner
CIDEPINT, Centro de Investigaci´on y Desarrollo en Tecnolog´ıa de Pinturas (CICPBA-CCT CONICET-La Plata), Avenida 52 s/n entre 121 y 122, B1900AYB La Plata, Argentina
Facultad de Ingenier´ıa, Universidad Nacional de La Plata, Avenida 1 esq. 47, B1900TAG La Plata, Argentina

P. R. Sere and A. R. Di Sarli
CIDEPINT, Centro de Investigaci´on y Desarrollo en Tecnolog´ıa de Pinturas (CICPBA-CCT CONICET-La Plata), Avenida 52 s/n entre 121 y 122, B1900AYB La Plata, Argentina

Injoon Son
Department of Materials Science and Metallurgy, Kyungpook National University, 1370 Sangyeok-dong, Buk-gu, Daegu 702-701, Republic of Korea

Hiroaki Nakano, Satoshi Oue, Hisaaki Fukushima and Zenji Horita
Department of Materials Science and Engineering, Kyushu University, 744 Motooka, Nishi-ku, Fukuoka 819-0395, Japan

Shigeo Kobayashi
Department of Applied Chemistry and Biochemistry, Kyushu Sangyo University, 2-3-1 Matsukadai, Higashi-ku, Fukuoka 813-8503, Japan

R. E. Núñez-Jaquez, J. E. Buelna-Rodríguez and C. P. Barrios-Durstewitz
Facultad de Ingenier´ıa Los Mochis, Universidad Aut´onoma de Sinaloa, Fuente de Poseid´on y Prol
Angel Flores S/N, C.P. 87223 Los Mochis, SIN, Mexico

C. Gaona-Tiburcio
Centro de Investigacione Innovacion en Ingenier´ıa Aeron´autica CIIIA, FIME-Universidad Aut´onoma de Nuevo Le´on, Avenida Universidad S/N, Ciudad Universitaria, C.P. 87223 San Nicol´as de los Garza, NL, Mexico

F. Almeraya-Calderon
Centro de Investigacione Innovacion en Ingenierıa Aeron´autica CIIIA, FIME-Universidad Autonoma de Nuevo Le´on, Avenida Universidad S/N, Ciudad Universitaria, C.P. 87223 San Nicol´as de los Garza, NL, Mexico
Grupo Corrosi´on, Centro de Investigaci´on en Materiales Avanzados, Miguel de Cervantes No. 120, C.P. 31109 Chihuahua, CHIH, Mexico

Mustafa Ö. Öteyaka
Department of Mechanical Engineering, Dumlupinar University, 43100 K¨utahya, Turkey

Edward Ghali and Real Tremblay
Department of Mining, Metallurgical and Materials Engineering, Laval University, Quebec City, QC, Canada G1V 0A6

Osama M. Alyousif
Department of Mechanical Engineering, Kuwait University, P.O. Box 5969, Safat 13060, Kuwait

Rokuro Nishimura
Department of Applied Materials Science, Osaka Prefecture University, 1-1, Gakuen-cho, Sakai, Osaka 599-8531, Japan